O9-CFT-146

Praise for
First Mothers

"Bonnie Angelo has written a superb book depicting and defining the profound influence of mothers on so many presidents. Angelo found the key to presidential personalities and ambitions—their mothers. Her book is beautifully written and long needed."
—Helen Thomas, UPI White House bureau chief

"*First Mothers* offers fresh and unique insight that is simply and gracefully written."   —Hugh Sidey, *Time*

"*First Mothers* gives us the human beings, in their attributes and deficiencies, with an understanding narrative voice."
—Carl Sferrazza Anthony, historian and author
of *First Ladies* and *America's First Families*

"Why hasn't anyone done this before? Bonnie Angelo's *First Mothers* is an important piece in the puzzle of why presidents are the way they are."
—Stephen Hess, Brookings Institution, author of
Presidents & the Presidency

"Angelo writes with a reporter's eye for detail and the telling anec-dote."   —*Boston Globe*

"This is an enthralling look at the women who've raised the men who've run the country."   —*Booklist*

"Angelo has written a rich and beautiful book whose message runs deeper than history."   —*Rocky Mountain News*

"Aside from being fascinating reading just for the stories and the history, *First Mothers* holds powerful lessons for today's parents. . . . Truly inspirational."   —*The Tennessean*

"Tantalizing bios. . . . Whets the appetite for more information about these First Moms."   —*Kirkus Reviews*

"*First Mothers* offers the story of each [mother] in rich detail."
—*Louisville Courier-Journal*

"The profiles are well written and lively."
—*San Antonio Express-News*

"Angelo is superb at weaving seemingly unrelated vignettes into a multicolored quilt that helps explain, in part, where the nation's leaders came from."   —*Denver Post*

"Angelo's book will surely be a great treasure for historians. For anyone who doubts the power of women, *First Mothers* should be required reading."   —*Greensboro News & Record*

"Succinct and highly readable."   —Amazon.com

"[An] engaging portrait of the mothers of eleven modern-day presidents."   —*People*

"Angelo turned up fascinating details about first mothers."
—*Philadelphia Inquirer*

"An absorbing, anecdote-laced account."
—Knight Ridder newspapers

"Filled with humorous and revealing anecdotes."
—*Winston-Salem Journal and Sentinel*

Diana Walker

About the Author

In more than twenty-five years with *Time*,
BONNIE ANGELO has reported on the
White House and the presidential families
during eight administrations. As a Washington correspondent and bureau chief in London and New York, she has covered
newsmakers and major events in all fifty
states and around the world. She lives in
Bethesda, Maryland, and New York City.

First Mothers

★

THE
WOMEN
WHO
SHAPED
THE
PRESIDENTS

BONNIE ANGELO

WITH A NEW CHAPTER ON FIRST MOTHER,
BARBARA BUSH

HARPER

NEW YORK • LONDON • TORONTO • SYDNEY

HARPER

First Perennial edition published 2001.
Reprinted in Harper 2006.

Designed by Jo Anne Metsch

Library of Congress Cataloging-in-Publication Data

Angelo, Bonnie.
    First mothers : the women who shaped the presidents : with a new chapter on
First Mother, Barbara Bush / Bonnie Angelo.—1st Perennial ed.
        p.   cm.
    Includes bibliographical references and index.
    ISBN-10: 0-06-093711-4 (pbk.)
    ISBN-13: 978-0-06-093711-9 (pbk.)
    1. Mothers of presidents—United States—Biography.   I. Title.

E176.3 .A543 2001
973'.09'9—dc21
[B]   2001026484

06 WB/RRD 20 19 18 17 16

To my mother,
Ethel Hudgins Angelo

CONTENTS

★

INTRODUCTION

★

$\mathcal{T}$HIS is a journalist's book, born of years of covering presidents and hearing them again and again pay tribute to their mothers and, noticeably, not their fathers. While much is written about the wives of presidents, I discovered that the stories of the mothers, the women who shaped the boys who became the men who became the presidents, are little known. Information about them is finite and varies greatly—a considerable amount about those who lived to see their sons in the White House and little about those, equally crucial to their sons' success, who had died long before.

Most meaningful were the personal reminiscences given to me by three president-sons, Gerald Ford, George Bush, and Jimmy Carter, who talked lovingly about life with Mother. Two younger sons, Senator Edward Kennedy and Edward Nixon, and two daughters, Eunice Kennedy Shriver and Nancy Bush Ellis, enlarged this portrait of their family lives. Daughters-in-law Lady Bird Johnson and Nancy Reagan shared with me their unique perspective in what is always a delicate human equation, and Emily

Gordy Dolvin recalled girlhood days with her irrepressible sister Lillian. Half a dozen grandchildren who knew their grandmothers well added spice to the family lore, especially the two who edited their fathers' memoirs, Margaret Truman Daniel and John Sheldon Doud Eisenhower, both distinguished authors in their own right; Luci Baines Johnson provided thoughtful introspection about her father and his mother; and Governor George W. Bush took time from his own presidential pursuit to reflect on his much-loved grandmother. I cannot begin to express my gratitude—and pleasure—for those interviews.

A word about why this book of presidents' mothers begins with Sara Delano Roosevelt: the Roosevelt era was a watershed in history, the beginning of contemporary America and the modern presidency, the prize that can now be won only by men of supreme self-assurance who are willing to withstand the grinding process and microscopic examination not required of candidates in the bygone era of nominations decided in smoke-filled rooms.

What was the wellspring of that confidence, toughness, and resilience? Here is my answer: twelve remarkable women, not stained-glass mothers but lively doers who funneled boundless energy into these sons, mixing praise and discipline in equal measure. And through him each put her stamp on American history.

First Mothers

1

TO THE MANNER BORN

A man who has been the indisputable favorite of his mother keeps for life the feeling of a conqueror, that confidence of success that often induces real success.

—SIGMUND FREUD

A RESTLESS young Franklin Roosevelt, under strict quarantine for scarlet fever in the Groton school infirmary, was startled by the scraping sound against the red brick wall, followed by a gloved tap on the window—and then the apparition of his mother, the regal Sara Delano Roosevelt, peering into the room. She was perched on a workman's ladder, risking her safety and shattering her dignity to circumvent the no-visitors edict. From that precarious roost, she talked with him each day and even read to him. When she learned of his illness she had rushed home from Europe to comfort him; from the day he was born her son had been her total concern.

When he was a student at Harvard, she rented an apartment in Boston to oversee his social life. When he and his young wife needed a larger house, she provided it. When he was stricken with polio, she pampered and cosseted him, against his wishes. When he was president, she schemed to bring the White House up to her standards. And when he was contemplating divorce, she threatened (so it was whispered) to cut off his funds from the family fortune. Whether Franklin wished it or not, Sara Roosevelt was determined to do what she deemed best for him. Not that she always won. After all, she had shaped this son in her own mold—confident, determined, and pleasantly stubborn when it came to getting his own way.

From the beginning she and her husband, James Roosevelt, created a world of privilege and principle, a secure little universe in which Franklin could grow up on the family estate at Hyde Park, covering hundreds of acres in New York's beautiful Hudson Valley. Shielded by his family's position, he was exposed to only a small circle of family, servants, the local gentry, and a few deferential shopkeepers in the village that bordered on Roosevelt land. The values and lifestyle of his parents were emblems of the world of a disappearing landed aristocracy.

At first their relatives in the other branch of the family—the Republican Roosevelts—regarded Sara and James as an odd couple. It is unlikely that Mrs. Theodore Roosevelt, whose son Teddy was away at Harvard, had harbored any intention of matchmaking when she invited her daughter's good friend "Sallie" Delano to a dinner party that included, among other guests, cousin James, a widower of fifty-one, a rather formal man with muttonchop whiskers—and a son Sara's age. Sara was twenty-five, tall and graceful at five-foot-ten, and world-traveled. By the end of the evening, Sara had accepted an invitation to visit him at Springwood, his country home, properly chaperoned, of course.

In May 1880 Sara arrived at Springwood; Hyde Park was abloom, and before the visit ended, so was love. Years later Sara wrote a nostalgic letter to her son, then the governor of New York:

Darling Son:

Just 51 years yesterday, the 7th, I came to visit. If I had not come then, I should now be "old Miss Delano" after a rather sad life!

That Sara was still unmarried when she met James was astonishing. She was one of the five "beautiful Delano sisters," as New York society called them, daughters of another family in the Hudson Valley aristocracy. Numerous young men paid court to her, only to be rejected as not suitable by her father, Warren Delano II. (One such was Stanford White, a budding architect whom she found charming, but her father did not, ordering her to return his flowers with a cold letter that would end the friendship. Warren Delano was a good judge of character, at least when it came to this young man, who became a great architect—and a notorious womanizer who, in the great scandal of the times, was killed by a jealous husband.)

★　★　★

WHEN Sara fell in love with James Roosevelt, she chose—perhaps subconsciously—a man who was very much like the father she adored, a protective man, a gentleman of substance. When James proposed, her answer was a prompt yes. Then came the awkward moment when he had to ask his old friend and contemporary Delano for his daughter's hand in marriage. James gently reminded the reluctant Warren that when he married Sara's mother he was almost twice her age. With that, he won the point and the maiden. Only six months after their first introduction, Sara and James were wed under a canopy of flowers at Algonac, the Delano estate across the Hudson, downriver from Hyde Park.

From that day forward, theirs was a happy life. A year after that first visit to Springwood, Sara was pregnant. She was a healthy young woman of twenty-seven, yet her labor was long—more than twenty-four hours—and harrowing. When the pain became more than her body could endure, the doctor administered chloroform, in much too great an amount. She lost consciousness. When the baby was finally delivered, he was limp, blue, and unresponsive. Only when the doctor desperately breathed into the tiny, quiet mouth did the infant begin to breathe. "It was nearly fatal to us both," Sara recalled as the son she had almost lost became president. Happily, mother and child came through the ordeal in good shape: "At quarter to nine," the relieved father recorded in his diary that evening, "my Sallie had a splendid large baby boy. He weighs 10 lbs., without clothes." To the new mother's eye, he was "at the very outset, plump, pink, and nice."

Sara Roosevelt was never pregnant again. Perhaps the near-death experience was too frightening to risk repeating. Or perhaps she was so confident that hers was the perfect child, there was really no need to have another. Whatever the circumstances, she was never heard to express regret at having only one child. This child would be the work of her life, her monument. In the years to come she would say that she had no grand design for her son, she simply wanted to bring him up to be an admirable Dutchess County squire like his father. However, she was adamant that her son would be worthy of the two distinguished family names he bore; she would make her father proud of him, and proud of her, for raising him so well. She gave no thought to his having a career in politics, which she considered beneath them.

And she was quite determined that her son would outshine his half brother, James, her husband's son by his much-admired first wife. "Rosy," as he was called, was thirty years older than Franklin, a nice man best known for marrying Helen Astor, whose mother, Mrs. William Backhouse Astor, was the reigning queen of New York society in America's Gilded Age. ("The Four Hundred," the defining term for that elite, stemmed from her invitation list to the grand ball she gave in 1888, limited to that number by the size of the ballroom in the Astors' Fifth Avenue mansion.)

Every second wife wants to win the inevitable comparison with her predecessor, and a successful son was the primary way for a woman of Sara's era to triumph. In addition, with the Roosevelts there was the implicit rivalry with the family's Long Island branch, who regarded themselves as the more distinguished members. Sara would have denied such motives, but they were all there, pressing her to push young Franklin. Her unreserved dedication to her son would be the essential element in forming the character of a future president of the United States. Her devotion shaped her life and his; it gave her an extraordinary degree of influence over him, a power she would never relinquish.

Sara had been the middle child in a large and loving family of nine children (two others were lost in infancy), but she maintained that an only child enjoyed important advantages. "So often I have noticed," she observed, a shade defensively in that time of large families, "that where there are many children, either the older child bullies the younger or climbs down the intellectual ladder to the younger one's mentality." In her view, an only child could develop more fully "because he has more chance of assimilating adult knowledge."

Certainly her only child was given that opportunity. She and James took him everywhere—eight long visits to Europe before he was fourteen, which put him at ease in adult society and, more important, gave him valuable insights into Europe and its politics. This worldview strongly influenced his foreign policy as he struggled to make an isolationist America comprehend the menace represented by Nazi Germany.

★   ★   ★

SARA addressed the challenge of a new baby with her usual confidence, following her own common sense. Along with the baby's nurse, she took personal care of the infant, which was not customary then for wealthy young mothers, and she continued to nurse him a year longer than the usual practice. "I felt [then] as I do now," she declared years later, "that every mother ought to learn to care for her own baby, whether she can afford to delegate the task to some one else or not." Her remark can only be seen as an oblique criticism of Eleanor, the daughter-in-law who seemed unable to be close to her children, to cuddle and comfort them, a distance her children never forgot or quite forgave. Eleanor acknowledged her detachment, in retrospect, blaming it on inexperience—"I knew nothing about handling or feeding a baby." Given that she bore six children (one, sadly, lost in infancy), her explanation sounds a bit thin.

Sara, the first-time mother, was thrilled with her son's every advance, meticulously recording each event. "Baby went to his first party yesterday," read her diary entry for November 11, 1882, when Franklin was ten months old. "He wanted to dance and I could hardly hold him. He manages to say a semblance of 'Papa' and 'Mamma.'" On Christmas Eve the following year, "Franklin gave out the presents for the first time. . . . Today James gave Baby his first coast [sled ride]." The intensity of her mothering, her attention to every detail of her son's young life, seems excessive, but to Sara it was nothing other than natural.

★   ★   ★

IT probably came as a surprise that this boy had a mind of his own, that he did not shrink from confrontation with her. And any mother can guess what the issues would be: clothes and haircuts. Young Franklin had a crown of beautiful blond curls, which Sara cherished; she dressed him in ruffled Little Lord Fauntleroy suits, which Franklin hated. Again, any mother could guess who won the argument. "We mothers," Sara wrote in *My Boy, Franklin,* her story of the family, "had daintier ideas [for] apparel than was popular with our sturdy offspring." She then "tried to induce him to wear Scotch kilts," but again he balked. They struck a compromise: sailor suits.

Reluctantly, she allowed his curls to be cut off, "oh, long before they should have been," she lamented, still regretful nearly fifty years later. But she conceded that "waving locks do not really go hand-in-hand with the qualities of self-reliance and responsibility that we tried to foster." She saved a few of the curls, tenderly storing them with other mementos of his babyhood.

"We never tried to influence him against his own inclinations or to shape his life," she insisted, with a mother's blindness. But it was a two-way relationship, for along with her edicts, she listened. When Franklin was about five, he suddenly turned melancholy. "I am unhappy," he announced. "Oh, for freedom!" "It seems funny now," said Sara in retelling the story, but "his voice had a desperate note that made me realize how seriously he meant it." She and James agreed that they—she—may have overregulated Franklin's life. He was rebelling at his mother's rigid schedule and, perhaps, her overcontrol.

The next morning she allowed him to do as he pleased—no rules, no schedule. "He proved his desire for freedom by completely ignoring us! That evening a very dirty, tired youngster came dragging in. We did not ask him where he had been or what he had been doing. The next day, quite of his own accord he went contentedly back to his routine." Sara was pleased with her experiment: Franklin discovered that "freedom" wasn't much fun if it meant being ignored, and Sara had the satisfaction of proving that Mother Knows Best, a conviction she held for the rest of a long life. And her son, for the rest of his life, expected constant attention—and criticism—from her.

The parents neatly divided the responsibilities of raising their son: Sara was in charge of the life of the mind and his father taught him the pleasures of the out-of-doors, shooting small game in Springwood's forest, swimming and sailing in the Hudson, riding his pony along wooded trails. (When Franklin was only eight, he and his father rode down the Hudson Valley to Grandfather Delano's even grander estate, twenty miles each way, just to spend the day.) James was the very model of an English country gentleman, following the example set by his titled British friends at their stately homes in the shires.

James spent minimal time managing the investments and income that flowed from his inherited wealth. (The Roosevelt money originated several generations earlier in dry goods, multiplied in real estate and the West Indies sugar trade, and expanded in railroads.) He turned down business opportunities that might have made him a very rich man, preferring the aristocratic pattern of living off one's inheritance. The elderly father found more pleasure in introducing Franklin to the joys of sport than in playing the market, and in return he was rewarded with his son's love.

When Franklin was six, he lobbied for a horse of his own, pleading his case in an earnest little essay: "If I bought a horse I would put him in the stable and have him groomed and washed every day as a well-groomed horse is always more healthy than a dirty horse but I would be careful not to let any soap get in his eyes because he would scream if I did." On the morning of his seventh birthday he was led to the stables where he found his most-wished-for present. Sara exacted a promise in return for the pony: he must personally take care of "Debby" as he did his red setter, Marksman. Taking care of others, human beings as well as animals, he soon learned, was a duty that came with wealth and privilege.

★   ★   ★

BRINGING up her son was, for Sara, a full-time vocation. She organized his day, hovering over him, praising and challenging, supervising a succession of governesses and tutors. The schedule was fixed: breakfast at eight. Lessons in the improvised schoolroom from nine until one. Time out for lunch. After a free hour for play, the schoolday continued. Drawing and dancing lessons, which he enjoyed, alternated with piano lessons, which he found tedious. After his classes were finished for the day, he was allowed to ride over and see his friends on the neighboring estate.

Franklin's young life was "sequestered," Sara acknowledged, but she dismissed the criticism of those "who pitied him for a lonely little boy and thought he was missing a great deal of fun." To the contrary, "I do not believe I have ever seen a little boy who seemed always to be so consistently enjoying himself . . . but being a rather introspective little fellow, he may have taken delight in forms of amusement that might not have a wide juvenile appeal."

As a child Sara had saved the colorful Chinese stamps that brought letters from her father, half a world away in Hong Kong, and she passed on to young Franklin the magic of stamp collecting. The hobby became his lifelong passion: on the last day of his life, April 12, 1945, FDR had worked on his stamp album at his Warm Springs, Georgia, retreat, and after his death his collection was sold at auction for more than $230,000, the equivalent of more than ten times that much today.

Sara loved to recall the evening when Franklin, about ten, was lying on the floor of her "snuggery" (her small sitting room), sorting stamps from his uncle Franklin in Egypt, while she read to him about the Pharaohs. "I don't believe you are hearing a word," she protested. At that, Franklin recited the last two sentences she had read, with the cheeky comment "I'd be ashamed if I couldn't do at least two things at once." In his White House years he would do the same, sorting his stamps (the State Department saved exotic ones for him) while enjoying lively conversation with his inner circle at cocktail time.

Most important, his mother encouraged him to read, widely and voraciously. Books were favorite gifts—his first, *The Kindergarten Children*, was under the Christmas tree in 1885, when he was only three. At thirteen the young bird lover pored over *Birds of Eastern North America*. Then came volumes of history, especially naval history, which very likely presaged his love of the navy and his early appointment as assistant secretary of the navy.

Not until Franklin was fourteen, two years later than was customary, could Sara and James bring themselves to send him away to boarding school. Years earlier they had selected Groton, a preparatory school north of Boston that patterned itself on England's renowned Eton and Harrow. Groton focused on making its students sturdy Christians of exemplary character, more athletic than intellectual, reflecting the values of America's ruling class—and measuring up to Sara's own rigorous standards.

This was a difficult time for her as Franklin was beginning the process essential to the development of boys: breaking away from the mother. After settling their son into his spartan quarters and bidding him a reluctant farewell, Sara wrote in her diary: "It is very hard to leave our darling boy. James and I feel this parting very much."

As she always would, she soon looked on the bright side, crowing that "almost overnight he became sociable and gregarious and entered with the frankest enjoyment into every kind of school activity." In reality the two-year delay in entering Groton made Franklin an outsider to classmates who had already formed close friendships; for the first time in his Franklin-centered universe, he was the "other." It didn't help that his classmates found him a bit of a prig with an affected accent—and that he was, of all things, a Democrat. But in his letters home, he gave no indication that things were going other than "finely." He had been brought up to avoid emotional display, especially any hint of self-pity, a toughness that would stand him in good stead as he coped with crutches in later years.

Before long his engaging personality and charm won over his classmates, and as he settled into a less isolated existence he began to assert the independence that came with separating himself from his mother. Although Sara attempted to influence him from a distance, she faced increasing resistance from her son, especially after he advanced from Groton to Harvard. Once she wrote from Europe to instruct him to decline a party invitation she deemed unsuitable. Franklin brusquely wrote back, "Please don't make any more arrangements for my future happiness." The mother must have been stung by such sharp resistance, but she was learning that her son had willpower to match her own. Now she would have to acknowledge his increasing independence, which sharply conflicted with her need to control.

Casual about his studies as always, Franklin threw himself into extracurricular activities. Hasty Pudding, sports, the glee club—the gamut. He was elected editor of the *Harvard Crimson* and president of his class. He was seen as a natural leader, born for success. Still, not everything came his way: he was rejected as a member in Porcellian, Harvard's most elite social club; both his father and his distant cousin, Teddy—by then in the White House—had been members. To be blackballed was a crushing blow, and with much loss of face he could only settle for second-best, Fly. He would never forget or forgive the slight by Porcellian.

Three months after Franklin entered Harvard, his father died of heart failure, both of his sons at his bedside. His will spelled out his trust and admiration for "my Sallie": "My wife [shall be] sole guardian of my son Franklin Delano Roosevelt, and I wish him to be under the influence of his mother."

Sara moved into an apartment in Boston to be closer to her son and to take her away from the echoing loneliness of her beloved Springwood. In every way she would fulfill the express wish of her late husband.

* * *

ON the deck of the clipper ship *Surprise,* Sara Delano, petticoats flying, was a great little sailor. Rounding the Cape of Good Hope, she watched the crew reef the sails, scanned the sea for monsters, and looked forward to celebrating her eighth birthday on the Indian Ocean. She was quite at home on *Surprise,* understandably, as the Delanos of New York were its sole passengers. Warren Delano II had chartered the ship to transport his family (plus nursemaid, personal maid, Uncle Ned and Aunt Sarah Delano, and cousin Nancy as tutor) to Hong Kong. In June 1862 they set sail on the four-month voyage, a glorious adventure. Years later her grandchildren were enthralled when their dignified grandmother sang sea chanteys she had learned from the sailors.

Young Sara—"Sallie"—found it all "incredibly exciting." And

the best part was knowing that wonderful Father, who had been away in China for three years, would be waiting for them at the journey's end. She was a papa's girl who, seventy eventful years later, still declared lovingly, "There was no one like my father." Again and again that sentiment is echoed by these presidential mothers, most of them self-declared "papa's girls." The relationship between father and daughter was clearly the significant factor in Sara Delano's development; she rarely mentioned her mother. In Franklin's case, there was a real struggle to separate from his mother. For both Sara and her son, those bonds, which were never loosened, influenced the character, behavior, and fate of two strong and remarkable people.

When the Delano family sailed for China, Warren Delano II was on his second sojourn in Hong Kong, rebuilding the fortune he had made, and lost, in his earlier years there. He traded in tea and silk, beautiful Chinese objects of all kinds—and opium. Many British and American merchants engaged in that lucrative business, which Warren justified as "honorable and legitimate." Legitimate, possibly, in the strictest legal sense. Honorable? If it was brought up, the family would explain that opium was needed for Union soldiers wounded in the Civil War. Battle casualties and millions of Chinese addicts quickly restored his fortune. As one member of a large family, Sara would inherit more than a million dollars, which suggests that the fortune was indeed considerable.

Sara's storybook life began in Hong Kong, where she lived in oriental splendor in a mansion set among palms and strange, lush plants; inside, houseboys stirred a breeze with a punkah, a large muslin panel. In 1864 her parents sent Sara and their three older children back to America to grow up in their own culture, entrusted to an uncle, Judge Samuel Lyman, in Northampton, Massachusetts.

In another two years, when Warren was again rich enough to retire, he reunited the family in Europe, where Sara celebrated her twelfth birthday in Paris. She studied in Dresden for a year, then in

1870 happily returned home. To Sara, the castles of Europe were no match for Algonac, the Delanos' forty-room Italianate mansion set on sixty acres with a painter's view of the great Hudson River. Algonac was a warm and lively world, filled with children of various ages, unmarried aunts and uncles, and visiting friends. The children, the house and gardens, and the horses required a staff of at least ten. Sara once said that her grandfather, a wealthy New England merchant, "was rather hoity toity. He had a large family and lived for himself and for them." She could have been speaking of her own father, her own family.

* * *

AT sixteen Sara was a slender, calm beauty, who wore her chestnut hair upswept and had lively dark eyes. Adding strength to beauty was a strong chin—the same jutting chin that, on the face of the thirty-second president, would be the hallmark of confidence, the jaunty chin that would be the delight of political cartoonists. An active girl, she loved riding in the spring, boating in summer, ice-skating and sledding in winter. She played the piano well, spoke German and French fluently, and had all the social graces. But it would be some years before she met the man who captured her heart and measured up to her standards, who fulfilled her ideal of a husband, an ideal based on her father. It seems, in retrospect, almost inevitable that the man she chose was old enough to *be* her father.

In no way did Sara feel that she was marrying "up" into the Roosevelt family. The Delanos predated the Roosevelts in America, and in Sara's view had an equally distinguished name. A great believer in heredity, she attributed Franklin's love of the sea and boats "directly to my own love of ships and distant horizons," which had come from her seafaring Delano forebears, owners of great sailing ships. She gave her son the Delano name, a public way of honoring her father (regretting that her brother had pre-empted the full name), and always insisted, "Franklin is a Delano, not a Roosevelt at all." Franklin's father, in the proud redundancy of the intertwined cousinhood of old-line families, had named his son by his first wife James Roosevelt Roosevelt.

The Delanos traced their line straight back to William the Conqueror. (One eager genealogist traced the Delanos back to the patrician Actii family of ancient Rome, which must have made even Sara skeptical.) The records are clear that in 1621 a French Huguenot named Philippe de la Noye (the name later elided into Delano) arrived in the Plymouth Colony in passionate pursuit of Priscilla Mullens, of "Speak for yourself, John Alden" legend. Philippe may have spoken for himself, but the answer he got was no. Still, his son Thomas married Priscilla's daughter—scandalously. On October 30, 1667, Thomas was fined ten pounds by the Plymouth Colony court "for having carnal copulation with his now wife before marriage." A Delano family historian with a sense of humor claimed this to be "the first shot-gun wedding in America."

The first of the Roosevelts, Claes Martenszen van Rosenvelt, left Holland with his wife (believed to be English) in the 1640s to settle in New Amsterdam, where he bought a farm near what is now Wall Street. After the Dutch ceded the colony to England, his son Nicholas anglicized the name, dropping the "van," a Dutch mark of status. The newly minted Roosevelt then ventured north on the Hudson River, prospered in the fur trade, and returned to Manhattan as a man of consequence. The Roosevelts and politics began with him: in the early 1700s he was elected alderman.

The Roosevelts then forked into two branches—the Oyster Bays and the Hyde Parks—and flourished. Franklin's great-great-grandfather Isaac, a wealthy Hudson River squire, staked his life and fortune on the cause of independence, served in the New York Provincial Congress, was one of the first state senators, and fought tirelessly for the adoption of the Constitution. For all of their material success, the Delanos couldn't match Isaac. The family called him, deservedly, Isaac the Patriot, and his portrait by

Gilbert Stuart hangs now, as in Sara's time, over the mantel in the library of the mansion in Hyde Park.

★   ★   ★

SARA and James embodied the values and attitudes of eastern, upper-class Americans of their era. Their life was much like that of other Hudson River patricians, the select who fill the pages of Edith Wharton novels. In the seasonal migrations of the rich, winter was spent in Manhattan, to join the social whirl and sample just enough culture; high summer found them in their "cottage" at Campobello, an island between Maine and Canada; spring and autumn drew them to Hyde Park; trips to Europe were sandwiched in between. But on the compass of Sara's heart, Springwood was her true home. It had been that from the day she moved there as a bride.

In this world that Wharton and Henry James portrayed from an insider's vantage point, a world that stretched from the New back to the Old, the Hudson River gentry occupied a special niche. The valley was dotted with sprawling estates that belonged to some of America's oldest and grandest names, alongside newer, more ornate houses built by a new breed of successful men—industrialists, railroad barons, shippers, financiers, the best of them nation-builders, the worst, ruthless scalawags—each outdoing the other in a show of magnificence fueled by new money and the need for social self-assurance. Lacking ancient lineage, they could purchase, at great cost, a heritage for themselves, works of art and entire rooms dismantled piece by piece from chateaus and castles in Europe. Just north of Hyde Park, Frederick Vanderbilt erected such an edifice, a fifty-four-room Renaissance palace in nineteenth-century America.

In 1898 this rampant manifestation of new money had even the members of the old, self-assured social order curious to have a look at the Vanderbilt establishment with its full-time staff of seventeen, every newfangled convenience, enough paneling and pictures for a museum and a dining table that could seat thirty guests at dinner, with a footman hovering behind every group of four silk-upholstered chairs. When the first invitation to dinner was hand-delivered to Springwood, Sara was ecstatic—the chance to see the gilded excesses firsthand. But by Roosevelt standards the Vanderbilts were nouveau, only two generations away from a farm on Staten Island. James, looking down his pedigreed nose, disabused his wife of any notion of attending. "If we accept," he explained patiently, "we shall have to have them to our house."

In comparison to the Vanderbilt showplace, the house at Hyde Park was human-scale, large though not overwhelming. But by the standards of ordinary Americans its inhabitants lived in a world apart. Even on a journey by train, the Roosevelts traveled in their own cocoon, thanks to James's grandfather, who had shown great prescience by investing early in railroads. His foresight became the main source of the Roosevelt money and led to lucrative and undemanding positions for James, and the use of a private rail-road car. Parents and son delighted in this perfect little home on wheels, its handsomely appointed sitting room and bedrooms, and its own cook/porter. The marvel is that the Roosevelt world, as enclosed as a snow-scene paperweight, could produce a president who would be the champion of those forced to spend their lives looking in from the outside.

★ ★ ★

IT was Thanksgiving of 1903. The Delano family was gathered at "The Homestead," Sara's grandfather's spacious house at Fairhaven, on the Massachusetts coast. During the afternoon Franklin, in his final year at Harvard, took his mother aside for a private moment. He had something important to tell her: he had fallen in love with Anna Eleanor Hall Roosevelt—his fifth cousin once removed, niece of President Theodore Roosevelt. He had asked her to marry him, and she had said yes. Sara, who thought she shared her son's life completely, was devastated: she knew nothing of this courtship.

Of course, she had known Eleanor from the time this rather pathetic girl, motherless from the age of eight, was a mere child. While Sara could easily understand how Eleanor, excruciatingly shy and certainly no beauty, would fall in love with her son, she found it quite impossible to understand why her handsome, charming Franklin had fallen in love with Eleanor. Of course, she belonged to a fine family with an aristocratic name, but her Franklin already had both. Worst of all, he was only twenty-one, far too young to think of marrying. Her father, she pointed out, had waited until he was thirty-three to marry, "when he had made a name and a place for himself, who had something to offer a woman."

Sara could not reveal the severity of the shock, even to her own diary, entering for that day only the terse line: "Franklin gave me quite a startling announcement."

On his return to Harvard, Franklin, hoping to cushion the blow, wrote his mother a loving and reassuring letter:

Dearest Mama,

I know what pain I must have caused you and you know I wouldn't do it if I could really have helped it—mais tu sais, me voila. That's all that could be said—I know my mind, have known it for a long time, and know that I could never think otherwise: Result: I am the happiest man just now in the world; likewise the luckiest—And for you, dear Mummy, you know that nothing can ever change what we have always been & always will be to each other—only now you have two children to love & to love you—and Eleanor as you know will always be a daughter to you in every true way. . . .

Eleanor, in the first of many rather saccharine letters, also anxiously tried to reassure her future mother-in-law:

Dear Cousin Sally,

I know just how you feel & how hard it must be, but I do so want you to learn to love me a little. You must know that I will always try to do what you wish for I have grown to love you very dearly during the past summer. It is impossible for me to tell you how I feel toward Franklin. I can only say that my one great wish is always to prove worthy of him. . . .

It was not going to be that easy. Apart from her possessive love, Sara had reason on her side. At twenty-one, Franklin was still a student with law school ahead of him, with no way to support a wife reared in affluence, and—seen in retrospect—the two were not well-matched. Why, then, did Franklin wait until after the fact to reveal to his mother this most important life decision, to her as well as him? Surely Sara deserved more consideration from the son who had been her life.

He faced a dilemma: his mind was made up, yet he knew all too well that his mother would object, because the inescapable fact was that there was no girl his mother would willingly accept, not yet, when he was still a boy, still her comfort and her life. He doubtless reasoned that announcing his news as a fait accompli would be a swift, surgical way to avoid a mutually lacerating confrontation. He almost certainly knew she would not readily accept his decision, and might never fully accept it.

The shock of her son's surprise that Thanksgiving was heightened by her impression that "he had never been in any sense a ladies' man." In his chatty letters he spoke lightly of girls he knew; he would have sensed that revealing any deeper feelings would not sit well with his mother. But a year earlier Franklin had fallen in love for the first time, with Alice Sohier, a popular young belle of seventeen, an episode he had also not shared with his mother. Alice's parents prudently sent their pretty daughter abroad and the romance was over: first love, first rejection.

When Eleanor returned to the Roosevelt circuit after several years of school in England, Franklin was attracted to this distant cousin's interesting mind, her concern for the poor, and her lustrous blue eyes. She in turn was awestruck that this golden boy would court her. He liked the fact that her uncle, his distant cousin, was the president he much admired—and would use as his model in his path to the White House.

Was there a deeper reason that led Franklin to this decision? He was moving through a necessary stage of separation from his mother, probably overwhelmed by her enveloping and possessive love. Now he was impelled to sever the silver cord, to take command of his life. Marriage would provide a measure of distance from control by the mother he so dearly loved, who, until now, had been the most important person in his life. Serious, self-deprecating Eleanor, the reluctant debutante whose glamorous mother, a brilliant figure in elite New York society, had cruelly tagged her "Granny," would never try to dominate him.

Ironically, if Franklin took this momentous step—even subconsciously—as a way of putting distance between himself and his mother, in reality he only extended her sway, subjecting his new wife to his mother's hegemony. At Fairhaven, Sara won one concession: Franklin and Eleanor agreed to keep the engagement secret for a year. This strongly suggests that Sara thought—hoped—that Franklin's ardor would fade, and an engagement that had not been announced could end without social awkwardness.

And what about Eleanor, a sheltered nineteen-year-old from a painfully dysfunctional family? "I never even thought that we were both rather young and inexperienced," she reflected in later life. When she told her grandmother that she wished to marry Franklin, "She asked me if I was sure I was really in love. I solemnly answered 'yes,' and yet I know now that it was years later before I understood what being in love was or what loving really meant." She explained no further. There is no answer, only speculation.

To smooth things over with his mother, Franklin suggested a Caribbean cruise. Sara was delighted; it might serve as a variation on the curative European tour, a long-established remedy for a daughter's ill-chosen attachment or broken heart. It would give Franklin, a young man with everything any girl might yearn for, time to rethink his drastic step. Franklin and his mother set sail, accompanied by his Harvard roommate.

The cruise remedy didn't work. Franklin returned home more eager than ever to marry Eleanor—after six weeks of his mother's constant company, the silver cord might well have felt like a bridle. In her memoirs Eleanor fairly crowed, "Franklin's feelings did not change." But Sara was not yet ready to give in. She took Franklin to Washington to meet an old family friend, the American ambassador to Great Britain, and boldly suggested that he take Franklin to London as his secretary. The diplomat could not oblige her—the job wasn't open and, anyway, Franklin was too young. Sara had run out of tools to derail, or at least defer, the engagement.

In a letter to Franklin, she made what was, for her, a rare admission of dejection and defeat: "I am feeling pretty blue. You are gone. The journey is over & I feel as if the time were not likely to come again when I shall take a trip with my dear boy . . . but I must try to be unselfish & of course dear child I do rejoice in your happiness, & shall not put any stones or straws even in the way of it. . . . Oh, how still the house is."

★ ★ ★

THE wedding date was set for March 17, 1905, chosen to accommodate Eleanor's Uncle Teddy, who wished to combine his role in giving the bride away with New York's politically important St. Patrick's Day parade. (Eleanor learned early that politicians' families come to accept such compromises.) Two hundred socially prominent guests attended the ceremony in the fashionable town house of her cousin Mrs. Henry Parish Jr., and scores more arrived for the reception that followed. The couple received 340 wedding gifts, their cards noted carefully by Sara. Eleanor was a lovely bride; features that had never been described as beautiful were transformed by rapture. Her gown of white satin and lace had the Gibson-girl shape, puffed at the shoulders and nipped at her narrow waist; her rose-point lace veil had been worn by her mother and her Grandmother Hall. Franklin gave his bride a gold pendant watch with her initials set in diamonds; Sara's gift was an impressive choker of pearls. The bridegroom (who was involved in such lively conversation that he was almost late at the altar) displayed his natural charm and buoyancy.

Every bride is the star at her own wedding, but at this particular ceremony the bride and groom were upstaged by the overwhelming presence of the president of the United States, who thoroughly enjoyed the attention. (His daughter, the tart-tongued Alice Roosevelt Longworth, quipped that her father wanted to be the bride at every wedding, the corpse at every funeral.) After the benediction was pronounced, TR flashed his Mount Rushmore–sized smile and welcomed the bridegroom into the Oyster Bay branch of the family: "Well, Franklin, there's nothing like keeping the name in the family."

The newlyweds spent a quiet week at Hyde Park—without Sara—then Franklin returned to his studies at Columbia Law School. In June the couple embarked on a three-month honeymoon in Europe, and by the time they returned to New York, Eleanor was pregnant. Sara had rented a town house for them near her own and furnished it completely, from cooking pots to parlor chairs, and hired three servants. Two years later Sara came up with the perfect Christmas present for the young couple—she was arranging to build them a town house of their own. What a generous gift! It was to be a six-story house on stylish East Sixty-fifth Street, an imposing thirty-five-feet-wide double dwelling, with a joint entry serving both sides. Sara, of course, would occupy the other half and in an architectural Siamese-twin arrangement, several rooms opened directly into both homes. In 1908 she presented it to them, completely decorated, according to her own taste.

Now and then Eleanor made a little effort on her own, only to be belittled or rejected. The home is the heart of any wife's life, and Sara's total control of it was too much for Eleanor. "I sat in front of my dressing table and wept," she wrote in her memoirs. "And when my bewildered young husband asked me what on earth was the matter, I said I did not like to live in a house which was not in any way mine, one that I had done nothing about and which did not represent the way I wanted to live. He thought I was quite mad and told me so gently, and said I would feel different in a little while and left me alone."

Did Franklin really not comprehend his wife's despair—or did he choose not to understand? Throughout his life he kept his eyes averted from the destructive tension festering between his wife and his mother. FDR biographer Arthur Schlesinger called it "jovial obliviousness," a euphemism for what many wives would call an insensitive, even spineless, lack of support for his beleaguered wife. Even in an age when women expected considerable deference from a daughter-in-law, Sara's naked assertion of territoriality and possessiveness was breathtaking. It would take a terrible toll on the marriage.

Despite the friction between his mother and his wife, there was a point at which Eleanor turned to her mother-in-law for help. In 1918 she made a shocking discovery: her husband was in love with her secretary, Lucy Mercer, charming and well-born, of Maryland's distinguished Carroll family. Her smile was radiant, her voice low and throaty, sensual attributes Eleanor sadly lacked. Confronting her husband, Eleanor offered him a divorce—if that was what he wished.

Sara did not take this development calmly. Eleanor might give up without a fight, but Sara would not. By all accounts there was a tense discussion with her son, in which she made it clear that the scandal of divorce would foreclose any thought of a political career and would poison his entire life. "Well," she wrote to him, "I hope that while I live, I may keep my 'old-fashioned' theories and that at least in my own family I may continue to feel that *home* is the best and happiest place and that my son and daughter and their children will live in peace and happiness and keep from the tarnish which seems to affect so many."

These were private family discussions, but there seems to be no question that Sara applied pressure: as keeper of the purse strings, she would cut off Franklin's funds and deny him his beloved Hyde Park home if he did not end the affair. For whichever reason, Franklin buckled. Eleanor complied, and shut her bedroom door, permanently—no sacrifice for her, as she had once confided to her daughter that "sex was an ordeal to be borne." On that subject, Sara agreed. Franklin obviously did not. (Though he promised to give her up, more than twenty-five years later Lucy was with him surreptitiously at his Georgia retreat when he died.)

Some years later Sara looked askance at another intimate relationship in her son's life, this one between Franklin and his devoted and beautiful secretary, Marguerite "Missy" LeHand. She shared his life completely, as part of the family, as hostess in Eleanor's frequent absences, partner on his outings, fun-loving companion over cocktails, relaxed in her robe in his bedroom. Particularly at his getaway cottage in Warm Springs, Missy was in effect his surrogate wife. The Roosevelt children—with the notable exception of son Elliott—insisted that there was no physical intimacy between the two. Since Eleanor didn't make an issue of this highly unusual arrangement, Sara couldn't. And besides, charming Missy made her boy so happy. Eleanor, in her remarkably candid memoirs, mused that her husband "might have been happier with a wife who was completely uncritical [which] I was never able to be. And so he had to find it in other people." Still resentful, she added, "I was one of those who served his purpose."

But he also served her purpose. As wife of the president, she could shine her light on the dark corners of poverty and racism. Her unflagging commitment to the have-nots brought both praise and criticism, and in time took her to the United Nations and worldwide recognition as a humanitarian.

★   ★   ★

"YOUR mother only bore you," Sara told her young grandson Jimmy. "I am more your mother than your mother is." It was a cruel statement, one that Sara did her utmost to fulfill. To Eleanor, Sara may have been an autocrat, a domestic despot, but to her grandchildren, the grown-up James said lovingly, "She was sort of a fairy godmother."

In effect, Sara co-opted her son's children—and neither Franklin nor Eleanor did much to stand in her way. The grandmother paid attention to them in a way that the mother did not. Sara sent them generous checks on their birthdays, took them on trips to Europe, showered them with expensive gifts. Often she flouted the parents' exercise of discipline—she gave the boys horses when the parents took away their pony; she gave teenage Franklin Jr. a new convertible after the parents refused to replace the car he wrecked. "Now what am I going to do?" the father lamented to a friend. This was the governor of New York speaking, the man who would stare down labor leaders and big business, control Congress, and take on Hitler. But he was unable to say "No!" to his mother.

Grandson Elliott confessed that the children "recognized the hostilities existing among the three grown-ups and we exploited the cold war to the hilt." In his view Sara "would never acknowledge that she continued to use her grandchildren as a weapon against Mother . . . she now chose to ignore her daughter-in-law's protests that we were being sadly spoiled."

"Granny's ace in the hole," James observed, "was the fact that she held the purse strings in the family. For years she squeezed us all—Father included—in this golden loop." That insight does much to explain FDR's lifelong subservience to his mother—purse strings entwined with apron strings.

In an article that was never published, Eleanor gave an unvarnished explanation of that internecine struggle: "She determined to bend the marriage to the way she wanted it to be. What she wanted was to hold onto Franklin and his children; she wanted them to grow as she wished. As it turned out, Franklin's children were more my mother-in-law's children than they were mine." So it seemed, when a submissive Eleanor stood by as Sara engaged a series of nannies—some of them nannies from hell—to take charge of the children. But Eleanor, admittedly, lacked a natural instinct for child-rearing. In one instance, to give little Anna more fresh air, Eleanor placed her in what seems best described as a cage, which was dangled outside a rear window of their New York town house. The experiment ended when a concerned neighbor, alarmed at the child's wailing, threatened to notify the Society for the Prevention of Cruelty to Children.

Sara didn't hesitate to belittle her daughter-in-law in front of guests—"If you'd just run a comb through your hair, dear, you'd look so much nicer"—turning Eleanor once again into an awkward, unloved little mouse. Though she was all too aware of what her mother-in-law was doing, Eleanor never openly challenged the autocratic matriarch, and as Sara constantly crossed the personal boundaries of both her son and his wife, Eleanor seemed to acquiesce and retreat. In contrast to Sara's possessive love for her son, Eleanor's emotional bonds with her own children became increasingly tenuous.

Her ultimate response to Sara's destructive intrusion into an always shaky marriage was to reinvent herself, to become the independent woman that the world remembers. Only in this very unusual marriage could she find the strength to stake out a place

for herself, to become her own person and find emotional fulfillment in other close relationships with intimate friends she could
trust and who recognized that the ugly duckling was indeed a
swan. She literally built on her new independence—a simple cottage that she named "Val-Kill." It was not far from Springwood,
but far enough to be completely apart, to be her own, where she
could entertain the many friends Sara disapproved of.

With Eleanor's self-liberation, the marriage evolved into a
working arrangement between a crippled husband and a wife who
walked out into the world, going to places he could not go, reporting to her husband what was going on in the unseen America. By
disengaging sufficiently from both her husband and his mother,
she saved her selfhood and her dignity. Eleanor found larger battles to fight, but she—and her children—would pay a high price
for retreating rather than directly confronting Franklin and Sara.

The conflict between mother-in-law and son's wife was exacer-
bated by Eleanor's transition into personhood. As cipher-wife,
producer of grandchildren, she was quite acceptable to Sara, but
the relationship became further strained when the submissive
daughter-in-law broke out of her cocoon, metamorphosing into
the new Eleanor. She came to realize that unless she took charge of
her life she would remain "a completely colorless echo of my husband and my mother-in-law and torn between them. I might have
stayed a weak character forever." As Eleanor the advocate raced
from project to cause to public meeting, a whirlwind of action,
Sara could only comment disapprovingly, "My generation did not
do those things."

The children, however, were defensive of "Granny." Elliott
pointed out his mother's ambivalence; in the early years Eleanor
"leaned on her mother-in-law for help, advice . . . she welcomed
many of the things, always intended as kindnesses, which Granny
insisted on doing for her." It was only later that Eleanor began to
resent her puppet role. She exonerated herself on grounds of inexperience—as though countless young women before and since had

not gone into marriage never having held a baby or fried a strip of bacon. In later life she reflected that it would have been better had she and Franklin not had servants as they began married life, which would have forced her to learn, to develop confidence. But Sara also had had servants all her life.

Eleanor attributed her children's unstable lives and multiple marriages to the fact that "they were not really rooted in any particular home," which put the blame, however obliquely, on Franklin. "My husband," she commented, "would never quite decide to make the break with his mother." He made no objection, apparently, that Sara had doors cut leading from her large bedroom through Eleanor's very small one, to give her direct access to Franklin's room, an unthinkable invasion of privacy. The mama's boy willingly made his wife second fiddle to his mother.

Yet, Eleanor acknowledged in her memoirs, in contrast to her own sense of alienation at Hyde Park, "the children loved the place, felt it belonged to them, and were completely at home there." Long after the children were grown and gone, on Sara's eightieth birthday they gave her a scroll inscribed, "Although we are now scattered . . . Hyde Park has been, and always will be, our real home, and Hyde Park means you and all the fun you gave us there." This must have pricked Eleanor's heart.

But as she traced everything that went awry in her life to Sara, Eleanor overlooked her own tendency, as a young wife, to retreat in sullen silence, resentful of her husband's magnetism, annoyed at his failure to back her up, frustrated by Sara's domination. Nor did it occur to her that the children might have been affected by what son Elliott termed "a marriage between two hopelessly incompatible people"—who lacked the will to escape the deceptively comfortable trap woven by Sara's manipulation of financial and emotional strings.

To the end Sara would not turn over to her son, outright, any portion of the fortune he would inherit, preferring to dole it out as a gift, as a means of control; thus her son, victor over paralysis and political foes, remained her hostage. This was a woman who loved to exercise power in her own way as much as her son did in his.

The one great gift Sara gave them outright was a summer place on Campobello, the small Canadian island between Maine and New Brunswick, where FDR spent so many happy summers as a boy. The sprawling shingle "cottage"—the colloquial term for a house of thirty-five rooms—had no electricity, but its big windows looked out to the water and sunlight flooded the house. The young Roosevelts loved Campobello, and though it was next door to Sara, there was breathing space between them. Eleanor at last felt that she had a home of her own, a place where she was indisputably Mother.

Franklin was relaxing in this happy and vigorous outdoor life when he was felled by polio in August 1921. At thirty-nine, life for Franklin Roosevelt, the golden boy of Hyde Park, could never be the same again. He fought its crippling legacy valiantly, drawing on every ounce of physical strength and newly discovered inner resources. This time Sara, in a reversal of her iron determination, wanted him to give in, to return to the quiet, solicitous life of Hyde Park. That way, she could lavish total attention on him. She would have him back, her boy once again depending on her.

But at this crossroads in their life, Eleanor finally stood up to Sara. She encouraged Franklin to keep fighting, to stay in public life. This tragedy forced both partners in this troubled marriage to redefine themselves and their destinies. Eleanor began her long journey from passive, put-upon wife to her place as humanitarian icon. She could reach out to people all over the world in a way that she found impossible with her own family.

★   ★   ★

ENTERING the voting booth in the village of Hyde Park on November 8, 1932, Sara Delano Roosevelt was stepping into political history: she would be the first American mother to cast her ballot for her son as president. In 1920, only three months after women at last had won full citizenship, she had voted for Franklin when he ran, unsuccessfully, for vice president on the Democratic ticket, but that could not begin to equal this exhilarating moment. And by midevening her pride was ratified by the results—a total triumph. In her self-possessed way, she remarked calmly, "I am never surprised by anything he does because he always succeeds. He has had a great triumph. Of course, he must expect to encounter opposition."

To all who asked the inevitable question—had she dreamed of her son as president?—Sara replied unequivocally, "Never, oh never! That was the last thing I should have imagined for him, or that he should be in public life of any sort. I know that traditionally every American mother believes her son will one day be President, but much as I love tradition, that is one to which I never happened to subscribe. What was my ambition for him? Very simple—to grow to be like his father, straight and honorable, just and kind, an outstanding American." Implicit in her firm answer was a question of her own: How could politics, even the presidency, compare to life in their personal paradise of Hyde Park?

Despite her arch dismissal of any suggestion that she had pictured Franklin as the second Roosevelt in the White House, Sara must have entertained such a thought, privately, after Governor James M. Cox of Ohio picked Franklin for his running mate in 1920. He was thirty-eight, the assistant secretary of the navy and a rising star in the Democratic Party, with a reputation for bucking New York City's Tammany Hall political machine. Congratulating her son on his nomination, Sara wrote, "If and when you are elected, you will belong to the nation." Though the ticket was trounced, Sara was confident that there were no limits on Franklin's future. One year later he was an invalid.

The presidential gleam had been in Franklin's eye long before James Cox looked his way. As a Harvard student, dining with friends, he had blithely remarked, "I think I might be president some day." "Who else thinks so?" someone parried, producing good-natured laughter all around. And in 1908 Franklin, then a young lawyer, confided to skeptical colleagues his plan to run for office—with the presidency as his long-range goal. His cousin Teddy had mapped out the way to do it: state legislature to governor to vice president to president. Which, with a shortcut, is what Franklin did. Perhaps his interest began in Washington in the winter of 1887, when his parents took him, not yet six, with them to call on their close friend President Grover Cleveland, in the White House. "My little man," the weary Cleveland sighed, "I am making a strange wish for you. It is that you may never be President of the United States." Franklin liked the big office; he ignored the advice.

In 1910 Franklin started, according to plan, at the modest level of the New York State Senate. When a local party leader asked him to run, he replied, "I'd like to talk to my mother about it first." The old pol rolled his eyes and explained to this rich young greenhorn that the boys back at party headquarters wouldn't much like that he had to ask his mama. Recovering from his faux pas, Franklin quickly said, "I'll do it." But the scion of Hyde Park had a lot to learn about grassroots politics: the pince-nez on a twenty-something was already an image problem, then he turned up at his first big meeting in riding togs. He kept the pince-nez, put away the twills, didn't mention his mother, and won.

Politics literally came home to Sara in the 1920 campaign. "Granny got her first real taste of the democratic process in action," her grandson Elliott recounted in his book about the family, *An Untold Story*. "Eight thousand people trampled the lawns at Hyde Park and five hundred more crowded into her house." Sara's sanctuary had never been subjected to such intrusion—but it was worth trading a day's calm for the place of honor next to her son, to see this throng cheering him. (Eleanor meekly stood off to one side.) Still, Sara would go just so far in this unseemly business of politics; when photographers asked her to kiss her son the candidate "one more time," she starchily refused. She contributed three thousand dollars to the Cox–Roosevelt ticket, which lost by seven million votes, setting a negative record for the time, but in the years ahead Franklin was to deliver a string of victory speeches, with his mother at his side, from Springwood's broad veranda.

By the time he ran for governor of New York in 1928—against all advice—Sara had become interested in the actual vulgar details of politics, in spite of herself. After midnight on that election night, as dispiriting returns piled up, everybody accepted defeat and went to bed—except the candidate's mother. It was 4 A.M. when the last votes were tallied and only Sara, along with a few election-night workers, was still there to cheer the upset election of Franklin Delano Roosevelt as governor, the underdog who carried New York against a Republican tide. Dawn was breaking when she retired to her bedroom, jubilant.

She found that she enjoyed the stream of political visitors to her house. Frances Perkins, whom FDR appointed secretary of labor, the first woman cabinet member, noted that Sara was "on friendly terms with people the like of whom she had never met." With genuine interest she asked about their families and homes, "which endeared her to them. . . . She remained completely maternal in her attitude." Sara's simple explanation: "I have always believed that a mother should be friends with her children's friends."

Much had occurred in their lives by the time the Democratic convention in Chicago nominated Franklin for president in 1932. "I suppose I should have been mightily thrilled over his nomination," she admitted. "But I cannot, in all honesty, pretend to having felt any elation at the prospect of Franklin's being chosen to run. The presidency, I have always felt to be a most taxing and trying occupation. None the less, I did sit up the better part of the night waiting to hear the verdict over the radio, for realizing how much it meant to Franklin, I could not help but be somewhat anxious." Note her word "verdict" rather than "victory." Sara looked on his choice of a career in politics not as a desire for power but as fulfilling his obligation to the American people, an exercise in her philosophy of noblesse oblige.

Reluctant as she was about his running for president, once he was in the fray, she felt he must win. In her elaborate language she wrote, "While I had not been enthusiastic over Franklin's seeking the nomination, once he became the candidate I felt it to be a great point that he be elected." As the campaign got under way, she found herself caught up in the excitement, aware of how much of her son she would be missing "had I not been able to attend the big eastern rallies, and I wouldn't for anything have given up my visits to Headquarters where I actually grew to know every eager face that was raised to me in welcome. . . . On that memorable Election Night, instead of remaining quietly in a chair receiving congratulations, as I suppose I should have, I fluttered up and down the crowded corridors part and parcel of the excitement engendered by Franklin's victory."

Sara was climbing yet another ladder to share an important part of her son's life. On the night of his victory, the mother who had shaped his life had "but one devout wish—a wish that amounted to a prayer—that as President of the United States he be a good one!" In 1936 the people had decided that indeed he was "a good one" and gave him a triumph like nothing the country had ever seen. At Hyde Park Sara and a bevy of grandchildren once again thrilled to the spectacle of a brass band and torches and flares and cheering throngs encircling the house. Politics can be heady stuff.

★   ★   ★

EVEN with Franklin's dazzling success, Sara continued to feel that politics attracted venal and crude men, not the sort who should be her son's peers, and made little effort to conceal that view. Shortly after he was nominated in 1932, FDR invited the crass, demagogic, and politically important Senator Huey Long of Louisiana to a make-friends luncheon at Hyde Park. Long ("the messiah of the rednecks," in the phrase of historian Arthur Schlesinger Jr.)

arrived wearing what Roosevelt's longtime secretary Grace Tully described as a "loud suit, orchid-colored shirt and watermelon-pink tie" and was seated on FDR's right. During a lull in Long's free-flowing opinions, Sara hissed in a voice that, Grace said, "could be heard out on the Post Road," "Who is that *awful* man sitting next to my son?" As the other guests chattered nervously to cover the embarrassment, Grace noted that Sara "had the expression of a willful child who had been caught raiding the jam pot, and felt guilty but satisfied."

Few presidents have stirred such extremes of love and hate as Franklin Roosevelt. His most vicious critics were the rich and conservative elite who branded him a traitor to his class. Not least among his detractors were the Oyster Bay Roosevelts, who deplored Franklin as a maverick, a usurper. The charmingly venomous Alice Roosevelt Longworth, daughter of *their* President Roosevelt and widow of a Speaker of the House, snapped, "Franklin is 80 per cent mush and 20 per cent Eleanor." Asked what provoked this family warfare, Sara sniffed, "I can't imagine, unless it's because we're better looking than they are."

When her son was the target of vituperation, Sara was as thin-skinned as any mother. "I have not minded enemies standing up for their principles," she declared in the 1936 campaign, "but I do mind the lies and *inventions* of jealous people. I have always been my son's severest critic, but I really do not see how people can say bad things about him." Severest critic? A politician should wish for such criticism.

Her reward was the role of Mother of the President. She looked the part with her queenly bearing, gold lorgnette, and perfectly coiffed white hair, and she played it to the hilt. She loved her extended stays in the suite called the queen's room and reveled in taking the ever-traveling Eleanor's place as hostess. Her special ally was Assistant Chief Usher J. B. West (White House parlance for manager), who shared Sara's raised-eyebrow appraisal of the First Lady. In cozy chats with her coconspirator, Sara mimicked

Eleanor's high-pitched voice and derided her unconventional con-
fidantes, most of them social activists of leftist persuasion, women
with unsettling attitudes and mannish suits. "Where *does* Eleanor
get all those people?" she grumbled. Sara was free with suggestions on running the Executive Mansion, but when she recom-
mended replacing the black butlers with whites, an indignant
Eleanor ruled that Sara could run Hyde Park, but the First Lady's
principles would be maintained in the White House.

Together the president's mother and West tried to do something,
anything, to improve the truly awful White House food her
daughter-in-law allowed to be served even to their most distinguished guests. Celebrated writer Ernest Hemingway described
the dinner he attended in 1937: "The worst I've ever eaten. We
had rainwater soup followed by rubber squab, a nice wilted salad
and a cake some admirer had sent in. An enthusiastic but unskilled
admirer." FDR adviser Harold Ickes, a gourmet, grumbled, "On
only one other occasion have I ever tasted worse champagne."

The problem was the housekeeper, a dragon named Henrietta
Nesbitt, whom Eleanor had brought down from Hyde Park, a
domestic tyrant who instructed the cook to stick to the prescribed
menus regardless of the president's requests, even when he was ill.
Franklin, the nation's most powerful figure, could not get rid of
the employee he thoroughly despised. In his third term FDR
declared to daughter Anna and an astonished Grace Tully, "I
really want to be elected to a fourth term." Big news, thought
Grace, he's made up his mind! Then he grinned mischievously. "I
want a fourth term so I can fire Mrs. Nesbitt." Instead, he hired
his own upstairs cook. (Ultimately, Bess Truman slew the dragon,
after she barred the First Lady from getting late-night snacks for
Harry.)

Franklin himself contributed to the sorry state of White House
meals with his constant worry about costs. He complained to
Eleanor that luncheon guests were being served two eggs each
when he had insisted on only one, and at an informal dinner Sec-

retary Perkins was amazed to see the president carve seventeen slices from one side of a small turkey. He also tried to ban second helpings, noting that "nobody takes a second help—except occasionally when I do." There was an obvious message in this observation that the cost-conscious president missed. But who goes to the White House for the food?

★ ★ ★

IN the family quarters of the White House, the staff privately called the president's mother "the queen." She wouldn't have minded the impertinent backstairs title—that was rather the way she thought about herself. "I don't think that she felt anyone was her social equal, except maybe the Queen," chuckled a family friend, "and she wasn't sure about that."

Not surprisingly, she was a royalist at heart, and her son, Arthur Schlesinger wrote, "had an odd weakness for European titles, however seedy." Sara's awe began, perhaps, in Paris in 1866, when twelve-year-old Sallie watched Empress Eugénie drive by in her royal coach. In her early years at Hyde Park she could almost feel the touch of monarchy when she and James would glide through snowy lanes in their red velvet–lined sleigh, which had been a gift from Czar Alexander II to Napoléon III.

As war broke out in Europe, displaced royalty took refuge in America, often visiting Hyde Park. When Crown Princess Martha of Norway was a guest in September 1940, Franklin, ever hospitable, invited the comely princess and her children to stay at the White House while getting settled in Washington. She accepted with alacrity and moved into the Rose Suite, where weeks stretched into months. The president didn't mind at all. He found her delightful, and the gossips made the most of this warm friendship. When the princess at last rented an estate in Washington, the president would personally pick her up for dinner at the White House. His son James surmised in retrospect, "There is a real possibility that a true romantic relationship developed between the president and the princess." Sara most certainly did not intend the contacts with royalty to be that close.

In May 1939 her ultimate fantasy was realized when the grandest of Europe's "face cards," the British monarch, King George VI, and his charming Queen Elizabeth (who in time became Britain's beloved Queen Mother) came for a weekend. Sara set out to show them how American aristocrats—the James Roosevelts—lived, reflecting the manner and manor of their British counterparts.

She was scandalized when Eleanor and Franklin decided to entertain the royal couple at an old-fashioned American picnic, replete with hot dogs and baked beans, strawberry shortcake, and tribal music by two American Indians, nothing at all like the champagne flutes and smoked salmon that the English deploy at the Derby or the Henley Regatta.

For their get-acquainted meeting in Springwood's handsome library, Franklin had a small bar set up, which brought a frown of disapproval from Sara. "The King," she stated with the assurance of one who knew the personal habits of royalty, "would prefer tea." Welcoming the royal guests, the president confided to His Majesty, "My mother does not approve of cocktails and thinks you should have a cup of tea." "Neither does my mother," replied George VI conspiratorially, and happily accepted a whiskey.

Sara's formal dinner did not go smoothly: first, her English butler abruptly took leave rather than work with black butlers from the White House, and then in mid-dinner a serving table collapsed with a crash, demolishing a stack of her fine Limoges china. "Mama tried in the best-bred tradition to ignore it," Eleanor reported in her newspaper column, "My Day." "Just exactly what happened to our well-trained White House butlers that night I shall never know. After dinner the butler carrying a tray laden with decanters, glasses and bowls of ice fell and slid right into the library, scattering the contents of the tray and leaving a large lake of water and ice cubes at the bottom of the steps."

To Eleanor the social disaster was "really funny," but to Sara, the embarrassed hostess, it was not at all a laughing matter. "My mother-in-law," Eleanor said later, "was very indignant with me for telling the world about it and not keeping it a deep, dark family secret." A hint of schadenfreude could be detected in her recounting the grande dame's discomfiture.

* * *

NO single portrait could capture the full measure of Sara Roosevelt; a tapestry of great size and rich colors would be required. Evaluating her grandmother's impact, Anna Roosevelt Boettinger conceded that Sara "was a martinet," but insisted that her positive influence on Franklin made up for the flaw. In Anna's view, Franklin Roosevelt, as president, "stood for a self-assurance of a past . . . he spoke from security." His inbred confidence and optimism lifted the spirits of a frightened nation that had lost its way—and, said Anna, it was Sara who imbued that confidence in him from the cradle. She shaped the man who would rally America. So, Anna was asked, Sara actually had an influence not just on her son but on the country? The granddaughter would go even further: "Oh, she helped *save* it. I don't think anyone's given credit to Granny—the credit that she really is due."

That credit, even taking into account a loving granddaughter's bias, was all the more remarkable for a woman of her background. Sara, born in 1854, during Queen Victoria's long reign, was never quite sure that she approved of the twentieth century as it unfolded in her grandchildren's era. Sara's values and attitudes reflected the emphasis on propriety and strict rectitude that the widowed queen imposed as standards for her court, her subjects, and her era. In matters great and small, ranging from respect for authority to a concern for good handwriting, the mistress of Springwood inculcated the values of her fading century in a son who was racing into the next.

Unlike Eleanor, Sara had no political agenda and, though she had strong opinions, rarely attempted to influence her son's policies—many of which were, in her judgment, much too liberal. (Her father had declared that until he met James Roosevelt he did not know a gentleman could be a Democrat.)

Now and again she passed along opinions: "My friends don't think you should recognize Russia" or, as Grace Tully remembered, "Franklin, darling, why is everyone opposed to so much of your program?" The president would smile benignly and say, "I think I know who you've been talking with," and let it go at that. Her ideological misgivings were more than compensated by her praise—after an early speech he gave to Congress, she wrote him, diplomatically avoiding content, "I am very proud of the way you keep your head."

But in matters of style and conduct she didn't hold back. No detail escaped her. When Franklin was appointed assistant secretary of the navy in 1913, then a major post, Sara offered a specific instruction: "My Dearest Franklin, Try not to write your signature too small as it gets a cramped look. So many public figures have such awful signatures and so unreadable." Thus his mother's influence is written with a flourish—in a signature of equal parts force and equanimity—on thousands of documents that are markers of history.

In her old age, Sara ruminated in a letter to her son in the White House, "I have been an unusually fortunate woman. First I had the love and protection of your grandfather, then of your father, and in my old age you have made possible for me the interesting life I am now leading." She took no credit for having dedicated her life to each of them, in succession—adoring daughter to Warren, caring wife to James, and totally devoted mother to Franklin.

In May 1941 she delivered a Mother's Day address on the radio, her one regular public performance, then was off to Toronto to lend her prestige to a favorite charity. The following month, in Campobello, she suffered a slight stroke, yet refused to engage a private nurse until her son begged her to do it for his "peace of mind." At summer's end a fading Sara returned to her beloved

Springwood. She rallied briefly and put on her prettiest bed jacket and a bright blue ribbon in her hair when her son came from Washington to be at her bedside. On September 7, two weeks before her eighty-seventh birthday, she breathed her last. Eleanor later stated, coolly, candidly, that as she looked at the woman who had been so much a part of her life, she felt no emotion: "It is dreadful to have lived so close to someone for thirty-six years and feel no deep affection or sense of loss."

The funeral was conducted in the Springwood library, where she had never given up her place of primacy at the fireside, and she was buried in the graveyard of St. James's Church beside James, who had waited for her there for forty-one years. The anguished president stood beside his car and never looked toward the grave.

Later, going through her many boxes, he came upon his lace-trimmed christening gown, neatly tied bundles of letters, hundreds of them, from schoolboy and college student, a few little toys, a number of gifts he had chosen with greatest care for Mama—and a lock of those blond curls she had so reluctantly cut. Until that moment he had kept his sorrow within himself, as Sara had taught him. Now, coming upon her cherished remembrances from the idyllic boyhood she had given him, the president gave way to his grief.

After Sara's death, he resisted any changes in the house, except one: his mother had asked that the room in which he was born be restored to the way it was on that most wonderful day of her life. This he did.

*　　*　　*

A few minutes after Sara quietly slipped away, a great noise was heard outside. Springwood's grandest, oldest oak tree had crashed to the ground—it had simply lived out its time.

2

BE A GOOD BOY, HARRY

Mothers' expectations for their sons are thought to be different from all other earthly ones. As many men see it, life itself is a Sisyphean task of trying to meet their expectations.

—LINDA R. FORCEY, *MOTHERS OF SONS*

MARTHA Young Truman settled herself in a comfortable chair to enjoy the impromptu concert for an audience of one. Her son was playing well that day, and if his technique was a bit rusty, that was of little importance considering the setting: the piano was in his personal office in the family quarters of the White House. It must have seemed like a dream to her. But there on the wall was her portrait—a famous New England artist had painted her in her best navy blue with the white jabot she liked—hanging right where Harry would see it whenever he glanced up from his work. It was only a month since the job and its wartime responsibilities had, as he put it, made him feel "like the moon, the stars, and all the planets had fallen on me." Amid national mourning, Harry Truman, his wife, Bess, and daughter, Margaret, had moved into the mansion only a week before, and already so much had happened. He had sent his big presidential plane, dubbed "The Sacred Cow," to fetch her and his sister, Mary, for their first visit to Washington, Martha's first time on an airplane in her ninety-two years. But so much was a first for her these days.

The week before, back in Grandview, Missouri, Martha stopped packing for her trip to sit close by her radio. Harry was announcing the news the country had waited so long to hear: the end of World War II in Europe, victory over the Nazis, the first step toward final triumph in history's most devastating war. For all of America's years to come, May 8, 1945, V-E Day, would still be celebrated. And added to the national joy was his mother's private pleasure that this triumph had occurred on her son's sixty-first birthday. Now, as he played her favorite Chopin waltz in A-flat and Beethoven's haunting Sonata Pathétique, Martha Truman may well have been remembering that earlier May 8, in 1884, when Harry was born on a pleasant spring day in Lamar, Missouri, a state where the horrors of an earlier conflict, the Civil War, were still living memories. She and her husband, John, had set up housekeeping there when they married—John had a hankering to be a mule trader, and Lamar seemed a good place to start.

It is a strange feeling to look at a wedding picture from so long ago, when you know the end of the young couple's story. On their wedding day, December 28, 1881, Martha—"Mattie"—and John Truman posed for posterity. John, in a new frock coat, was seated in an armchair, to obfuscate the fact that at five-feet-four he was a good two inches shorter than his bride. By the measure of Jackson County, he was also a good cut short of her social standing. His face was agreeable enough, but unmemorable. Martha, standing with her hand resting lightly on his shoulder, was twenty-nine, considerably older than most brides of her time. Her hair, in Victorian fashion, was parted in the middle and drawn back severely into a bun. Her wedding dress was elegant—black taffeta ornamented with elaborate handwork and jet beading and topped by a wide puritan-style white lace collar, perhaps worn by her mother as a bride back in Kentucky. The dress and her matched bracelets suggested that she was of a prosperous family, which indeed the Youngs were. But it is her face that won't let you go: she looks

straight through that primitive lens, directly at you. Her gaze is
level, the blue eyes are interested in you, a smile flickers at the corners of her.mouth. Martha Ellen Young Truman was stepping into
her new life with easy confidence.

In Lamar they bought a little house for $685—four tiny rooms
downstairs, two under the high-pitched roof. Measuring twenty
by twenty-eight feet, it would fit into the spacious room she was
sitting in, on the second floor of the White House. When the baby
was born, healthy and hollering, it was some recompense for the
sadness of a stillborn son the year before. John marked the event
by planting a pine tree in the front yard. They called him Harry, a
good plain name that was short for Martha's brother Harrison.
When it came to a middle name, they couldn't decide which
grandfather to honor—Anderson Shippe Truman (who lived with
them from the time they were married) or Solomon Young. So
they settled on simply "S." This Missouri compromise would
cause considerable confusion in later years, but allowed Martha to
know, in her heart, that the S was for her beloved papa.

Within a year, hoping to expand John's tenuous business of selling whatever had four feet, they moved to nearby Cass County.
Martha gave birth to another son, whom they named John—no
disagreement there—with Vivian (for a Confederate cavalry officer) as the middle name he would always use. Before two years
had passed, Solomon, getting on in years, asked his son-in-law to
come back to the Youngs' fine homeplace at Grandview and help
run his large farm. For John, who was thirty-six by then and
hadn't made a go of his livestock trading, Solomon's request came
as a godsend—and maybe that's what Solomon had in mind, to
bring Mattie, his favorite child, back under his protection. For
Martha, no decision was needed. She would have run all the way
to Grandview. Back to her sprawling family, to her feisty red-
haired mother, and her awesome papa. Back to the big two-parlor
house that crowned Blue Ridge, with its view of just about the
whole world—they could see Kansas City, more than fifteen miles

to the north, and on a sparkling day (so Harry would say) they could see clear to Lawrence, Kansas, forty miles in the distance. Best of all, she could share the home she loved with her husband and their two small sons.

When Harry was five, a baby sister, Mary Jane, arrived. Though she came as a surprise to the little boys, there was not an ounce of sibling rivalry. Harry adored her. Like a little mother, he would braid her pretty hair and sing to her and rock her to sleep, the two of them in a big wicker rocking chair. His tender concern never ceased, even when he was president and Mary constantly turned to him for help and advice about the most petty problems of her routine life. "I sometimes think that's why I never did get married," Mary Jane concluded in later life. "I just never met anybody who was as nice as Harry."

Harry's early childhood on the farm and his extended family, with three generations and three children, gave him a lifetime supply of happy and consoling memories. "I had just the happiest childhood that could ever be imagined," he reminisced in his memoirs. It was a cousinly world, loving and secure. The boys would roam free over the five hundred acres, searching for birds' nests, catching tadpoles in the brook that ran through the bluegrass pasture, losing themselves in fields of towering corn. There was the barn to explore, the long porch for a racetrack, and mud holes to romp in. (He never forgot the spanking Martha gave him as the ringleader in a grand mud escapade.) Active imaginations invented games and built things, and in those pre-TV days when children relied on themselves for entertainment, Harry could never remember being bored. Out there on the prairie little Harry Truman romped in a life as sunny and secure as that of young Franklin Roosevelt's world at Hyde Park and as solidly Middle American as Dwight Eisenhower's Kansas boyhood.

Life on the farm can be austere, but not on Martha's farm. Resilient and buoyant, she brought up her boys to see the fun side of things, as well as respect the eternal verities. She would toss two-year-old Harry out of a second-story window into the arms of big, strong Uncle Harrison, a drop of probably no more than three feet, but to a toddler it was a great, squealy game, one that most mothers would have frowned on. Cousin Fred Truman remembered how Aunt Martha cleverly turned her hands, by lamplight, into animal shapes silhouetted against a white wall like their own puppet show. And she was always ready to play lively tunes on the piano and teach them songs. She made their life merry—and used her considerable intelligence and imagination to stimulate theirs.

The fun was in counterpoint to the ceaseless work of the farm, where even the small children had chores, which Harry would call the source of his relentless work ethic: "We went to bed early, got up early, and were always busy." That's not to suggest that it was all work, not at all. There were picnics and church socials and, when the setting sun called it a day, parlor dances. Harry treasured those years as an idyllic life, the kind of wholesome American lifestyle that Broadway musicals and Hollywood mythologized in the 1950s, while ignoring the harsher realities of running a family farm.

Martha was the hub of this large family's life and work, a woman determined never to forget her pioneer-parents' example of hard work and hardy spirits. The Youngs were resourceful people who had carved out a good life on America's advancing frontier and would always be the yardstick Martha used to measure her own life. There was never a time when she failed to measure up.

★   ★   ★

THE Fourth of July was always a big event for the Trumans and the Youngs. They would gather at the farm, thump the ripest watermelons in the patch, and hand-crank peach ice cream. The ladies brought their special deviled eggs and what everybody said were the best pies in the county. At the end of the festive midsummer celebration, when it was about dark, they would eagerly await the fireworks. As the children jumped and squealed at the boom and sizzle, Martha noticed that when everybody looked up and oohed and aahed as the multicolor fireworks arced into the sky in spectacular streaks, her Harry lost interest. Other such moments stuck in her mind, like the times he couldn't see the horses at the far end of the pasture.

The next day she took action. On her own—her husband was away cattle trading—she hitched two horses to the farm wagon, settled Harry alongside her, slapped the reins smartly across the broad flanks, and was off to Kansas City, a good fifteen miles away. She wanted a specialist to check her six-year-old's eyes—right away. (The question arises: since this was not an emergency, why not wait until John returned? Did she act to forestall grumbles that she was fussing over the boy too much?) The doctor diagnosed "flat eyeballs"—hypertropia was the medical term. He was given glasses that were thick and expensive, and they worked. On that day the course of his life was changed, thanks to the timely intervention and decisiveness of his mother.

Martha, perceptive and wise, understood what was at stake with Harry's poor vision: his ability to read, to learn, to get ahead. "She was very smart to realize that his eyesight was not good," said granddaughter Margaret, grateful that correcting his sight opened up the wider world to her father and proud that the boy who became a voracious reader of stories of great men and events grew up to be as famous and powerful as the heroes he had read about. "Those expensive glasses cut him off from the rough and tumble boys' world," Margaret said, "but they gave him a vision that stretched back across history and forward to tomorrow." Years later, Dean Acheson, the erudite secretary of state, observed that Truman, unlike presidents who want important issues boiled down to a page, "read the documents themselves and acted on them. It was the habit of reading, and of understanding, that followed all through his life, from his boyhood on."

"The habit of reading" came straight from his mother. As Harry neared school age in 1890, Martha decreed that the family move to Independence, about fifteen miles away, for its better schools. It was a pleasant courthouse town of about six thousand people, with tree-lined streets and imposing Victorian houses, a southern kind of town, not at all like booming Kansas City with its Yankee background and its boasts that everything was up to date. The Trumans bought a nice house set on three acres, a miniature farm in town that allowed John to get back into cattle trading and raise vegetables for the family. At one time he kept five hundred goats, not an asset calculated to endear him to neighbors on South Crysler Street, but the kids loved to hitch up a pair to a cart and trundle through the neighborhood.

The house, with its acres and barn and Martha's welcoming presence, was the headquarters for a congenial young gang. Since his eyesight kept Harry on the sidelines, the others sometimes had him umpire their baseball games (perhaps those thick glasses were a pretext for claiming "the ump is blind" when Harry made a disputed call). In later life Harry would always insist that he had not been popular. "The popular boys were the ones who were good at games and had big, tight fists. I was never like that," he told a group of youngsters visiting his presidential library. "Without my glasses I was blind as a bat, and to tell the truth, I was kind of a sissy." Easy to say when the world had seen tough Harry Truman in action, but he probably reassured the anxious schoolboy who asked the question about the president's younger years.

The comfortable Truman home, Martha's world, was filled with books—the complete works of Shakespeare, a red-bound copy of Plutarch's *Lives*, and a four-volume set of *Great Men and Famous Women* that she gave Harry when he was ten. She shared her love of Tennyson with him, helping him memorize long verses. (Her inscribed volume of the poet, which he treasured, is now in the Truman library.) "His mother would buy anything that they could enjoy, that would improve them," said Cousin Ethel Noland. "And that boy read!" Before Harry was five, Martha had used her large-print Bible to teach him to read, and with the help of his glasses he never stopped. "By the time I was thirteen or fourteen years old, I had read all the books in the Independence Public Library and our big old Bible three times through." He added laconically, "I suppose considering the fact that I became President of the United States, it wasn't time wasted."

But in those days wearing thick-lens, wire-rimmed glasses "was hard on a boy," Harry reflected. "It makes him lonely and gives him an inferiority complex." So he kept his nose buried in a book and lost himself in the grand stories of heroes and history. As a schoolboy he was diligent, responsive, the natural teacher's pet—but like the other modern presidents in their early years, not an exceptional student. At home, he was the attentive son, eager to please his mother, who loved all of her children, but Harry was always special.

Before Harry was able to reach the pedals of the Trumans' big Kimball upright, Martha, an able pianist whose natural "ear" was honed at the Baptist Female College, began teaching him to play little pieces. From the first time his mother showed him middle C, young Harry discovered the magic of music. In the Truman household, the piano was more than a piece of furniture in the parlor; it was at the heart of home life. With family and friends gathered around, Martha played and everybody sang current favorites and Stephen Foster classics. The children's request was always "Little Brown Jug," a lively jig that she played and sang with great gusto. John Truman was such an enthusiastic singer, said his daughter, "you'd always know where he was, out in the barn or in the field, because you'd hear him singing," As a result of her college training, Martha's accomplishment went beyond parlor songfests to classical music, a love shared by Mary (for many years the church pianist) as well as Harry. As for Vivian—who was so different from his brother in every respect—Harry chuckled that "Mamma couldn't get a lasso long enough" to get him to piano lessons. Music-making extended into the third generation in Margaret Truman, who made headlines as a concert singer, in part because she was the president's daughter, though her talent was not inconsiderable.

Spotting talent in Harry, Martha was determined to develop it, just as she had led him to literature. She signed him up with the music teacher next door, who used a new "shorthand" method for reading music—numbers representing notes—and in no time Harry was playing "Return of Spring," a favorite of Martha's, and Mendelssohn's "Songs Without Words." Such was his skill that the company marketing the numbers system featured him in an advertisement: "He will play at sight any piece set before him, seemingly as well as if had been drilled." That was catnip for Martha, who then placed Harry with a Vienna-trained professional in Kansas City, Mrs. Edwin C. White. Each morning he was up at 5 A.M. to practice for two hours (what a tolerant family!) and twice a week he made the long streetcar ride to his lessons. His performance in recitals—which terrify most pupils—won much praise, especially from Martha. By then he was playing work that required advanced technique and emotional range— Mozart, Chopin, and Beethoven.

A moment of fantasy became real for Harry in Kansas City's Shubert Hall, where Paderewski, the most famous pianist in the world at that time, was performing. Mrs. White wangled an after-concert backstage meeting with the maestro and her star pupil, who, she explained, was having trouble with a technically difficult passage in the maestro's own Minuet in G. Paderewski sat down at the piano with the near-paralyzed, jelly-fingered fifteen-year-old and showed him just how to do it. Martha, not easily intimidated, was indeed awed by this—Paderewski himself instructed her son! Surely this was an omen that a concert career lay ahead for Harry. She was beginning to see Harry as a professional musician, playing encores, taking bows, and making a living at something far more rewarding than farming. Sixty-one years later, at a White House dinner given in his honor by President and Mrs. Kennedy, the aging former president sat down at the grand piano in the East

Room and once again played Paderewski's Minuet in G from memory—negotiating the tricky part just the way the legendary pianist had showed him.

★ ★ ★

FOR the Trumans, the first eleven years in Independence were secure and untroubled. John Truman, an undistinguished man with the volatile temper that often accompanies a sense of failure ("He would fight like a buzzsaw," said his son, with a touch of admiration), was at last doing quite well. Martha was bringing up the family at the level she had been born to. An avid Democrat, John had also gained a foothold in local politics, which led to some minor appointments. Through those contacts Harry, at sixteen, got his first taste of big-time politics—the 1900 Democratic Convention in Kansas City, where perennial candidate William Jennings Bryan was at his silver-tongued best and Harry got to run errands for his father's important friends.

Everywhere John looked, men were getting rich by speculating in commodities and, determined to be one of them, he began rashly investing in wheat futures. In 1901 his modest luck turned. "He got the notion he could get rich," Harry said ruefully. "Instead, he lost everything at one fell swoop and went completely broke." Savings, home, all gone—along with any hope of college for Harry. That was a withering blow for Martha as well as for Harry; to her, education was the touchstone for a full life. Harry, entranced by military history, would have loved to attend West Point. (As a child he used to camp out in the yard, pretending he was a soldier, and as a student he built a scale-model of Caesar's bridge over the Rhine.) But his bad eyes ruled that out, and so he could only look for work.

On his deathbed in 1914 Harry's father whispered to his son, "I've been a failure," a lacerating self-judgment that Harry would always stoutly deny. Despite his financial setbacks, John Truman was respected as an honest man whose word was his bond. On the day of his funeral (long before Harry had become a personage) Grandview schools were closed in tribute, and the *Independence Examiner* declared him "An Upright Citizen Whose Death Will Be a Blow to His Community." He never made money, but to those who knew him he was not a failure.

High-school diploma in hand, Harry landed the job as a clerk in a Kansas City bank. In his same boardinghouse was a fellow from Abilene, Kansas, Arthur Eisenhower, who went on to become a vice president of the bank. He had a brother called Ike, whose only goal at the time was to shine on the Abilene High School football field. The young athlete would later figure significantly in Harry's life.

Harry took every advantage of cosmopolitan Kansas City's rich menu of cultural offerings, but best of all, K.C. was only a trolley ride away from Independence, and the girl of his dreams, Bess Wallace. She had golden hair and blue eyes, was a slugger on the baseball diamond and the best fencer in town, and could whistle through her teeth. And she loved books, too. From the moment he laid eyes on her in the Presbyterian Sunday school, when they were six, Harry was in love, and for the rest of his life he never looked at another girl. And even though Bess had paid scant attention to him as they grew up, he never let that discourage him.

★   ★   ★

BESS Wallace came from a proud family in Independence, but the Wallace family's story could not match Martha's. Few could. The basic facts—in 1852 Martha Ellen Young was born in Jackson County, Missouri, married a farmer named Truman, had three children, and died there—miss the real story. Martha's life was entwined with the story of America, from pioneer days to the first thrust of the space age; it spanned the greatest era of change the world has ever witnessed, compressed into just one hundred years. Her ninety-four years were bracketed by war—as a child she suffered through the Civil War, in her midyears she saw her son off to fight in World War I, at the golden close she celebrated victory in World War II with her son, the president. And she was one of only thirteen women in history, until now, who lived to see their sons become president of the United States.

Born in the age of horse and buggy, of steamboats and mule-drawn plows, Martha lived to fly in luxury and listen to Edward R. Murrow in London reporting the first, worst face of the space age—German V-2 rockets in their trajectory of destruction over Britain. The year she was born Millard Fillmore, who had a beautiful signature and not much else, was in the White House, the second vice president to assume office upon the death of the president (typhoid fever had felled Zachary Taylor); her son would be the seventh. Martha marveled at the arrival of electricity (not until 1929 at the Grandview farm) and indoor plumbing, radio and refrigerators, telephone and tractors. And in the minds of brilliant men the computer age was stirring. Oh, Mattie Young had seen it all. She exuberantly embraced the new—at least the good parts—but never stepped away from the standards and beliefs of an earlier time.

This woman of strong will and staunch values, which she passed along to her son, was the daughter of a legendary America. The Missouri-Kansas border was the country's westernmost settled edge in 1841, when Solomon Young, with his wife and two small children, left Kentucky for a new horizon, a new life. They came by steamboat, following the Ohio, the Mississippi, and then the Missouri, the great river that was the pioneers' pathway to the uncharted West. In western Missouri it was the rich topsoil, never disturbed by hoe or plow, that lured farm families like the Youngs from comfortable homes in Kentucky, Ohio, and Indiana, states that free-ranging spirits felt were getting a little crowded. Beyond Missouri and Kansas stretched vast spaces officially designated as "unorganized territories," where only the courageous, or unwitting, dared venture.

Within three years Solomon Young, a tall, burly young farmer,

had staked his claim (at $1.25 an acre) for choice land on Blue
Ridge. Then as westward migration swelled, fueled by the American zest for breaking free of old limits, Solomon, always a go-getter,
hitched his future to the migrants' wagons as a "bullwhacker." He
was a wagon master on the Conestogas, huge canvas-covered
prairie schooners pulled by six teams of oxen, that carried courageous families and their worldly possessions west to places yet to
be named by white men; he managed trains of as many as eighty
wagons, stretching three miles along the Santa Fe Trail. Between
trips he regaled Mattie with tales of his adventures, and in the
course of time little Harry listened wide-eyed to the same stories,
well embroidered with retelling.

The big, bearded Solomon could scare his little listeners with
stories of lurking Indians and the stunning natural forces of burning sun, fierce wind and torrential rain, enormous storms, and
endless parched badlands. Though the expeditions were harrowing, not for the fainthearted, Solomon went out repeatedly. They
tested a man's mettle—and they were lucrative. Historian David
McCullough, author of a distinguished Truman biography, estimated that Solomon was worth fifty thousand dollars (many times
that amount in today's dollars) in 1860, a tenfold increase over his
assets of 1850. He no longer called himself a farmer; he was a
"freighter," in the term of the day. The Youngs were, in their con-
text, wealthy. For Sunday dinners and special occasions, their
table was laid with sterling silver.

He was away as long as a year on those expeditions. And during
that time his spunky red-haired wife, Harriet Louisa, was left
alone on the farm with a growing brood of eight children. Back in
Kentucky her family had been well-off, with a gracious Greek
Revival house. In Missouri she joined the ranks of insufficiently
celebrated American heroines: the pioneer women. The promise of
land had brought them there, but once there they found that
nature, unwilling to give up her dominance, was the enemy, delivering blinding blizzards and scorching heat, torrents, drought,

insect scourges. The nearest doctor could be many miles away, and unreachable. And with their husbands away for long periods, isolated prairie wives were often driven mad by the aching loneliness.

A collection of pioneer women's memories, gathered by Joanna Stratton, whose forebears knew such trials, portrays lives of incredible hardship. Many lived at first in rude huts dug into a hillside and covered with moss until a log cabin could be built. Drawing water from the nearest well or stream, gathering buffalo manure chips for the fire, butchering, canning, making wool, sewing the family's garments . . . the work was endless and babies kept coming. Maybe worst of all were the grasshopper years—a monstrous one hit the Young farm in 1874—when suddenly "grasshoppers by the millions in a solid mass filled the sky" and it "looked like a great, white glistening cloud, for their wings caught the sunshine." The invaders came on wings of destruction, devouring not only crops in the field but everything made of wood—trees, fences, furniture, utensils, even a cabin's outside walls. They ate clothing, polluted wells, and made animal meat too vile to eat. Ruin was always just a hailstorm or an infestation away. Yet these women of grit and dreams found time to fashion quilts of such beauty that today they are museum pieces.

The Youngs were spared the worst of pioneer travails, thanks largely to Solomon's astute managing of his new world, but Harriet Louisa faced danger of a different kind. Harry proudly told how his grandmother, alone except for her children and a woman servant, routed a band of Indians who appeared at the house: "The Indians told her they wanted honey and if she didn't give it, they would take it and her too. They sharpened their knives on the grinding stone, and then she turned loose a large dog. Away went the Indians, some leaving their blankets." All in a day's life for a frontier wife. "There wasn't a thing in the world that ever scared her," said her admiring grandson. Nothing, at least, that she let on about.

Solomon Young, a shrewd man, sure of himself, was the prototype of the pioneer settlers. Many who headed west were escaping bad luck or themselves. Not Solomon. He was not running away from anything; he came to tame the West, to ride the country's expansion. He had a knack for buying and selling land and live-stock; he prospered and became a major figure in his community. Mattie, his next-to-youngest, and her boy Harry both thought he was wonderful.

Harry's other grandfather, Anderson Shippe Truman, also a Kentuckian, was a pale figure by comparison with Solomon. He had come to Missouri in 1846 for no better reason than to please his well-born new wife, Mary Jane Holmes, whose family had set-tled there. Both families owned several slaves, but Ethel Noland, tender of the Truman family tree, insisted, "They never bought one, they never sold one. They were inherited from the Holmes side of the family." When Anderson Truman's son John married Mattie Young, the older man, a widower, quit farming and moved in with the couple for the rest of his days, even when they returned to Solomon's farm. If Martha ever wished that this remote grand-father, who stayed mostly in his room, would spend some time with his numerous other children, she never voiced the thought. Family must be looked after, older folks cared for. Her sense of duty came first.

And Martha by then had lived through such momentous and extraordinary events that lesser domestic matters could not dis-turb her.

★   ★   ★

THEY came with a clatter, men and horses sweating from riding hard. Little Mattie Young, a wisp of a nine-year-old, and the other youngest children huddled under the kitchen table, trying to hide and yet stay close to their mother, the formidable Harriet Louisa. Mamma, they knew, could handle anything, almost. But these armed men swooping down on the Youngs' farmhouse, there in the heart of the Missouri slave territory, were the dread

Red Legs, led by the fearsome Jim Lane himself. Though they were not Federal troops, they wore the Union uniform, with red leggings that gave them their fierce name.

If only Papa was here, Mattie kept thinking. Solomon Young was a man people listened to, and he could have reminded Lane that he was a declared Unionist—not an easy decision in Jackson County in 1861, where there were many slaves and much support for the Confederacy. Indeed, Martha's family owned a few slaves—a cook, a housemaid, and one or two farmhands—but Solomon was dead-set against the war that had finally broken out in April, like a boil that popped and spewed its poison every which way. In his expeditions all the way to California he had seen a great continent-wide nation in the making and knew that the war could obliterate that future. But the day the Red Legs (also called Jayhawkers) drew a bead on the Young farm, Solomon was far away from home, somewhere out west, and in painful evidence of the deep divide within Missouri and Kansas families, Solomon's older son, Will, had left home to join the Confederate Army. Harriet was helpless.

The very name Jim Lane struck panic in a pro-South sympathizer's heart. In his turbulent life Lane was a terrorist in the eyes of many and hero to others, a senator (the first to be chosen by the new state of Kansas) and an abolitionist zealot who had contrived a friendship with President Lincoln. Gaunt, unkempt, hatchet-faced ugly, with a strange voice and wild manner, Lane was admired at home as "the liberator of Kansas," but on the Missouri side of the war's most brutal dividing line he was hated more than the devil himself.

Dismounting while sizing up their loot, the Red Legs demanded food. Harriet would tell her grandson Harry how she made biscuits "until her wrists were blistered from working the dough and rolling it out." Meanwhile, the raiders shot the Youngs' four hundred hogs, taking only the hams, and killed all the hens "out of sheer cussedness." They set fire to the barns and, Martha would say, "stole everything loose that they could carry," including a prized featherbed that had taken months of plucking goosedown and sewing. And they made away with the family's treasured silver. All of her life Mattie remembered, clear as a tintype in her mind, standing in the havoc of the barnyard, her face blackened by the ashes of war. It was a searing memory that toughened her, affected her priorities, and left her an unreconstructed Southerner for life.

A greater tragedy was narrowly averted. Jim Lane, determined to root out any family rebels, put a noose around fifteen-year-old Harrison's neck. "He was only a boy, but he wouldn't turn informer," his namesake, the president, would relate proudly. "They tried to hang him; time and again they tried—'stretching his neck,' they called it—but he didn't say anything." Harry was pleased that "people said I took after him." Finally Lane tired of his vicious sport, and his Red Legs, marking the Youngs' fine farm for a later raid, galloped back into "Bloody" Kansas.

The war raged for four years between the first shots fired on Fort Sumter in 1861 and the surrender at Appomattox Courthouse in 1865. But for the Youngs and other families along the Missouri-Kansas border that separated slave and free states, the fighting had begun in 1855 with skirmishes on both sides of the river, and went on for nine years. Guerrilla raiders, like the Missouri "Bushwhackers," were as savage as Lane's Red Legs, their massacre of the Kansas town of Lawrence foremost among their many heinous acts. In the space of two years more than two hundred people were killed and as much as two million dollars in property was destroyed on the two sides of that no-man's-land along the border.

Though Americans know a great deal about the Civil War, few have heard of the unspeakable General Order Number 11, the local Union general's retaliation against the Lawrence massacre. In August 1863 the Youngs were among twenty thousand residents accused of Southern sympathies and driven from their homes, forced to move to specified "posts." Given only fifteen days to pack up one wagonload of belongings, Grandmother Young— again on her own—piled as much as she could in an oxcart. Martha, brother Harrison, and four sisters trudged, bewildered, behind the cart as they made their way into three years of exile in a Kansas City "post." Chattels, furniture, the remnants of life had to be left behind as plunder for Federal soldiers. Numbers of houses were torched, a vengeance the Youngs luckily escaped. Many years later Harry, telling of this neglected chapter in American history, said bitterly, "They called them 'posts,' but they were concentration camps." While that was clearly an overstatement, without exaggeration the "posts" effectively functioned as detention camps.

All of this Mattie Young experienced before she was eleven years old. Such memories never fade, not even when one is well past ninety and content, and her unreconstructed stance has long been a family joke. Harry loved to tell his mother's reaction when he, then a senator, came home from a gathering of Civil War veterans in Kansas. "If you'd kept your eyes open," Martha said tartly, "you'd probably have seen your grandmother's silver and could've brought it home with you." Granddaughter Margaret chuckled that "Mamma Truman always talked as if, at that very moment, somewhere in Kansas they were all collectively dining off her mother's silver." When Martha came for her only White House visit, her son Vivian teasingly told her that the only unoccupied bed in the mansion was Lincoln's. "You tell Harry," she snapped, "if he tries to put me in Lincoln's bed, I'll sleep on the floor."

His mother's direct experience of the savagery of the Civil War, and the years immediately before and after it, as well as the destruction left behind, gave Harry Truman a visceral sense of the bitter crops of hatred and despair that could spring up in lands desolated by conflict. This inspired perhaps the greatest initiative of his presidency, the plan for the reconstruction of Europe's social and economic order, its infrastructure and industry, after World War II, the most generous policy ever extended to a defeated enemy. Because of it the conquered nations became pillars of democracy. (This aid was denied to the captive nations behind the Iron Curtain and to Russia itself because of Stalin's paranoia.) Though the plan bore the name of General George C. Marshall, then secretary of state, its principal author was President Truman. It would go down in history along with the Truman Doctrine, which pledged American support to any democratic nation threatened by the Soviet Union.

Years after the triumphant restoration of Western Europe, Truman was asked if his family's personal experiences had in any way influenced him. "Very much so," the former president replied. "You can't be vindictive after a war. Europe had suffered in the same way we had suffered after the War Between the States, and that made me think that Europe had to be rehabilitated by the people who had destroyed it. And that's what happened." (As an avid reader of history, Truman would have been well aware of how different this country would have been if Lincoln had lived to implement Reconstruction in the South in his spirit of "with malice toward none, with charity for all.")

The misbegotten policies of Republican Reconstruction, driven by the radicals in the party, were still much on his mind in 1952, when he hoped to withdraw troops from Korea. Nothing, he observed in his memoirs, creates greater antagonism than "the presence of unwanted soldiers, foreign or domestic. That was the way people in the southern states felt during the terrible reconstruction period, and when I was a very small boy I had heard much of southern reactions from my father and mother."

★   ★   ★

WHEN the Youngs were allowed to return home after almost three years, life on the Grandview farm seemed more precious. For Martha, it was not a hard life. "Papa told me if I never learned to milk a cow, I'd never have to do it," she confided to granddaughter Margaret. And so she didn't. She didn't much like to cook, either, and when she ran the house she always managed to have help in the kitchen, no small matter when as many as two dozen hearty appetites had to be fed at threshing time. In Margaret's candid opinion, "She only made one dish that was praiseworthy, and that was fried chicken." So Sunday lunches featured fried chicken, and everybody told her what a good cook she was.

This spirited daughter preferred to mount one of Solomon's fine horses—his trotters were the envy of the county—and ride carefree across the fields. When she was about eighteen, Martha, bright and strong-willed, insisted on going to college—and she went. In the 1870s it was amazing for a farm girl to go to college. As Margaret Truman said of her grandmother, "When she decided to do things, she just *did* them. She got what she wanted." So with her trunk packed with long black skirts and high-necked white blouses, her favorite books, and her Bible, Martha was off to the Baptist Female College in Lexington, Missouri. The college brochure boasted, "The building [there was only one] is well furnished throughout with all modern improvements. The regular course of study is as good as can be found in the best institutions of the west." For two years she studied art, music, and literature and looked forward to the dances put on each month by the college, which was surprising for a Baptist institution in those days. The girls were even allowed to meet thoroughly vetted cadets from nearby Wentworth Military Academy. And when she was back home at Grandview, there was more dancing, sometimes two or three evenings a week, in somebody's parlor. Martha's life was, at a very different level, a frontier parallel to Sara Delano's life in Hudson Valley society.

"I'm what you might call a lightfoot Baptist," she used to say, an impish challenge to "hardshell" Baptists who banned dancing, cards, and all things frivolous. That is not to suggest that Martha was any less religious; at eighteen she made a lifelong commitment to the Baptist Church, formalized with total immersion in the Little Blue River near the farm. Her faith was underscored by humility, as she explained when her son became president: "I believe in true religion but I do not believe in making out that one is better than he is. I think Harry is that way." Her belief was bedrock in her life, and it was passed on, undiluted, to her son.

★   ★   ★

IT was hard to distinguish Harry from the Missouri topsoil when he finished work in the fields after a long day, knocked the mud from his workshoes, and came into the kitchen to chat with Mamma. Sometimes she'd say, "Harry, how about a little Mozart before supper?" The dirt farmer could still draw Mozart's eloquence from the keys. He had come back to Grandview from Kansas City in 1906 to help his father run the farm, a spread of six hundred acres plus another three hundred across the road, not sorry to give up an unfulfilling couple of years as a bank cashier. He figured he wasn't cut out to be a banker, anyway. The farm he returned to was a different place from the magical place of his childhood. Martha's fine old family home had burned to the ground in the 1890s, and been replaced by a basic two-story white frame house as square and plain as Missouri. "This one," said Mary Jane Truman, "was just kind of thrown together." (Decades later Truman would sketch Grandfather Solomon's handsome house as he remembered it, as his model for his presidential library.)

And now, with John's financial burdens and without Solomon's skills, Harry found "about two bushels of debts to pay." Farming with mules and horses was backbreaking, and yet Harry enjoyed the hours of thinking-time—much of it centering on Bess Wallace—as he plowed the straight furrows he became famous for. "They had to be straight," he'd joke in political speeches, "or I'd hear about it from my father for the next year." That was not altogether a joke. Cousins and friends would say that John was much closer to his boy Vivian, the son who bore his father's first name and was a carbon copy of him, the son who truly loved farming. Though Harry wasn't a natural farmer, he conscientiously read up on agricultural matters such as crop rotation and water management. He was a farmer in his mind, if not in his heart.

This would seem a fairly sterile existence for a young man not yet thirty. Happily, that was not the case. In Independence, an easy fifteen miles away after he bought a secondhand Stafford touring car, there was always a party or a picnic and easy fun. "We just had a hilarious good time," Ethel Noland remembered. "And remarkably within the bounds of propriety." Sticking to the straight and narrow was easy for Harry because he was old-fashioned even for those times; raised by a nineteenth-century mother, he was a nineteenth-century man. As the twentieth century loosened up in its second decade, Harry continued to be the model of a young man on his way to sip a sarsaparilla with his best girl on the porch swing. His scrupulous morals—from which he would never stray—were rooted in a more solid time. It was prewar, small-town America, an age of innocence, of convention, of values set in cement, a world that lingered in the last century long after the calendar pages had flipped to the new.

And it was a world of generous hospitality: at the farm, there were always guests for Sunday dinner and relatives who visited for days, even weeks. The constant flow of visitors warded off the sense of isolation that makes a farm seem somehow cut off from the world. "I like to have company," Harry wrote to Bess, "for we'd die of lonesomeness if no one came." His parents were growing older, the farmwork never lessened, and he was past the age for catching bullfrogs in the creek.

To lure Bess out to Grandview, he had a brainstorm: build a tennis court and she will come. So he did, but she didn't. Although crestfallen and humiliated, he didn't give up pursuit. He courted her with letters, literally hundreds of them, written by kerosene lamplight. Not by any measure romantic, his letters described the

daily trivia of farm life: the calf that threw him and broke a leg
(his), the weather (never right), troubles with the hogs (smelly).
There were frequent warm references to Mamma and fewer, often
irritated ones, about Papa. At the heart of the letters was a shared
enthusiasm for books and magazines—Harry and Bess discussed
them, exchanged them, and were drawn together by them.
Martha's passion for reading had provided her son with his key to
happiness, and she truly liked this fine girl with whom she was
sharing Harry. His letter of November 4, 1913, reveals the
romance, in less than romantic phrases: "When it comes to the
best girl in all the universe caring for an ordinary gink like me—
well, you'll have to let me get used to it."

Before he could get used to it, he was off to war, a doughboy in
France, the perfect example of the wartime ditty, "How You
Gonna Keep 'Em Down on the Farm After They've Seen Paree?"
For nearly two years, much of the time in the thick of battle,
Artillery Captain Truman carried a picture of Bess in one breast
pocket, a picture of Martha in the other. Then he was back home
again, and his beautiful Bess was waiting for him. And so was his
mother, with a much more understanding heart than that of Sara
Delano Roosevelt.

★   ★   ★

IT had been a marathon courtship. Twenty-nine years elapsed
between the day six-year-old Harry was captivated by little Bessie
Wallace and the day of their wedding on June 28, 1919. The summer sun smiled down on them with such enthusiasm that flowers
and guests alike wilted in little Trinity Episcopal Church. Bess was
thirty-four, Harry thirty-five. She looked lovely in a tiered dress of
white georgette and a white picture hat; Harry was natty in a gray
suit custom-made for him by his best man, a tailor who had served
with him in France.

Out at the farm early that morning Martha and Mary had
picked daisies for the church with Harry and cooked dinner for

twelve threshers. Then they put on their new dresses and hurried to the wedding and the ice cream–and–punch reception on the lawn of the Gates-Wallace mansion. The occasion was bittersweet for Martha: Harry was marrying a fine girl, a town girl from a prestigious family. She knew that Harry had permanently kicked off his muddy workshoes; he could no longer be just a farmer. And though she might long for the old days with all of them together at Grandview, she knew it was well past time for her son to find his real calling, especially now that his beloved Bess was his wife.

Martha, aware but not unduly impressed by Harry's leap up the social ladder, could take satisfaction in knowing that she had made him a literate man, a history scholar, a musician—she had shaped the farm boy into the man who had won Bess. With marriage, he left the farm for good, charting a new course for his new life. He sold off the farm's equipment and livestock to raise money to open a haberdashery with a friend, Eddie Jacobson, a high-quality gents' furnishings store located in the heart of Kansas City. Truman & Jacobson did very well for a year after its opening in late 1919, but in 1921 a deep recession struck and men could no longer buy silk shirts and fine cravats. In 1922 the store went under, but rather than declare bankruptcy, Harry struggled for the next twenty years to pay off thirty-five thousand dollars in debts.

At the most propitious moment fate, in the guise of the Kansas City political machine, stepped in. Harry was picked to run for judge of the eastern section of Jackson County, an administrative rather than judicial post based in his old hometown of Independence. One factor in the Democratic bosses' choice was John Truman's unblighted record for honesty during his stint as road overseer—a tribute the son would deeply appreciate. Harry won the election and was on his way.

The new Mrs. Truman came appended to a formidable mother: Madge Gates Wallace. Bess's Grandfather Gates was a leading citizen of Independence, a flour miller who had become wealthy from his "Queen of the Pantry" flour that came in sacks featuring a pretty girl, modeled, they said, after his daughter Madge, who, they also said, indeed considered herself a queen. In 1867 Grandfather Gates built an impressive home on the town's finest street, a three-story house with gables and porches, gingerbread trim and stained glass windows, and a dining table that could seat thirty people. The Gates family was top drawer in Independence; Harry Truman was just another dirt farmer—and not a noticeably successful one. That he was widely read and played Bach counted for nothing in Madge Wallace's social ledger. "She was a very, very difficult person," said a longtime acquaintance, retired teacher Janey Chiles. "There wasn't anybody in town she didn't look down on. And Harry Truman was not at that time a very promising prospect." Nor would he appear to be for some years thereafter, except to his mother, who was confident that he would amount to something—after all, he was Solomon Young's grandson.

Bess asked that they live "temporarily" with her mother, and Harry graciously acquiesced. "Temporarily" stretched into thirty-three years. It shouldn't have been much of an adjustment for a man who had spent thirty years under his mother's roof, but it would be hard to find two women more unlike than Martha Truman and Madge Wallace. Martha was nurturing, hospitable to a fault, fun-loving; Madge was cold and critical—it was lemon zest compared to persimmons. So, from the time they were married, Harry and Bess lived their life with Madge. In their apartment in Washington she shared Margaret's bedroom; in the White House she occupied a spacious bedroom looking across Pennsylvania Avenue to Lafayette Square, and chose rarely to leave it.

Throughout her long and crotchety life, Madge Wallace was treated with utmost consideration by the world figure to whom she never showed respect. Truman's White House physician, Dr. Wallace H. Graham, spoke of "stolid and stoic" Madge Wallace's condescension to her son-in-law. "The president would always knock on her door. He'd say, 'Mother Wallace, may I come in?'

She would reply, 'Is this a social visit, or just what is it?' She was not always cordial. . . . I don't think she was overly fond of the president. She was, in a manner, giving the cold shoulder. The president treated her with royal respect at all times."

Despite all those years they had lived under the same roof as part of the same family, she still called him "Mr. Truman," and in 1948, when Harry was the struggling underdog in the race of his life, Mother Wallace, with monumental tactlessness, wondered aloud why Harry would run against that nice Governor Dewey. When the president fired General MacArthur for insubordination in the Korean conflict, a staff member overheard Mrs. Wallace churlishly criticize him to Bess—"Why didn't he let General MacArthur run the Korean War in his own way? Imagine a captain from the National Guard telling off a West Point general!" That time Bess snapped back in defense of her husband—Harry, she sharply reminded her mother, was commander in chief. The president's valet, who saw all of this up close, offered a wise insight for the mother-in-law's coldness: "Mrs. Wallace always thought Harry Truman wouldn't amount to anything. It galls her to see him in the White House running the country." What it all came down to was her implacable prejudice—Harry Truman was not in her class. She was right; she was not in his class at all.

In the Trumans' last year in the White House Mother Wallace, then ninety, suffered a stroke, lingered four months, and died in December, just a month before the Trumans returned to live in the old Gates home in Independence. Madge's fine house is now a favorite tourist attraction—because Harry Truman lived there.

Despite the fact that she always looked down her nose at him and his family, at her death Madge's gentlemanly son-in-law had only kind words for her, even in his personal diary: "She was a grand lady. When I hear these mother-in-law jokes I don't laugh. They are not funny to me because I've had a good one." With admirable grace, Harry had accepted Mother Wallace's attitude as just "her way." Martha would have said of her son's impeccable courtesy, What else do you expect? That's the way I brought him up.

★   ★   ★

WHEN his mother visited Harry and his family in the White House, he gave her a full tour of what he called "the taxpayers' house." A portrait of Andrew Jackson, the first log-cabin president, her favorite president, had been hung on his bedroom wall. In the East Room she praised Dolley Madison for saving the famous portrait of George Washington when the British burned the White House in 1814, and Harry the historian would have told her how British Admiral George Cockburn, taking possession of the White House, seized one of President Madison's hats as a trophy and ate the meal that had been laid out for Dolley. Admiring the desk Harry used, Queen Victoria's gift to President Rutherford Hayes in 1880, crafted of oaken timbers from a Civil War fighting ship, she would have noted that the name, the USS *Resolute,* was singularly fitting for this new president. Then on the South Portico, taking in the vista that sweeps across the lawn to the monuments to great presidents, she must have been overwhelmed at seeing all this with her son, who was now one of their number.

She tested White House bedrooms like Goldilocks: Lincoln's imposing, heavily carved bed was too full of history; the "queen's room" was too fancy, the bed too high—you needed a three-step stool just to climb in—but the small adjoining room, for a lady-in-waiting or a staff aide, had a neat single bed that Martha, who "never cared for fuss and feathers," deemed just right. Significantly, Sara Roosevelt had found the queen's room quite suitable—an insight into the very different characters of two mothers whose sons served a nation in crisis. Without affectations, without superior airs or attitude, Martha was the same confident self she had been since she stood in her black wedding dress with her hand on John Truman's shoulder.

Not in her wildest fantasies—if down-to-earth Missouri women were given to fantasy—would she have placed herself in the White House, the Truman White House. "I never wished for Harry to be president some day, as all American mothers are supposed to do. I never even dreamed that some day he might be vice president. I just raised all my children to know that they must always aim to do the right thing. They might make mistakes, but they'll be honest mistakes and that's only human." (She might have taken her cue for this answer from Sara Roosevelt, who had spoken virtually the same words in answer to this inevitable question. Sara, however, had not suggested that there could be mistakes.)

Martha could feel no joy at her son's unexpected ascent to the presidency. A Roosevelt devotee, she, like the nation at large, was devastated by his sudden death. She was perhaps less surprised than most, because Harry had confided to her his alarm about the president's obviously failing health. The American public did not share this knowledge; it was kept closely guarded by a small circle of people around FDR. "It was too sad, the way it came," she said to a reporter. "We even cried. I felt sorry for Harry and I felt sorry for Mrs. Roosevelt. I did not rejoice at all. . . . If he had been voted in, I'd be out waving a flag. But it doesn't seem right to be happy or wave any flags now."

Though he was anxious about Roosevelt's health, Harry set about being vice president with no thought that the worst would occur immediately. On the day before his life was irrevocably changed, he explained in a letter to "Mamma & Mary" just what a vice president did in that amorphous position. He described himself as the object of curiosity for "people who want to see what a V.P. looks like and if he walks and has teeth. I'm trying to make a job of the Vice Presidency and it's quite a chore. I'll be home one of these days and tell you all about it."

Within hours the world's most overwhelming job had fallen on his shoulders. After taking the oath of office in the White House Cabinet Room, Harry, still stunned, went home to his apartment and made his only telephone call. It was, predictably, to his mother. She knew instinctively that he needed to be bucked up. "You put on your fighting harness and do all you can," Martha exhorted him, then said good-bye with what had become her personal benediction: "Now, you be a good boy, Harry." In the weeks that followed, the denigration of Truman, a modest senator little known around the country, rankled her. "Harry is more able than many people give him credit for," she declared. Time would prove her right.

The politically savvy Martha had not wanted Harry to run as FDR's vice president. "I do not mean it wouldn't be a great honor," she explained to an interviewer, "but Harry [then a committee chairman] is doing good in the Senate—he ought to stay there." In fact, his leadership of the Truman Committee's investigation into corruption in war production had won the respect of Capitol Hill colleagues who had once dismissed him as a creature of Tom Pendergast, the ultimately imprisoned boss of the old Kansas City machine. Harry was also reluctant to leave the Senate behind. "I am going to the convention to defeat myself," he told his Aunt Ella. "I don't want to be vice president." When party leaders argued that this was the only way Roosevelt could dump increasingly leftist Vice President Henry Wallace, he couldn't refuse. Once Harry said yes, the fighting Martha emerged. "Since he's running, I want him to win. I'm not too old to campaign yet!"

When Harry was elected county judge, Martha was pleased that at last he had found his true calling, and from his first Senate race in 1934 he had counted on her help in his campaigns. After one of his victories, when a reporter asked the inevitable question, Martha, in a tone underlining its stupidity, repeated, "Am I proud of him? Say, I knew that boy would amount to something from the time he was nine years old. He never did anything by halves." Indeed he didn't; she had seen to that.

So, at ninety-one, Martha joined in the campaign, mobilizing Democratic women volunteers, reaching out to defense workers new to the area, charming the press. On Election Day someone asked if she had voted yet. "I certainly did," she replied saucily, "and I'm thinking about voting again." That night, avidly following election returns on her radio, she said philosophically, "If Harry doesn't win, he won't be disgraced. It won't kill him if he isn't elected."

★ ★ ★

IN a time when ladies tended to say little and smile much, Martha was outspoken, with quick humor and opinions spiked with Missouri show-me skepticism. Granddaughter Margaret, much like her, admired that exuberant spirit: "She always had a quick, snappy retort. If she didn't approve, she could decimate you. She was a woman with a glint in her eye. She had a mind of her own on almost every subject from politics to plowing."

On one occasion, ignoring the origins of many of Washington's leading lights, Martha declared, "If there are any good Yankees, I haven't seen one yet." After her unvarnished opinion of the atomic bomb—"I don't think they should ever have invented it"— she was muzzled by the White House. "The president's mother," it was explained, "is apt to speak a little too frankly for interviews." (However, she had the political wisdom to defend dropping the bomb because it had shortened the war and saved lives.)

In the White House Harry wrote letters to "Mamma & Mary" two or three times a week, an astonishing flow—"the demon letter writer," his daughter dubbed him. And he was a faithful telephoner as well, calling his mother every Sunday and after big events. Half an hour after Japan surrendered and World War II was over at last, the telephone rang in Martha's bungalow in

Grandview. It was Harry, calling to rejoice with her. "I knew he would call," she told a guest matter-of-factly. "He always does."

In his chatty letters Harry gave Mamma, always keen on politics, the inside story behind big events with the warning "Don't say anything about these things because there'd be headlines from Boston to Los Angeles. Don't ever let anybody talk to you about foreign affairs. It is a most touchy subject, especially in that part of the world." ("That part of the world" was the still "touchy" Middle East. Truman would recognize the new state of Israel eleven minutes after it would come into being, and his old partner, Eddie Jacobson, came to the White House as Israel's first, though unofficial, ambassador.)

Looking back over his life, Harry was cocky about his plain-spoken disdain for artifice and ego, an attitude surely inculcated by Martha. The accidental president would not try to imitate his illustrious predecessor; he was Harry S-for-nothing Truman, former dirt farmer from the Missouri sticks, failed businessman, and, much credit to his uncommon mother, completely confident in who he was. He saw no need to change. He had truly found his niche.

★   ★   ★

ON her ninety-third birthday Martha declared, "I'm going to live to be 100 years old." And why not? She came from a family that lived hard and long. With those hardy pioneer genes, she was heading for her goal until a second broken hip got in the way. It was so unlike Martha to be out of action. "She never, ever moved slowly," said granddaughter Margaret. "Even in her nineties she was going like the speed of light. And her head stayed sharp—right up until she died, her daughter had to read the newspapers to her." Every day she perused the *Congressional Record*, one of its few known voluntary readers, "to learn what all those senators and representatives say." Could there be any doubt that Harry's mother was a true political animal? As her sight faded, radio was her good friend: "I like newscasts and songs—cowboy songs and all kinds." She never switched off.

Cousin Ethel Noland, a lifelong teacher, mused that Martha "was just the type of mother that you'd think would furnish a President for the United States. She never deviated from her idea of what was right. Her principles were so sound—she had just about the finest set of values I have ever known." Martha was as surprised as everybody else that her son became president, but she had prepared him by raising him "to do the right thing." Politically "the right thing" may be as changeable as the moon, but morally "the right thing" remains as fixed as the North Star, and she taught him that. Among the boyhood mementos she had saved was a composition in which a serious young Harry wrote that "with a true heart, a strong mind and a great deal of courage" a man can achieve his goals in life. A schoolboy wouldn't have realized that he was distilling his mother's character.

Martha's influence on her son was not exercised in matters of public debate but made its mark on the president's tone, style, and his approach to his mission. "She set an example for him," observed Margaret Truman Daniel. "Her thinking was very direct and clear; his head was never turned by fawning or flattery." Truman said in retrospect, "I tried never to forget who I was and where I'd come from and where I was going back to. I just never got to thinking that I was anything special." And, sure, "I've had a few setbacks in my life, but I never gave up." Straight from a resolute mother.

In late May 1947, when Martha's broken hip was not mending and death's door was opening, Harry rushed again to her bedside. She wanted to talk politics. "Is Taft going to be nominated next year?" (Senator Robert A. Taft, said Margaret, was her least favorite Republican.) "He might be," Harry reckoned. Looking hard at her son, his mother asked the question the whole country was wondering about: "Are you going to run?" "I don't know, Mamma." Martha, never one to pussyfoot, was a bit sharp with him: "Don't you think it's about time you made up your mind?" When he finally decided to go for it, his mother was gone, but he knew she would have approved—and would have campaigned from her bed if necessary.

At his triumphant inauguration in 1949, after one of the great upsets in the annals of politics, the president's joy was not entirely complete. "I wished my mother had lived long enough to see me sworn in as an elected President," he wrote in his memoirs. "When I succeeded Franklin Roosevelt, my mother so wisely said it was no occasion for her to rejoice. But now that I had been elected President in my own right, it would have been a great thrill for her to be present as her son took the oath." And then she would have waved flags.

★   ★   ★

MORE than once Harry Truman told of his profound experience on the day his mother died, July 26, 1947. For two weeks she had been sinking inexorably, her son keeping vigil at her bedside. He then had to return briefly to the White House. Racing back to her in the presidential plane, he was dozing when his mother's face suddenly appeared in his dream "with amazing clarity." He sat up, terribly shaken. A few minutes later he was handed a radio message just received by the pilot. Without reading it, he knew the contents: "I knew she was gone when I saw her in that dream. She was saying good-bye to me." What she had said to her son in his strange dream was "Good-bye, Harry. Be a good boy."

★   ★   ★

CHARM was never Harry Truman's strong suit. Neither handsome nor charismatic, a man much given to plain speaking, he would not have fared well in current image-driven politics. It is almost impossible to imagine Harry Truman standing before a microphone defending his "character."

While not an uncomplicated man, he was something of an open

book. What you saw was what you got—he knew exactly who he was and felt no need to defend it. No setback, disappointment, or failure could shake the confidence that came from his being, as the French put it, at home in his own skin. The presidents who preceded and followed him were enormously self-assured, but few would call them "humble," and Truman was, in the best sense, humble. His strengths derived, to a great extent, from his being fully aware of his own weaknesses and limitations.

Harry Truman, who knew history as well as any president in the twentieth century, understood what Aristotle meant when he said that action defines character. His study of the past led him to insist, "men make history; history doesn't make the man." And he did. Like his mother, he would never shrink from action; he was not a passive man.

He was very much a "mama's boy" and his country benefited from all that his mother did to implant the beliefs and attitudes that kept him humble and made him memorable. It might come as a surprise to Harry—but not to Martha—that historians increasingly rank Harry Truman among the strong and successful presidents.

The burdens his mother imposed on him were not those impossible demands that require the labor of Sisyphus, forever pushing his boulder up the steep hill only to see it roll down again. He would fulfill her high expectations; he would surpass the initial low expectations that had greeted him on taking office.

His success, in no small measure, stems from the wise parenting of a stalwart and uncommon woman, one who had the good sense not to seek perfection in her son or in any human being. She knew that perfect is all too often the enemy of good.

3

―――――――

SIX BOYS AND A PIANO

The bearing and training of a child is woman's wisdom.
—ALFRED LORD TENNYSON

ORE than once Ida Eisenhower declared that her third son, Dwight David, was "my most troublesome boy." Pure sunshine when he wanted to be, but possessed of a temper that would flash quicker than lightning in August and a habit of scrapping with his older brother when he knew he couldn't win. After one such mismatch, he had hurled a brick at Edgar, and only his bad aim averted serious consequences. They were inseparable brothers, known as "Big Ike" (Edgar) and "Little Ike" (Dwight) to their friends, though never to Ida, who disliked nicknames. And as for their fights, well, much as she deplored violence of any kind, she knew that boys will be boys.

Especially in turn-of-the-century Abilene, an unpaved and unpolished Kansas frontier town that had been tamed for scarcely twenty-five years, a legendary place whose brawling and shoot-outs were still the stuff of cracker-barrel yarns. Ida was inured to the noisy turmoil of a house over-full of boys—there was an even half-dozen now, with the new baby in the crib. And for Ida there would always be seven sons; she could never leave out little Paul, the baby who died of diphtheria when he was still at her breast. Still, six sons—Arthur, Edgar, Dwight, Roy, Earl, Milton—were enough for any woman to handle.

It all went into making a lively family, a warm home, the joy-filled life that Ida had yearned for when she was a motherless little girl, turned over to stern grandparents back in Virginia. Even though her life as wife and mother in Abilene was not easy for her, not at all, it satisfied her heart, and that was what mattered to Ida.

Rising above the family's relentless money worries and the burden of putting 189 meals on the table every week (the number included grandfather Jacob, patriarch of the Eisenhower clan, who lived with them), Ida was a pillar of comfort and cheer. Her unaffected grace was imprinted on every feature of her amiable countenance, brow unfurrowed, hazel eyes alight with interest and open to the world around her; the square set of her shoulders in their leg-o'-mutton sleeves bespoke the confidence to face come what might. Most of all, there was her smile, bright as a prairie sunrise, a generous, encompassing smile that, when replicated on the face of her third son, would reassure an anxious world as bombs fell and battles raged across Europe. The warm and easy grin that brightened front pages and television screens and became the hallmark of the celebrated general and popular president was Ida's smile, once removed.

Dwight, the son most like her, writing about his race for the White House in 1952, shared a secret behind the famous smile: "The candidate . . . steps blithely out to face the crowd, doing his best to conceal with a big grin the ache in his bones, the exhaustion in his mind." His mother would have understood that little admission all too well. A sparkling smile can mask private pain and disappointments, inner emotions that one must never reveal, and in the Eisenhower household Ida had to supply the smiles for both parents, because her husband, David, was a somber man who rarely smiled at all.

Ida believed there was always too much that had to be done to waste time chewing over worries, especially after the family acquired, from her husband's brother, the house on Abilene's South East Fourth Street that would be her home until the end of her long life. She was thirty-six years old, married almost fourteen years, and for the first time there was a yard big enough for her boys to play in. It was a white frame house with high-soaring gables and a porch meant for neighborly rocking chairs, a house with enough bedroom space for the boys to be shoehorned in, two to a bed. In the parlor Ida's cheerful spirit expressed itself in a bright jumble of patterns and colors on every surface—carpet, wallpaper, upholstery—and every chair wore a fancy antimacassar, Ida's handiwork. Most treasured was the handsome coverlet on the "fainting couch" (known more prosaically as a daybed), the work of her husband's grandfather, Frederick Eisenhower, a skilled weaver who had brought it with him from Pennsylvania. (It is still admired in the Eisenhower home museum in Abilene.) Like all modest houses of that time, the Eisenhower home had no running water or indoor plumbing; baths were taken on Saturday nights in big galvanized tubs filled with well water hoisted steaming from the black iron stove.

Best of all, the house was set on three bountiful acres, a boys' kingdom for games and tussling, with nearby woods for roaming and streams for fishing. A big barn built in the Pennsylvania Dutch style housed an assortment of farm animals, a smokehouse, an orchard, grapevines, and a strawberry patch. Three acres might not sound like much by Kansas standards, but three acres, carefully cultivated, constituted a small truck farm, a cornucopia of vegetables for the family table and for the boys to sell for much-needed cash. Those three acres meant security for Ida, the prudent ant of the fable. Not that she did the stoop labor—that's what boys were for. The Lord gave her the sons and loaned the land, and what she did was bring these two gifts together. Had it not been for her industry and wise managing, what her neighbors called her "get up and go," the Eisenhower boys might have gone to bed hungry, and much worse, they might never have fulfilled their potential.

"Tutor and manager" was how Dwight described his mother. Her role as tutor shaped every aspect of their family life; her role as manager was perfected in her distribution of chores. Scrupulously fair, she showed no favoritism toward one son over another and rotated the tasks, taking into account each boy's individuality. With Ida in charge, even slicing a pie became a Solomonesque exercise. She would have one boy cut the pieces, who would then give a brother first choice, a system that made for geometric precision in the size of the slices, if not spontaneous generosity.

In an era when discipline, rather than child psychology, was emphasized, Ida was a genial commanding general marshaling her troops. Before and after school the boys fed chickens and pigs, milked cows, curried the horse; they worked the fields, chopped wood, cured meat. Though farm chores were unremitting, the boys generally preferred them to such dreaded indoor tasks as dragging out of bed before dawn to light the stove in a freezing house, or lugging heavy, sloshing buckets of water for the primitive washing machine. While some of the six disdained working in the kitchen, Ida made Dwight into an enthusiastic cook, a skill he would enjoy throughout his life. He realized, in retrospect, that for him, even as a boy, cooking was a "creative outlet."

For the Eisenhowers, supper was the family gathering time. It began with a long blessing, invoked by David in his undisputed role as head of the family, and, impossible as it would seem, the six rowdy boys were expected to be restrained. (Granddaughter Peg Eisenhower Bryan, whose father, Roy, the fourth son, was a pharmacist in nearby Junction City, laughs now at her embarrassment when "my brother Bud and I were noisy—and we were dismissed from the table. We had to go sit on the stairs while the rest of them ate. My grandparents weren't mean—they never yelled, but they were strong disciplinarians.") After supper there was Bible reading by kerosene lamp light at the dining table. If one of the boys slipped up in reading, he lost his turn and had to pass the Good Book to the next brother. Dwight could boast that his mother had given him a watch for reading the Bible all the way through, from Genesis to Revelation. Later all would kneel for family prayers, which always included, General Eisenhower later remembered, "a plea for the hungry, the weary, and the unfortunates of the world."

Some evenings, after the Bible reading, Ida settled herself at the cherished ebony piano she had bought before she was married and had clung to as she and David moved from place to place. It was more than a musical instrument; it was the talisman that brought the magical sounds of beauty and cheer into their home and softened the struggle that had not been part of her dreams. The family would gather around to sing the hymns she had known from childhood in the Shenandoah Valley—"Bringing in the Sheaves," "Rock of Ages"—hymns that comforted and refreshed when body and, occasionally, spirit flagged, familiar songs praising the Lord, the very personal God she depended on to help her keep this family going. She played very nicely, and the Bible, she'd observe, bids us make a joyful noise unto the Lord.

Bedrock to Ida's standards was her unwavering commitment to serious education. Back in 1882, in an improbable decision for an orphaned girl of twenty, she made up her mind to go to college, and she did. As a mother and exponent of William Butler Yeats's aphorism "Education is not the filling of a pail, but the lighting of a fire," she provided the spark that inspired her sons to make something of themselves. She saw futures for them that would take them far beyond their father's disappointing lot as a mechanic at the creamery and later in charge of pensions at the utilities company.

Of course she wouldn't put it in those words—it would be unthinkable for Ida to criticize or speak ill of her husband. But something, somehow, had planted a rare spark in Ida Stover. She would do everything in her power to encourage her sons to stretch their minds and ambitions, as she herself had done from the time she was seventeen. In his odd way David was an intelligent man but didn't, or couldn't, seem to reach out to his sons, leaving Ida to do it for both.

Though times were hard for the Eisenhowers—by any measure other than the one set by their own pride, they were poor—neither Ida nor David would tolerate any thought of their sons dropping out of school to help support the family, no matter how desperately the money was needed. That meant that the boys would go to school wearing hand-me-downs that were patched and patched again. Years later, a friend still remembered that "the younger ones especially were little better than ragamuffins." Ida's great-nephew, Wes Jackson, who is a water management specialist in Salina, Kansas, feels that "surely it must have been difficult, even humiliating, for Dwight, wearing his mother's old high-button shoes to school." Even so, Ida saw to it that each boy had one proper outfit kept nice (and passed down) for Sunday meetings.

She always encouraged the boys to read, yet when Dwight's schoolwork and chores began to suffer because his nose was always in a book, she locked the books in a closet. He found the key, of course, and as soon as his mother was out of the house, out came his books. His favorites were tales of the exploits of ancient military heroes, particularly Hannibal, the Carthaginian general, and his brilliant strategy of using elephants to transport his troops across the Pyrenees and the Alps in his march on Rome. Ida's key, it seems clear, opened doors to more than the closet.

★   ★   ★

IN the flickering lamplight, David Jacob Eisenhower pored over his Bible, its text rendered in Greek, which, he liked to argue, conveyed more precise meaning. There may have been a trace of pride in his choice, a hint of vanity that here on the rough Kansas frontier, he was proficient in classic Greek. He provided evidence for an observation by his grandson, John, Ike's son: "The Eisenhowers were poor, but educated."

That David and Ida were deeply devoted to each other for a lifetime is unquestioned, and baffling. Though they came from similar backgrounds, husband and wife were like two sides of a coin: Ida outgoing, buoyant, resilient, David withdrawn and given to melancholy. Grandson John, a distinguished military historian, remembers his grandparents well from his summer visits with them as a boy. David, he said, "was morose, detached—maybe broody." Milton, the son who made his mark as an academic, underscored the contrast: "Mother had the personality. She had the joy. She had a song in her heart. Dad had the authority."

Ike described his father as "breadwinner, Supreme Court and Lord High Executioner." There was no closeness between sons and austere father, and only the requisite filial respect. Even by the stern man-of-the-house standards of the times, David's manner was brusque and antisocial. Nettie Stover Jackson, Ida's niece, told of a visit with them as a girl—two weeks during which Uncle David never spoke to her. "I don't know whether he knew how to smile," she said. "I doubt he ever laughed out loud in his life. He was distant, just distant." John as a boy was quite proud to buy his grandfather a pair of low-cut shoes to replace his heavy hightops. "His reaction? I got scowls. Ike had respect and sympathy for him, I suppose, but nobody but Grandma was close to him. She would go along with whatever he said."

Though David worked long days as mechanic at the creamery, there was never enough money. If the boys wanted a bit of cash in their pockets, they had to earn it. "Swim or sink, survive or perish," said John. "That was the philosophy." Dwight, at the ripe age of seven, proudly brought home a nickel a day for delivering newspapers, and a few years later, he and Edgar turned their garden plots into a profitable enterprise, selling their vegetables door to door. In the off season Ike added a new item: hot tamales, which Ida taught him how to make. The tamales, at three for five cents, were best-sellers—and if any were left over, Ike and Edgar happily gobbled up the surplus.

The young entrepreneurs targeted the affluent part of town, north of the Union Pacific tracks, which divided Abilene geographically, economically, and socially—and on all counts the Eisenhowers came from the wrong side. Edgar found it humiliating to stand by deferentially as haughty ladies pawed through his produce, criticizing the quality and complaining about the price. It was even worse when they sold produce door-to-door because often the stuck-up kids they knew at school lived in those big houses. But the boys needed the cash, so they swallowed their prickly pride and knocked on the doors. Edgar detested being put in this embarrassing position, but Ike insisted, in long hindsight, "I never had any such feeling."

Not so, declared John, who worked with his father on his memoirs. "All of them had a chip on their shoulders about being from the wrong side of the tracks." Some of the boys' old schoolmates recalled that Ike had been openly resentful and contemptuous of North Side snobbishness, and his reputation as a hotheaded scrapper indicates that there was a lot of getting even. It's easy to forget such feelings when you have outshone the whole North Side, when they have all embraced you as their hero.

Ida was well aware of the family's socioeconomic status, but such differences in wealth meant nothing to her. She was leading her family on a different path, a straight line laid out by the faith that was the compass of her life, a path in which character sets the direction and hard work gets you to the finish line. Character, of course, didn't just happen; it was Ida's handiwork, work that she never put aside.

David, "the czar" of the household in Ike's memory, focused on obedience and discipline. It is not unusual for a man who has been stamped a failure to compensate for his own lack of power by exercising the strictest control over his children. "He was very Pennsylvania Dutch. Very authoritarian—very!" said John. "He was a man of discipline who believed in beating his kids until they were in their teens. Then my dad turned on him and said, 'No more!' "

In his life story Ike recounted that memorable episode in painful detail. David, told that Edgar had been skipping school to work for the town doctor, stormed home from work. "His face was black as thunder. He reached for a piece of harness, at the same time grabbing Ed by the collar. He started in." Twelve-year-old Dwight, witnessing the merciless beating, began to cry as loudly as he could, hoping to alert his mother, and clutched at his father's arms. "Father stopped his thrashing and turned on me. 'You want some of the same?' " At that, the son did the unthinkable—he stood up to his father. "I don't think anyone ought to be whipped like that," he cried, "not even a dog!" To his astonishment, his father did not punish him, which suggests that David, a God-fearing man, realized that in his fury he had completely lost control of himself and was abusing his sons.

In marked contrast, Ike remembered that his mother "talked more of standards, of aspirations and opportunities." Ida was the balance wheel between severe father and resentful sons. A whipping might have given the man who delivered it a certain satisfaction that he was training his sons in right and wrong—and less admirable psychological satisfaction—but it was Ida's values that guided her boys, each of them, to grow into upright, productive men. "She did so much," said John, "but she demanded no credit."

"Grandma was very warm, very friendly, very companionable—but really very impersonal," he said of the grandmother he much admired. "She was very direct and straight." As a historian he regrets that a lively and interesting woman, with her own imperfections and contradictions, has been turned into a plaster saint. "She was put on such a pedestal by her sons. No flaws, no quirks, no idiosyncrasies." He likes to remember other moments, like the time his saintly grandmother, in an Ike-sized outburst, dispatched a visiting cat from her porch. "I hate cats!" she declared with uncharacteristic vehemence. "She kicked that cat off the porch," John chuckled. "It was a place-kick—about three steps out into the yard."

Underlying every aspect of the Eisenhower home life was the parents' religious fervor, which exceeded even the rigid demands of the River Brethren, an offshoot of the Mennonites that had defined the Eisenhower standards for generations. It was a deep disappointment to them that none of their sons shared their religious commitment and none felt uncomfortable with the Brethren. As the parents grew older, they too broke away from the River Brethren, electing to move to a group that was even more rigidly fundamentalist. Ida ultimately joined a small independent sect that affiliated with Jehovah's Witness, and in her late years she could be found passing out pamphlets on the street. Several members of her family were disturbed that she was being exploited—the pacifist mother of the topmost general passing out antiwar tracts—but the general himself defended her right to do whatever added to her happiness. The episode was no more embarrassing than Ida's peace protests during World War I—with a German name like Eisenhower, whose elders spoke German at home, she would have been arrested but for the intercession of a well-placed friend. Running afoul of the law in this issue did not trouble Ida, who would take the risk, certain that she was standing up for what was right.

With her unblinking faith and natural optimism, Ida created a cheerful and supportive world for her household of men. She gave her boys a belief in themselves and shared her gift for laughter. Nurturing, she knew instinctively, was more than baking eight loaves of bread every other day and ironing that pile of shirts. No accolade would have pleased her more than the observation by her "most troublesome boy"—the nation's thirty-fourth president— that his mother was "the happiest person I ever knew."

★   ★   ★

THERE was little in Ida's personal history to suggest this sunny turn of mind. At the time of her birth on May 1, 1862, in the Virginia hamlet of Mount Sidney, the Commonwealth was racked by the turmoil of the Civil War, and the once-tranquil Shenandoah Valley was scarred by skirmishes between Confederate and Union troops fighting to control that key north-south route. To the strongly pacifist Stover family, both sides were anathema.

The Stovers were peaceful people, staunchly independent farmers of Swiss-German stock, like the Eisenhowers, members of the River Brethren, who baptized members in local streams. Zealously nonconformist, the sect rejected both the customs and civic duties of mainstream society, shunning contacts outside the Brethren community and turning its back on worldly temptations. To show that they were separate and apart from the "others," they dressed in plainest black and recoiled from such manifestations of the devil as tobacco and alcohol, cards and entertainment. Their faith and their lives were based on literal interpretation of the Bible, the Good Book that was both conscience and solace, guide and daily sustenance, and at the heart of the Brethren's belief was unshakable opposition to war for any cause.

Within the Brethren there was no formal hierarchy; members chose to communicate with their God face-to-face, so to speak, without intermediaries. But insofar as they had a leader, Jacob Eisenhower, David's father, filled the role in their community in Pennsylvania and in Kansas, preaching (in German) and presiding over services in which the faithful would testify, occasionally speaking in tongues. He was the revered patriarch, a solemn, bearded figure with the massive dignity of a Michelangelo fresco.

In about 1730 the Stovers (originally spelled Stoever) settled in the beautiful valley that is mated to the Blue Ridge mountains in western Virginia, seeking to carve out a life removed from the distractions and temptations of town. For well over a hundred years they farmed, stayed quite separate from the cavalier Virginians, and multiplied. One Simon Stover built a plain, sturdy farmhouse set amid rolling farmland and orchards, and fathered seven sons and one daughter—Ida Elizabeth. Suddenly the peaceable Stovers found themselves in the path of the most brutal of brotherly wars, in which they supported neither side, though they believed slavery to be against God's will.

Their valley was home territory for General Stonewall Jackson's Confederate troops, and as the war raged soldiers in both blue and gray thundered up and down the contested road that ran past the Stover farm, Southern forces battling to hold their ground and General Philip Sheridan's troops implementing the Union leaders' scorched-earth policy. "Not even in Sherman's march through the deep South," Ida's general-son would write, "had war been more cruel than in this valley of small farms."

To avoid fighting, Simon paid one thousand dollars to a substitute, but as terrible losses in battle required more and more replacements, he was drafted for the second time. Rather than violate his conscience, he went north into Union territory, leaving behind his wife with their brood of eight. When Ida was three years old Confederate soldiers stormed into the Stover home, searching for her fighting-age brothers who had not signed up with the army. The harrowing experience of being literally in the line of fire took its toll when, two years later, her young mother died; Ida always felt that her mother was a casualty of the war. Years later, passing on these stories to her grandson John, Ida declared that she became a pacifist as a result of her family's suffering. For one of her persuasion, war brought a grim symmetry: her life began amid the horrors of battle and her last years were marked by fears for her beloved Dwight, in command of Allied forces in Europe, in history's greatest conflict.

At her mother's death, her father parceled out the children among relatives, placing Ida, who was six, and the younger brothers with their strict maternal grandparents, the Links, just down the road. She lived with them for eleven years, a childhood that was hardly a childhood at all, spent in cooking, cleaning, quilting. "It would never do to burn any food or take it out underdone,"

she recalled when she was an old lady. "That was sure punishment!" And she never forgot laboring over a quilt on winter days:
"My fingers got so cold I often botched the work and had to pull
it out and do it over again—and not cry about it, either."

During those years she was active in the local Lutheran
church—her only attachment to a mainstream denomination—
and won a citation for memorizing 1,365 Bible verses. (She was
just as proud when her Dwight was rewarded for reading the Bible
from cover to cover.) Sometimes Ida and her brothers would ride
horses over the wooded hills. "I was a tomboy," she would say,
half laughing, half boasting. "I was tough and wiry and could gallop bareback as heedlessly as any of my brothers." Then they'd
rest in the shade of a tree and dream about their future. West, they
agreed, we'll go west to Kansas! And I will go to college, Ida
insisted. She had another wish, which she may have kept to herself: she would have a piano of her own. What strange notions, in
the 1870s, in backwoods Virginia.

"Because I was a girl, I was told I must listen, not talk, and not
expect to go to school much," she recalled. Since she had no one
to help her to carry out her plan, her solution was—as it always
would be—to do it herself. She took herself to Staunton, the nearest sizable town, got herself room and board as a mother's helper,
baked pies and raised chickens for sale, put herself through high
school, then went back—as a teacher this time—to the one-room
country school where she had been the star pupil. Such a self-
starter would be bound to pass on her drive and determination to
little Dwight and his brothers.

Turning twenty-one, Ida came into a goodly legacy from her
father. At once she headed for Kansas, in the company of an aunt
and a group of new homesteaders, to join her brothers, who had
settled near Topeka in a colony of Dunkards, another small fundamentalist sect. They were eager to be part of the westward drive
that was stirring ambitious young Americans, lured by the promise of opportunities as unlimited as the vast dimensions of frontier

America itself. The railroads, thrusting deeper into land rich in untapped resources, until then inaccessible, opened up new possibilities for commerce and industry, and each arriving family spawned a need for stores selling household goods and every kind of service. Now that homesteaders no longer feared attacks by hostile Indians, who had been driven farther west, farmers—and their courageous wives—from long-tamed towns and the gentle countryside of the East could envisage an almost normal family life on the prairie.

Ida had arrived in a land of endless skies and unbroken vistas, as open as the people who had come there to fashion new lives. For this girl who had never ventured beyond the Virginia valley where she was born, setting out to find her future—in a state only one year older than herself—was a courageous undertaking. She had left behind Virginia meadows as verdant as the green pastures of her beloved Twenty-third Psalm, sparkling streams that tumbled clear from hidden sources high in the mountains, creamy dogwood and purple-red Judas trees dappling the countryside in spring, chestnuts and beech blazing saffron and gold in autumn. Never again would she feel the land gathered close around her like a shawl or be sheltered by the forested slopes of the rumpled old Blue Ridge mountains.

She would soon discover that in Kansas weather was a more fearsome force than the passing summer storms and puffy snows that as a girl she could forecast from the color of the sky. In Kansas she would welcome any small splotch of shade when the summer sun scorched crops and eyeballs; she would huddle against winter cold that bit deeper than anything she had experienced back east. She would fear the funnel cloud skipping capriciously, delivering destruction, and in any season of the year she would brace against the wind that whined, loudly, tiresomely, staining the wash hanging on the line. This was Kansas, and Ida Stover accepted the differences with the equanimity that was programmed into her spirit.

Before she left Virginia, Ida made the most extravagant purchase of her life: she paid the immense sum of six hundred dollars for the finest piano in the store, a Boston-made Hallett and Cumston of gleaming ebony, with panels of pierced carving backed by bold red cloth. This fine instrument was loaded in the freight car along with the group's washstands and cradles destined for Kansas, and she would never let it go.

★   ★   ★

O N E of Ida's brothers had already arranged for her to attend Lane University, a Brethren-run institution in nearby Lecompton, a "university" consisting of three buildings and about one hundred students. Ida took courses in literature, history, and—most eagerly—music, and soon found herself pursued—most eagerly— by a fellow student, an earnest young man named David Jacob Eisenhower.

David had been fifteen in 1879 when his family, along with some three hundred other members of the River Brethren community, sold out their considerable holdings in Pennsylvania's Susquehanna Valley to make the trek to Kansas. There were tears in some eyes and trepidation in all minds as the Eisenhowers sold their fine nine-room house, capacious barn, and a hundred fruitful acres. They were headed for Abilene, which, a few years earlier, had gained notoriety—for good reason—as the sin city of the frontier, full of gamblers and fancy women and gunmen waiting for the arrival of lonesome cowboys fresh off the Chisholm Trail.

Then, just like in the old Saturday-morning westerns, legendary Wild Bill Hickok, the fastest draw in the West—surely wearing a white hat—took over as marshal in Abilene in the 1870s and, with his similarly talented sidekick, Martha Jane Burke, aka Calamity Jane, began to clean up the town. Since Wild Bill had not tolerated much in the way of breaking the law, Abilene had become a more respectable place by the time Jacob Eisenhower and his wife, Rebecca, arrived from the East with their five children.

A local historian called the Eisenhower group the most meticulously organized operation ever seen: they came with half a million dollars in cash and fifteen railcars packed with every kind of
farm, construction, and household equipment. Jacob bought 160
acres not far from town in a fertile, well-watered spot called
Smoky Hill River (in the area labeled "the Great American
Desert" on mostly guesswork early maps) and set about con-
structing a life of faith and frugality and discipline. All the while
he and his good wife were raising a family that steadily increased
to a total of thirteen offspring, who would be brought up to live in
the same time-honored way.

The new country planted new notions in the head of David,
Jacob's first son and namesake, who persuaded his father to allow
him to attend Lane University. For an Eisenhower, that was a radical break from his family's tradition of farming, and a serious step
away from his powerful father's expectations. From the time they
settled in Pennsylvania in 1741, the Eisenhowers had been farmers. (Originally from Germany, several generations earlier the family had fled to Switzerland to escape religious persecution, then
migrated to Holland and eventually to America.) With industrious
husbandry of their fertile land, five generations of Eisenhowers
had prospered; they were prudent citizens who thought in acres
and turned the fruit of their labors into hard cash. But David, his
grandson John would say, "hated farming, he hated to get his
hands dirty—and there he was in Kansas."

When David approached him with his plan for college, Jacob
gave it deep thought. Working the land was God's own labor, but
the Brethren believed an individual must make his own decisions.
He consented. David, in his second year at Lane, met and was
beguiled by Ida, this self-possessed, vivacious girl with a smile that
made a man's heart sing. Equally important to Jacob's son was the
fact that Ida was a devout member of the River Brethren—rejecting the farming tradition did not signify David's apostasy. To this
girl who fulfilled both his dreams and his inbred requirements, he

confided his ideas for a different future: in this bustling new world of the West there were so many things a fellow could do instead of tying himself to the land. With his courses in mechanics, along with the classics required at Lane, perhaps he would be an engineer or maybe go into business.

They say opposites attract, and Ida was soon captivated by this young man. He was different from most students, tall, with strong dark coloring, serious beyond his years, uncomfortable around people, and given to dark moods. He was determined to marry Ida Stover, and he won her. On September 25, 1885—his twenty-second birthday—they were wed. Ida wore a dark silk dress that was far from Brethren plain, all pleats and draping and tucks and buttons. Her smile was even more radiant than usual as they stepped confidently into a future that could only have looked bright. Jacob, while surely heartsick that this Eisenhower son was abandoning the blessed land, abided by the Brethren precept that each individual is responsible for his own life. And so the patriarch gave the young couple his blessing and the same generous wedding gift he had given his more dutiful children: 160 acres and two thousand dollars as a nest egg—along with a handsome Bible that was to be read daily and heeded. Through the years it would serve as the archive of the newly formed family's history.

* * *

DAVID and Ida left college behind them to hurry into their new life. David, the confident new businessman, took out a mortgage on the farmland and with the two-thousand-dollar wedding gift bought a general store in a nearby village called Hope. He found a partner for the enterprise, a man named Milton Good. What auspicious names: Hope and Good. The partners, with Ida's input, enthusiastically planned an emporium that stocked a bit of everything, a place to buy chicken wire and candlewicks and teacups and swap opinions over the knitting yarn or pickle jar. In sad reality, this would be the high point of David Eisenhower's career as a businessman. From the outset, he was unsuited by temperament to fill the role of small-town merchant, an endeavor that calls for ease with people, for an outgoing geniality that was not in him.

Then the years that had brought plentiful rain to the prairie and enabled its farms to turn comfortable profits came to an end. Withering heat and merciless drought scorched Kansas, driving farmers—David's customers—into debt and bankruptcy. The slump, which would infect the entire country, was of such severity that even the well-fixed Jacob was helpless as the bank in Hope failed and his part ownership in it evaporated. Amid the general panic David was rocked by personal betrayal: his Good partner absconded with what little money there was, leaving David to face financial ruin alone.

So, a mere three years after David and Ida had married with such optimism about their future, he was bankrupt. Both Good and Hope had abandoned him; his tenets of self-denial and hard work had not been enough. Worsening his plight was his responsibility for a growing family. By then their first son, Arthur, was two years old and another child was on the way. Desperate for work, David went south, to Denison, Texas. Reaching back to his student days at Lane—those untroubled days seemed so distant, how could things have gone so badly so quickly?—he drew on his courses in mechanics to land a menial job as "engine wiper" with the Cotton Belt Railroad. His pay: ten dollars a week.

As David sank, his young wife grew stronger, more self-reliant, and always supportive of her husband. When he left for Denison, Ida, twenty-six years old and pregnant again, had to stay behind in Hope to deal with the demoralizing aftermath of bankruptcy, worsened by lawyers who, she would always be convinced, cheated her. Seeping into her mind was the realization that the burden of managing the family would increasingly rest on her. Many weeks later, in January 1889, without husband or mother for moral support, she gave birth to their second son, Edgar, then had to pack up their few belongings for the move to Denison. Her treasured piano was left with a friend until things got better, as surely they would. Even though they had no funds, to sell her piano was out of the question.

In Denison they found a small house situated on a dismal, cinder-grimed plot facing the railroad tracks, a shabby dwelling that would be proudly marked, sixty-odd years later, as the birthplace of America's thirty-fourth president. In less than two years, on October 14, 1890, Ida gave birth to their third son. They named him David Dwight, for his father and, according to family lore, for famous evangelist Dwight Moody, who was drawing enormous crowds at his revivals. (On growing up, the son reversed the order of the names.) A wild Texas storm raged on the night this son was born, which, the story goes, foretold his stormy temper. The augury might have been fanciful, but the temper was beyond dispute. Throughout his life an explosive nature lay just beneath his irresistible grin and genial air, and from his early childhood Ida was bound and determined to teach him to curb his unquelled anger.

Even with three healthy sons and a loving wife, David was beset by his failure. His headstrong decision—hubris, perhaps, in his merciless self-judgment—to turn his back on the Eisenhower tradition had proved disastrous; his bankruptcy in Hope had sapped his spirit and scarred his psyche. In the long history of the Eisenhowers, he was the first husband unable to take care of his wife and children. Always a moody man, he retreated ever deeper within himself, reading his Bible and poring over his strange astronomical charts that supposedly related various stars to the pyramids of Egypt.

Ida, meanwhile, was managing three babies under age four. No matter how frugal she was with David's niggardly wages, she faced the frightening reality that they could not make ends meet. In those dark days the sunlight seemed to fade even within Ida's resilient spirit. Their dreams lay in shards. Somewhere deep within that spirit a determination was taking shape: she would bring up her boys with the confidence to reach out to people, to take on the world. David was her dear husband, but she wanted their sons to do more than spend time mulling over mysterious connections between stars and pyramids.

Suddenly a shaft of light broke through the gloom of Denison—as Ida, with her trust in the Lord, knew it would. Jacob, aware of his son's humiliation and despair, called him back to Abilene. The Brethren had started a cooperative, Belle Springs Creamery, to increase profits on milk from their prospering farms, and they needed a mechanic. The fact that Jacob remained a commanding figure in the Brethren and David's brother-in-law managed the creamery was not irrelevant, yet it seemed that the maker of great patterns was at work: the patriarch had come to the rescue of the prodigal son mired in hard luck. A fully lived life was nearing its finish and a struggling son was given a new chance.

For Ida the return to Abilene was the answer to her prayers. She could enjoy once again the comforting companionship of family—the Eisenhowers by then numbered in the dozens—and the close-knit community of their church, a network of friendship that meant much to a young woman barely twenty-nine years old. Once again she was packing up to start anew, this time renting a tiny, one-story frame house, once again on the wrong side of the tracks, in a rough section derided as "Hell's Half Acre." The little dwelling measured 818 square feet—for a family soon to be seven. The older boys were squeezed into one room, while the baby of the moment slept in the parents' room. Many years later, looking around his enormous office, General Eisenhower observed that it was bigger than his family's first Abilene house: "I don't know how my mother jammed us all in."

That early experience had a lasting effect on Ike. In contrast to most men on the rise, who equate office footage with power and prestige, whenever he was offered a choice of offices, he went for the smaller one, which, he surmised, "may reflect a subconscious effort to test my own capacity for the use of space against my mother's." In that contest, Ida Eisenhower would have easily bested her famous son.

After about eight years and three more babies—and still not the daughter she longed for—in that primitive, overcrowded cottage, fate stepped in, again with a nudge from the kindly Eisenhowers, to improve David and Ida's lot. Brother Abraham had a calling to put aside his work as a veterinarian, to go forth and spread the word of God. He and his wife, in their specially fitted-out "gospel wagon," were pushing west to California, where the ratio of souls needing fixing was greater. Abraham had come up with a plan: David and Ida should rent his house on South East Fourth Street for a modest amount and take over the care of their father, a widower well into his seventies. Clearly, Abraham was already doing God's work—Solomon himself couldn't have come up with a wiser solution to the problem of David and Ida's outgrown house, at the same time entrusting the old patriarch to the care of capable Ida, and giving David a face-saving way to keep his head above water.

At last, a home of their own—with space in the tiny parlor for the ebony piano, which soon resounded with hymns of thanks. For fifty-four years, until her death in 1946, this would be Ida's home, the center of her life. This modest home was to become a national showplace, evidence that great leaders can spring from unlikely beginnings—when a woman like Ida is in charge.

★   ★   ★

THE military academies have long been a magnet for confident American boys. A six-year-old Jimmy Carter was already determined to go to the United States Naval Academy at Annapolis; Douglas MacArthur never considered any college but West Point; only bad eyesight kept Harry Truman from his dream of the United States Military Academy. Ike's choice, however, was more practical than idealistic: with no money for even modest tuition, he saw the chance for a free college education, and as an outstanding high-school football player, he hoped to play on the famous Army football team. His first choice had been the Naval Academy, but he was over its entry age limit. So the difference of a few months meant the nation lost a probable admiral but gained the commanding general of World War II—and Dwight David Eisenhower gained a preeminent place in history that was not possible for a naval leader in that war.

On the day he got his appointment, in the spring of 1911, he was walking on air. "This was a good day in my life," Ike wrote in his memoirs. "The only person truly disappointed was Mother. She believed in the philosophy of turn the other cheek. She was the most honest and sincere pacifist I ever knew, yet at the same time she was courageous, sturdy, self-reliant. It was difficult for her to consider approving the decision of one of her boys to embark upon military life even though she had a measure of admiration for West Point because one of her instructors at Lane University, a favorite of hers, had been a West Point cadet."

And so twenty-year-old Dwight Eisenhower was on his way to the U.S. Military Academy at West Point, clear across the country. With his new suitcase neatly packed, a very small stash of money pushed deep in his pocket, and exhilarated by this first step into his future, he had said good-bye to his father before he left for the creamery, and now his mother and brother Milton were with him on the front porch. For Dwight the moment was freighted with guilt; he knew all too well that in choosing a career in the army he violated his mother's deepest principles, her unyielding pacifism. Still, she managed a smile, along with the usual motherly cautions to a boy leaving home for the first time. With a struggle, she kept her composure, but as her beloved son strode toward the depot, she dissolved in tears. Never before had her sons known their indomitable, resolutely cheerful mother to weep.

Given her lifelong stance against war and all things military, it is surprising that Ida did not plead with her son not to choose a military life. Yet, from the time he sent his application to the Kansas senator who would make the appointment, she never tried to change his mind. In this instance her pacifism ran up against her religious belief that each individual must chart his own path. "I never saw Grandmother try to impose her thoughts on anybody else," says her grandson John. And besides, didn't her David go against his father's wishes when he turned his back on his family's nearly religious farming heritage to go into business? However she rationalized Dwight's decision, all she said was "It is your choice."

By the time her son had achieved the pinnacle of wartime leadership, Ida the pacifist could justify her pride in him: supreme commander of Allied Forces leading the greatest army ever assembled. To her, his was a modern crusade to safeguard all of the verities she lived by. As he achieved an unassailable place in history, Ida declared, "My highest hope is that Dwight may be an instrument in bringing peace to this troubled world." Looking at it that way, she saw his military role as a plan "ordained by God."

But that triumph lay years ahead. At West Point, Ike showed early promise in his deepest interests, football and baseball. Academically he was only a middling student, except in English, where he was in the top ten. Then the young athlete was permanently benched by a serious knee injury, which left him despondent, with passing thoughts of leaving the academy and abandoning his future in the army. But he soon regained the optimism that had come to him from Ida and was on his way; once he found his footing among cadets who had come from better schools and far more privileged backgrounds, Ike began to shine as a leader, in pranks as well as cadet training.

After his second year, he raced home to Abilene for a long summer furlough, strutting a bit, his brothers thought, in his snappy uniform that turned every girl's head. Ida, notwithstanding her feeling about the military, spent hours lovingly ironing his summer whites, creased to knife-edge perfection and starched until they could stand unoccupied. She had to admit it: her Dwight—tall, blond, with an easy laugh and a smile to match—was about as handsome as they come.

Among the girls whose heads swiveled was Gladys Harding, the prettiest thing in Abilene and a singer of considerable talent. Back at West Point, Ike, smitten, wrote letters to her pulsing with restrained passion and declarations of his devotion—"Girl," he called her, and somehow it sounded as romantic as Keats or Shelley. (Many years later, after the deaths of both Ike and Mamie, Gladys turned over to an Eisenhower biographer a small chest of many neatly ribboned letters.) In New York, where she was hoping to make it as a singer, they met for a wonderful, dreamlike weekend, doing the city, being in love. But when Ike, soon to graduate, talked of marriage, Gladys talked of her career on the stage. She thought he would wait for her. He didn't.

As a young lieutenant, with those shiny gold bars on his broad shoulders, Ike was breaking into the placid by-the-numbers life of the peacetime army at Fort Sam Houston, just outside of San Antonio, Texas. There, life was centered on the officers' club, with a little work on the side. And there his own life was about to take a most important turn.

★   ★   ★

THEY met on the lawn of the officers' club on a warm October Sunday. It may have seemed unplanned, but a mutual friend, playing matchmaker, had set up the ostensibly casual encounter. Miss Mary Geneva Doud—Mamie—wearing a pink outfit studded with pink roses and a romantic straw hat with a pink band relaxed in a canvas chair. She was a few weeks from nineteen, with china blue eyes the size of a full moon and a sparkling smile, a flirtatious belle whose date book was always crammed to the margins. Lieutenant Eisenhower, invited to join the little group on the lawn, accepted most readily, then asked Mamie to walk his assigned rounds with him. She accepted most readily, high heels and all. Before they had finished that pleasant duty, Mamie's seven o'clock date was fuming that she was so late and Dwight Eisenhower had decided to marry her.

It would take a lot of courting. Mamie thought he was fun, but "fun" officers clustered around her every season in San Antonio and numerous other young officers had fallen under her spell. This pretty butterfly did not have settling down in mind; she relished her social whirl. In a letter to a friend back in Abilene, Ike—not revealing his intentions—mentioned Mamie's avid pursuit of the social life as her only fault. She had been born to it, brought up in a family with two homes, smart cars, and lively parties. Her every whim was indulged. (When she was six, she asked for a diamond ring for Christmas, which dutifully appeared beneath the tree; it was a very small stone, to be sure, but a diamond nonetheless.)

Ida had once commented with satisfaction that Dwight always got what he went after, and within months he got Mamie. On Valentine's Day he slipped a ring on her finger and on July 1, 1916, they were married in the Douds' Denver home, with only Mamie's family present. Travel was daunting in those days, and David was not one to take time off from the job. A Presbyterian minister performed the ceremony, and a harpist played ethereal music. The bride was a picture in a long gown of chantilly lace with a pink satin sash—her trademark color, which would come to be known, when she was First Lady, as "Mamie pink." Friends came in for a champagne reception, a number of them harboring the thought that Mamie, still so young, could have done much better. "After all," one commented, "he was only a soldier and a poor boy, and everybody felt he was marrying above his class."

Everybody but Ike. Their backgrounds, their upbringing, were poles apart, but Ike was not at all bothered by the gap between the Eisenhowers' pinched circumstances and rigid mores and the Douds' affluence and easy lifestyle. His mother had imparted a different set of values and boundless confidence. Insecurity was not part of the Ike package—and an officer's spiffy uniform is a great equalizer.

★　★　★

IT was four in the morning when the Union Pacific from Denver lumbered into the sleepy Abilene station and the newlyweds stepped out into the steamy night. On the platform David Eisenhower was there to greet them. Mamie's first impression was surprise that her new father-in-law was in shirtsleeves—the first of the many differences, small and large, that she would encounter between the Eisenhowers' simple life in Abilene and her social circles in Denver and San Antonio.

At the Eisenhower home, lights burned brightly at this odd time before night eases into day. Ida anxiously listened for the sounds of their arrival breaking the stillness. Since Dwight was the first of her sons to marry, this role of mother-in-law was a new experience and she wasn't sure just what to expect. All she knew was that Dwight's bride was certainly pretty, judging by snapshots he had sent. There are few relationships more complex and delicate than that of mother- and daughter-in-law, all the more so when their first meeting comes after the marriage. Inevitably a feeling of comparisons made and judgments passed hang over the early encounters of the two women crucial to a man's life; they must deal with the tangled emotions of giving up the son or gaining the husband. Sara Roosevelt and Eleanor never sorted it out; Betty Ford and Rosalynn Carter weathered an uncertain beginning, Jackie Kennedy courteously tolerated the differences; and a rare few like Lady Bird Johnson and Barbara Bush converted the tension into lasting friendship. Understandably, young Mamie was nervous about what her new in-laws might think of her, but her social graces would see her through any awkwardness.

Meeting Milton, Ike's youngest brother, Mamie—one of four girls—exclaimed, "I've always wanted a brother!" and kissed him on the cheek, an unusual gesture in the seven-male Eisenhower household. Milton, who became a distinguished university president, would recount this introduction with the fond admission "I've been her willing slave ever since!" Dwight showed his bride the remnants of his boyhood—the big barn and the garden they had worked—and then it was nine o'clock, time for breakfast and another surprise for Mamie. "My mother was determined that we should have at least one fine meal in her house," Ike explained in his memoirs. So, instead of bacon and eggs or pancakes, Ida served "a monumental fried chicken dinner." Mamie, well aware that she was being judged like a prize Tamworth at the county fair, was "on," smiling prettily, friendly but not forward, as comfortable as a girl used to having fried chicken every morning for breakfast. Before noon she had them eating out of her hand. Then, as quickly as they'd arrived, she and Ike were off again, headed for Fort Sam Houston.

The differences between this mother and her daughter-in-law went far deeper than a breakfast menu or the fact that Ida deplored nicknames while the Douds used nothing else. Mamie grew up as the coddled daughter of a wealthy family; both her grandfathers had accumulated sizable fortunes. (Her Doud forebears came from England in 1639; her maternal grandfather was a Swedish immigrant.) John Sheldon Doud, Mamie's father, had built such a successful packinghouse business in Boone, Iowa, that by the time he was thirty-six he could retire comfortably to Denver and devote himself to enjoying life with his family.

Their year was divided between Denver, from late spring through fall, and San Antonio, to escape the rough Colorado winters. The seasonal transfer was a three-week procession made in the Douds' latest touring car, with two servants in the retinue. For the Eisenhowers, scrimping was a way of life; Ida and the boys handled all of the household work, watching every penny, while the Douds were a jolly, fun-loving family who liked dancing and cards and enjoyed having a drink or two.

Mamie's mother, Elvira Carlson Doud—in this family of nick-

names, she was "Min"—played the harmonica con brio and
whipped it out at parties to entertain with her repertoire of popular tunes. Mamie attended a fashionable Denver finishing school
(where she was considerably stronger in finish than schooling),
had a closet full of up-to-the-minute dresses, and was indulged in
any passing whim. "I was rotten spoiled," she admitted to her
granddaughter-in-law, Julie Nixon Eisenhower.

The Eisenhowers' religious scruples prohibited all of the above
and more. In a time when only a few daring ladies smoked,
Mamie did—on later visits to the Eisenhower home, she would
hang out of the upstairs window and surreptitiously light up a cigarette. Ida, a born homemaker who had taught Dwight how to
clean and sew, as well as cook, when he was a boy, must have been
bewildered that Mamie had mastered none of the housewifely
arts. "I'm not a cook," Mamie blithely conceded. "I was never
permitted in the kitchen when I was a girl." Nor is there any indication that she objected to that deprivation. (She would have
winced at the Eisenhower boys' favorite dish, called puddin', Ida's
special cornmeal mush with some dark concoction better left
unidentified spooned over it.) And when Mamie was pregnant, it
was Ike who altered her dresses to fit her expanding waistline. It is
funny, and telling, to compare the personal recipes the president
and First Lady contributed to a VIP cookbook published by the
Women's National Press Club in the Eisenhower years. Ike's dish,
"The President's Old Fashioned Beef Stew," featured hearty por-
tions for sixty; the First Lady shared her recipe for "Million Dollar Fudge."

Though Mamie respected her husband's parents and admired
Ida for the job she had done with their sons, she was never altogether at ease in the Eisenhower home. The two women had very
little in common, a gap that was made more difficult by David
Eisenhower's dour personality, which could cast a pall over the
evenings. (But for lively young grandson John, the countrified life
of the Eisenhowers was more fun than the fancier city ways of the

Douds: "In Abilene I was allowed to sit out on the lawn and eat fried chicken in my undershirt. The other grandparents were more formal—I had to dress properly.")

Mamie's close and loving family immediately drew Ike into their lives. He was a surrogate son to John Doud, father of multiple daughters, and Min adored him from first meeting. (In the White House years, Min visited for weeks at the time, a lively addition, in striking contrast to the tenure of Madge Wallace during the Truman years.)

One pleasure Ida and Mamie shared enthusiastically was the piano. Both played well, Mamie entirely by ear, while Ida had the benefit of professional instruction in college. In the gypsy life of an army wife (more than thirty moves, she guessed) Mamie rented a piano, no matter how cramped their quarters, paying five dollars a month even if they couldn't afford it. Ida understood that extravagance—she had only to look at her own piano to remember that she had done much the same. At parties young officers and their wives gathered around while Mamie played and Ike led the singing in no recognizable key, much as the Eisenhowers had done back in Abilene. There was, however, a marked difference in musical tastes: Mamie liked Tin Pan Alley's newest tunes, while Ida stuck with the favorite old hymns. Ragtime for Mamie, "Rock of Ages" for Ida.

★   ★   ★

EVEN as a junior officer, Ike stood out as a natural leader, skilled in dealing with people while exercising firm authority, abetted by his buoyant personality. The authority was honed at West Point, but the basic traits had been implanted by his mother. She did, however, fall short in her efforts at temper control—from childhood those lightning flashes were as much a part of Ike as the contagious smile.

In the heat of party politics, which he always found distasteful, his tendency to explode was never fully overcome. As president, when a news conference question got under his skin a vein in his forehead would stand out noticeably, signaling, "Caution: high voltage temper." Merlo Pusey, a senior White House adviser, wrote, "Sometimes his anger is aroused and it may set off a geyser of hot words. The president's emotions are close to the surface, and his irritations are registered on his face almost as readily as his general good humor." His storms of temper passed quickly and his rare moods of depression were also transitory. Following his mother's example, he could put aside personal conflicts and present a cheerful and confident face to the world.

At the height of the battle for Europe, Ike's old friend, General George Patton, was shocked by the icy ferocity of Eisenhower's tongue-lashing following the infamous incident when Patton slapped a young soldier suffering from combat fatigue. Ida had urged him to curb the heat of his anger, but he compensated by developing a coldly remote side, which perhaps reflected the dour authoritarianism of his distant father.

In private life, he was equally volatile. White House seamstress Lillian Rogers Parks, in her gossipy book about her thirty years backstairs in the executive mansion, tattled that President Eisenhower had a "terrible temper. . . . Poor Mamie lived in constant fear that Ike would burst out at the wrong time." Julie Nixon Eisenhower remembered playing bridge with her husband's grandparents when Mamie made a stupid play. "Why did you play that?" Ike demanded. "Just 'cause I wanted to," Mamie replied airily. At which point, said Julie, "Ike throws down his hand, leaves the bridge table in disgust, his face a flame color." There were still traces of the boy Ida called "the little pouter."

Ida's approach to dealing with his outbursts was the opposite of his father's method. Where David would wield a hickory switch and gruffly order the boy to bed, Ida would wait out the worst of a tantrum, then reason with him quietly. Sixty years later, in his

memoirs, her hero-son recounted the lifelong lesson he absorbed
from his mother's wise counsel following a particularly violent
episode when he was ten:

> Hatred was a futile sort of thing, she said, because hating any-
> one or anything meant that there was little to be gained. The
> person who had incurred my displeasure didn't care . . . the only
> person injured was myself. This was soothing, although she
> added that among all her boys, I was the one who had most to
> learn.
>
> I have always looked back on that conversation as one of the
> most valuable moments of my life. To this day I make a practice
> to avoid hating anyone. If someone's been guilty of despicable
> actions, especially toward me, I try to forget them. Eventually,
> out of my mother's talk, grew my habit of not mentioning in
> public anybody's name with whose actions or words I took vio-
> lent objection.

Recognizing his temper as a character flaw, Ike tried to restrain
it. He confessed in his wartime diary, "Yesterday I got very angry
and filled a page with language that this morning I've expur-
gated." He devised a practice, possibly at Ida's suggestion, of writing the name of the offending person on a slip of paper and
burying it in a box in his desk drawer—a physical and symbolic
act of banishing the individual from his thoughts.

He must have forgotten to place Harry Truman's name in the
box. As long as the Truman-Eisenhower relationship was between
president and general, it was cordial. Truman honored the general
at a White House dinner and praised him, in a letter to Mamma,
as "a real man." But when Eisenhower emerged as the Republican
president-elect, relations turned bitter; Supreme Commander
Eisenhower was not accustomed to the attacks that are part of
politics, and Truman, leaving office after almost eight years, took
umbrage at perceived slights to the presidency and to his efforts to

provide a smooth transition. Ike's antipathy to Truman ran so deep that he wasn't sure "if I can stand sitting next to him" in the short drive to the Capitol. Ethel Noland, Truman's wise cousin, had a theory about Ike's frostiness: "He never could forgive Harry for saying that military men never made good presidents, which he did say. He still says it."

Ironically, these two leaders had more in common with each other than with any other modern presidents. They stemmed from the same breed of hardworking, cash-pinched, God-fearing, small-town/farm families, Middle Americans by virtue of geography and philosophy. Both were open, artless, and conventional, hewing closely to the fundamental values instilled in them by their mothers, strong in their faith in the Almighty (though not demonstrably pious), respectful of the military, shrewd poker players, and connoisseurs of good bourbon. And they really hated each other.

Years passed before the two aging ex-presidents could put the past behind them, at least for the cameras. Their mothers, women blessed with common sense and uncommon wisdom—"mother wit," they call it in the rural South—might have said to their famous sons, You both honored the office, and history is big enough for the two of you.

★   ★   ★

AS a term of reference, "mama's boy" is freighted with negative baggage—the image of a wimp tied to his mother's protective apron strings. In reality the image is off the mark. Take, for example, three of the most famous generals of the twentieth century: Dwight Eisenhower, Douglas MacArthur, and George S. Patton, all as tough as they come. All lovingly credited their success to their mothers. Ida implanted values and made her Dwight toe the line; Mary Hardy ("Pinky") MacArthur was equal parts coach, cheerleader, and lobbyist on behalf of her Douglas. As he left to take the West Point entrance exam, she gave him a pep talk worthy of Vince Lombardi: "You must believe in yourself, or no one else will believe in you. Be self-confident, self-reliant, and even if you don't make it, you will know you have done your best. Now, go to it!"

While Ida shed tears as Dwight left for West Point, Pinky, the widow of a three-star general, moved from Milwaukee to West Point, settling into a hotel room overlooking the parade grounds for her son's four years, and afterward moved to every new post to stay close to him (wrecking his first marriage in the process). Through the years she used every connection, pulled every string, to advance his career—with considerable success. Douglas idolized her.

General Patton—"Blood and Guts," they called him—was as rough a soldier as ever roared into battle, but when it came to his mother, he was all devotion and sentiment. Visiting his boyhood home, long after his mother's death, he wrote this tender note to her, which he placed in a little box holding her dearest mementos:

Darling Mama, I had always prayed to show my love by doing something famous for you. . . . Nothing you ever did to me was anything but loving. I have no memories but love and devotion. . . . Perhaps this is foolish, but I think you understand. Your devoted son, George S. Patton, Jr.

And Eisenhower, his comrade and superior officer, would surely have understood. At his wartime headquarters in England, the only photograph on Ike's office wall was of his mother, a serene, white-haired Ida in profile, seated on her front porch back in Abilene; the resonance with Whistler's famous portrait of his mother was striking. In the most harrowing days of his life, her son could see something more in that portrait—Ida encouraging him to be his best.

In May 1944, amid final planning for the most massive invasion of history, the supreme commander took time to dispatch a top-secret order to Washington, making sure his Mother's Day message would be delivered to Ida in far-away Kansas. These were strong men who were grateful to their strong mothers and proud to be Mama's boys.

★   ★   ★

"BOYS are the best investment in the world," Ida declared in her waning years. "They pay the biggest dividends. The boys were our bank. We invested in them every spare penny we could save." And the investment paid rich returns. All six did well (some more so than others) in what they chose to do. Milton was president of three major universities; Edgar, a leading lawyer in Seattle; Arthur, vice president of a Kansas City bank; Earl, a businessman; Roy, a pharmacy owner. And Ike. No imagination could have stretched to include a supreme commander *and* a president. Yet throughout his brilliant career, power and success never went to his head, as gold braid and presidential trappings often do; he never abandoned Kansas common sense, plain talk, and commitment to old-fashioned values—those things he had absorbed from his mother.

"They were all self-made men," said John Eisenhower of his uncles and his father. "All prima donnas—and they were pretty tyrannical with their own kids. The whole family was very strong and self-contained, with utter independence of feelings. I think they got that from Grandma: you don't lean on people." John was amused that whenever these strong-willed brothers got together, even as self-assured men, "they would yell and argue—on the verge of fisticuffs." They were still the scrappy Eisenhower boys, just like the old days, with peace-loving Ida still sighing that boys will be boys.

Edgar and Milton had been the family stars until Ike was catapulted to the front pages, and his rise to power was met with very different reactions. Milton, the youngest brother, became Ike's closest confidant, while Edgar, in John's view, "had a very tough time adjusting to being overshadowed and to Milton being so close." The Abilene High School yearbook had forecast Edgar as a two-term president, Dwight as a mere professor of history at Yale. Didn't that make Dwight the usurper?

While Milton was supportive and discreet, the ultraconservative Edgar freely blasted his president-brother, untroubled by the stir he caused. Over lunch in the White House, Ike snapped to Edgar, "The only reason anybody is in the least interested in you is because you are my brother." So, as John tells it, "Edgar went straight out and told the press, 'He chewed me out!' " Which predictably produced more headlines. Ike laughed it off: "Edgar's been criticizing me since I was five years old!" In Edgar's mind, he was still "Big Ike."

★   ★   ★

DAVID, in retirement, suffered a series of small strokes and grew increasingly morose, lamenting that he had accomplished so little in life. "Don't be despondent about your career," Ida gently reassured him. "Your career was raising six good boys. That's your investment." It was a kind and loving thought for her husband, who had never regained his self-esteem after his early failure, and whose influence on those sons was minimal, even negative, compared to hers.

On March 10, 1942, Ike's diary began with a bare entry: "Father died this morning. Nothing I can do but send a wire." His next sentence dealt with his irritation at Admiral Ernest King. On the following day he regretted that he could not be there to comfort his mother. "I think my Mother the finest person I've ever known. She has been the inspiration for Dad's life and a true helpmeet." Not until the third day did he set down thoughts of his father: "He was a just man, well liked, well educated, a thinker." Modest, quiet, undemonstrative, uncomplaining—those were Ike's adjectives for the complex David Eisenhower. "My only regret is that it was always so difficult to let him know the great depth of my affection for him." A doleful coda for a good man bereft of the human touch. Among David's possessions, the one thing the son asked as a memento was a sample of the name they shared, written in his father's elegant, perfectly shaped handwriting.

"I feel so lost now," Ida told a reporter. "With all of them gone, nothing seems real any more." The old house that held so many memories felt empty, too quiet. Yet she maintained a cheerful countenance to the end. It was surely her sunlit soul—and the contrast with gloomy David—that led Ike to attach such value to the virtues of cheerfulness and enthusiasm. Up until her final years, Ida stayed busy working in her flower beds, and her fingers, though not as nimble as they once were, were still flying on her fancywork. Rocking on the front porch that was meant for just that, "Mother Eisenhower" would chat with passersby, many of whom claimed Dwight as their old best friend, no longer remembering him as that rowdy kid from the other side of the tracks. Her face, beneath its nimbus of white hair, reflected a glad heart and joyful spirit. She had steered her sons safely through trying times, curbed their wayward inclinations, inspired them to succeed, instilled in them her own true principles. She was content.

Late in his own life, Dwight spoke lovingly of "her sincerity, her open smile, her gentleness with all and her tolerance of their ways. And for her sons, privileged to spend a boyhood in her company, the memories were indelible. . . . Mother was by far the greatest personal influence in our lives."

Late in the night of September 11, 1946, Ida awoke with a pain in her stomach and asked for a glass of water. Insisting that she was all right, and still motherly at eighty-four, she told her nurse-companion, "Make sure you are covered up—you might catch cold." Then, as quickly as turning out a light, she was gone. To her last breath, Ida Eisenhower was thinking of others.

Just as Ida had approached her death with serenity, so did her Dwight many years later. John Eisenhower attributes his father's attitude directly to Ida. "I think the big thing Ida Eisenhower gave her sons," he muses, "was some inner faith that developed into serenity. My dad was not religious in any conventional sense. But I was around him in his last moments, as I was with Milton [the youngest brother] in his last days. Both accepted their impending deaths with admirable calm. I think that security came from Ida Eisenhower."

★   ★   ★

HOW gratifying it would have been to know that her Dwight had marched into history not only as a great general but as a beloved peacetime president, remembered as the military leader who deplored war. "More than any general in history, Eisenhower hated war," declared his biographer Stephen Ambrose, "and that's why the American people loved him. They wanted that kind of man in charge of their boys." In an early study, columnist Marquis Childs characterized Ike as "the war hero who spoke eloquently against war . . . a professional soldier who had always found war repugnant and who was convinced, after World War II, that another war would mean mass suicide."

Ida had been dead for six years when her son took the oath as president, his hand on the Bible opened to a verse she had often quoted. Her influence was woven into his inaugural address. Working with his speechwriter, Ike had set down four key points: "understanding, heart and determination, productivity, readiness to sacrifice." They were the essence of Ida's teaching, Ida's example, extended now to the country. She would have been pleased that Dwight, never a very religious lad, began his inaugural address with a "little private prayer of my own" and that the following Sunday, in Washington's National Presbyterian Church, he was, at last, baptized. And when he delivered his farewell address to the nation eight years later, Ida the pacifist would have applauded her soldier-son's warning against the might of "the military-industrial complex," a caution that has been cited again and again over the years. Had his mother lived to see him in the White House, she could have seen the clear mark of her influence.

★   ★   ★

JUST as he was a soldier who genuinely hated war, Eisenhower was a politician who disdained politics. All the same, he was a very effective politician—distancing himself from messy conflicts, avoiding confrontations, husbanding his popular support behind big decisions. Any man who could handle the massive egos and imperial wills of Churchill, Roosevelt, and de Gaulle simultaneously knew how to achieve his ends, more effectively, through compromise and amiability rather than confrontation—the way Ida would have done it.

He also was able to separate from his mother and her world, to remain a loving son while leaving behind the rigidity of her faith and the limits of Abilene. It required the true wisdom of a mother for Ida Stover Eisenhower to understand that all of her sons would break away from the old mores and traditions of the Stovers and Eisenhowers. She had seen her husband attempt to do just that—and fail; she was resolved that her boys' lives would not be blighted by hopes unrealized and dreams deferred. The ties of affection between mother and son remained strong because Ida knew how to let go, to let each one find the key to a door that opened into a wider future.

4

THE GLUE THAT HELD

IT ALL TOGETHER

Men are what their mothers made them.
—RALPH WALDO EMERSON

*T*HE crucial 1968 California presidential primary was only days away. As Robert Kennedy raced from one end of the state to the other, driving home his impassioned message, his brother Ted and his three sisters were everywhere. Once again the Kennedys were immersed in a family campaign. In a rare quiet moment between stops, Bobby Kennedy contemplated a question: what had led the Kennedys—the whole family—to throw themselves once again into the thick of politics, after it had brought the family such tragedy? He squinted from under his eyebrows and his face crinkled into that lopsided grin: "Have you met my mother?"

Conventional wisdom had it that the formidable patriarch, Joseph P. Kennedy, had propelled his sons onto the path to power, dipping deep into his fortune and connections to catapult his second son into the White House, and had he not been sidelined by a crippling stroke he would have been doing the same for Bobby. But what did Rose Kennedy add to the mix?

At seventy-seven, Rose was regal, even glamorous, as she entered the ballroom wearing a size six Paris outfit and the aura of a Personage. She delivered her short, unscripted remarks in a slightly Hepburnesque pitch and that hallmark accent that gave no quarter to TV's homogenization of American speech. She talked about her family and its dedication to public service and why this audience, any audience, should vote for Bobby: because he would make a difference.

Were her crowds, which cut across every demographic divide, won over by her message? Clearly, they were enthralled by this legend in their midst, this woman who with stoic grace had weathered tragedy of Shakespearean dimension. (And, within days, in the same Ambassador Hotel ballroom her family would be struck again.) They shook her hand, paid rapt attention, and many of them must have voted for her son, because Robert Kennedy won the California primary handily, too late. Rose was unstinting in her effort to help Bobby, the son most like her, win his race; she would work for him with all the zeal she had unleashed for Jack in 1960. She was on the campaign trail doing what she had always done and would always do: everything possible for her children's success. Would she do the same for Teddy, she was asked, when it was "his turn"? "Of course," she replied, with a slight frown, as if there could be any doubt. By her political calculations she would then be eighty-five (allowing two terms for Bobby—the possibility of defeat was not factored into Rose's thinking) and she assured her audience that she would still be in campaign form. (And at eighty-five, she was.) Time spent with Rose Kennedy made it easy to understand just what Robert Kennedy meant.

Rose Fitzgerald, daughter of the powerful mayor of Boston, had brought to her marriage to Joseph Patrick Kennedy a maiden name as illustrious in her world of Irish Boston as the one Sara Delano brought to her union with James Roosevelt. Both were matriarchs fiercely defending their children as a lioness would her cubs. And both women had powerful attachments to their fathers; their unassertive mothers, who dutifully yielded to the

strong personalities of their husbands—and their daughters—
recede into the background in the family histories.

The linkage of influence from grandfather to mother to son was
never so clear as in the relationship between Rose Fitzgerald and
her father. Her mother, Mary Josephine Hannon, striking and
quite stylish for a farm girl, was a reserved and deeply religious
woman who shunned the public life of her husband, the mayor.
Rose, in her late teens and remarkably self-possessed, was
delighted to to be her stand-in. "My love of politics came from my
father," she said more than sixty years later. "He conveyed to me
a great deal about history and politics. He made me think about
those things. And I in turn tried to do that with my own children."

Asked, in 1972, if she had ever wanted to be a senator, she
replied, after a few moments' reflection, "I think I'd rather be the
mother of a senator. You can form a child's mind, and soul." "As
the twig is bent," she wrote in her memoirs, "so the tree inclines."
Her energy and attention, unlike Sara's unwavering focus on
Franklin, had to be allocated among a family of nine. Her eldest
son, who bore the paternal name, Joseph Kennedy Jr., was the
brother the family expected to succeed in politics, but it would be
her second-born son, bearer of her proud Fitzgerald name, who
won the ultimate prize. The patriarch lived long enough to experience the terrible wisdom of the injunction "Be careful what you
wish for, because your wish may be granted."

★   ★   ★

ROSE Fitzgerald Kennedy, gathering her brood around her on
the waterfront at Plymouth, Massachusetts, told them, not for
the first time, about the courageous little band of Englishmen
and -women who braved all dangers and abandoned all comforts
to carve out a place for themselves in that wilderness. Her daughter
Eunice Kennedy Shriver remembers that her mother, in her school-
teacherish way, would toss out questions: "Why do we come here?
Who were the Pilgrims? Who was Miles Standish?" The children

were quick to come up with the answers because, Eunice laughed, "she took us there many times."

Plymouth Rock was only one of the expeditions Rose Kennedy led in her station wagon, crowded with nine young Kennedys. When they were at their Palm Beach house, she would round them up for a trip to a nearby Indian village. "We would drive into the reservation and see how they lived," Eunice recalled. "And when we were back at home, she asked us about what we had seen." It was her way of making history come alive. The children didn't mind her quizzes—they wanted to please Mother, and they wanted to get out in front of the others. Competition, coming in first, whether in the potato sack race at school or in presidential politics, was bred into the Kennedys and was as much a part of their makeup as the gleaming teeth and thick mop of hair.

The drive to win was instilled in each of them by example and expectation. Not only was their father a winner in the cutthroat competition of Wall Street, Hollywood, and Washington, but Rose, even as a schoolgirl, had been a formidable contender. Let the others be shrinking violets, she wanted to be the best at everything—and by and large, she was. She expected no less of her children.

A census-taker would have recorded "Rose Kennedy—housewife," which would be something like listing Bill Gates as "computer programmer" or Mark McGwire as "baseball player." To Rose, "mother" was more than a noun, feminine gender, it was an active verb, "the greatest career," which she pursued with prodigious energy. In her memoirs, *Times to Remember,* she elaborated: "I looked at child rearing not only as a work of love and duty, but as a profession that was fully as interesting and challenging as any honorable profession in the world, and one that demanded the best I could bring to it. . . . I have relished my role." And one thing more—her husband had entrusted her with the primary responsibility in bringing up their children. He demanded much of them, and of her; she would more than meet those demands.

To that end she was teacher, nurse, social adviser, lay preacher, drill sergeant, camp counselor, and, in later years, campaign worker. As her children arrived over a period of seventeen years— Edward Moore Kennedy, Senator Ted, was the last—she considered her role in their lives: "Whenever I held my newborn babe in my arms, I used to think what I did and what I said to him would have an influence, not only on him, but on everyone he meets, not for a day, or a year, but for all time and for eternity. What a challenge, what a joy!" She grasped instinctively what child-development experts impress on mothers today: literally from birth an infant begins to absorb, to learn, most of all from its mother. In that doll-sized head an active brain is already beginning to decode messages, to learn signals, to catch on to the world. Rose wanted the right messages stored in those receptive new minds.

The full force of Rose and Joe Kennedy's intentions was directed at ensuring a great future for their four sons. (Their daughters, they hoped, would marry well and have wonderful families.) Well before his ninth and last child was born in 1932, Joseph Kennedy dared to think of a Kennedy in the White House—if not himself (a notion he seriously entertained), then a son. Logically his eldest, Joseph Patrick Kennedy Jr., his namesake, his favorite, would carry the Kennedy banner. There are studies showing that fathers of highly successful men are primarily concerned with their sons' choice of a career, while the mothers envision for them a future without limits. Joe Kennedy's career aspirations were boundless.

President John F. Kennedy, reminiscing about his childhood, pointed out that his father "was not around all that much." The dynamic Joe was consumed with building his fortune on Wall Street and in Hollywood's nascent film industry, amassing the wealth and power to support his dynastic plans. There was, however, time for extramarital relationships that would set a powerful and destructive example for two generations of Kennedy males.

In her husband's absence Rose embraced the role she was born for. In the early years of the twentieth century, achievement for a woman of her background was defined by her skills as a mother and the success of her children. True, maiden ladies—spinsters—with a measure of intellect or ability could win respect in careers as teachers or librarians. Factory jobs were for working-class girls; a career in a profession like medicine or law was unthought of, except by the rare "bluestocking" intellectual. But gradually office jobs and telephone switchboards opened the door to new opportunities for young women whose families were edging up toward the middle class.

★   ★   ★

THE mayor's beautiful daughter had attracted a number of suitors including, it was rumored, Sir Thomas Lipton, the celebrated British yachtsman whose sleek craft floated on Lipton tea. In truth, Sir Thomas was her father's good friend who enjoyed a teasing relationship with young Rose (but was rather preoccupied with his French mistress). The family loved to retell how, on Britain's Isle of Wight, Honey Fitz once commandeered King Edward VII's royal launch, instructing the skipper to take him and his daughter out to the Lipton yacht, at anchor in the harbor. His Majesty, arriving a few minutes later, could only sputter as he watched the wake of his launch headed toward the *Erin*. With a story of such cheeky impertinence, the much-amused Sir Thomas and his guests forgave the mayor for not being the monarch.

Honey Fitz might hobnob with a baronet of the empire and pinch the king's boat, but he could never achieve social acceptance in the closed world of Boston's founding families. They had fiercely resisted the Irish from the first arrival; the immigrants and their American-born offspring fought back through tribal solidarity, political skills—and large Catholic families. In the nineteenth century the new Irish Americans lived as apart from the old Bostonians as southern blacks from whites before the civil rights movement. East Boston was closer to Galway than to Beacon Hill—two worlds separated by a common language, in George Bernard Shaw's phrase, by religion, and by every aspect of their cultures, from the way they comported themselves to the songs they sang and the food they ate.

The tectonic plates of Boston society were (and still are) grinding inexorably against each other, and though the Fitzgeralds lived on the fault line there was no question as to which side was theirs. On one side were the "Brahmins," whose old-line family trees were heavy with early colonial—better yet, Revolutionary—antecedents. These American patricians came to be identified in shorthand as WASP: white Anglo-Saxon Protestant, with bloodlines, distinction, and old money implied. In this social order, even more stringently than in Boston business firms, the rule was "No Irish Need Apply."

Boston's other society was led by Irish "aristocrats" who had made good by dint of their wits and their work, and who, through education and careful observation, had acquired a patina of social graces despite being only two or three generations away from dirt-floored cottages in Ireland. They embodied the American dream that was a magnet for Europe's daring young who could foresee no future in "the old country" and had the spunk to set out for a new life.

Rose and Joe's grandfathers had taken that chance, and their sons had done well, especially Honey Fitz, perhaps the most popular figure in Boston. By one account, in his first two years as mayor the peripatetic Honey Fitz attended fifteen hundred soirees, danced with five thousand ladies, and still found time for a thousand meetings. Rose, her Irish charm laced with innate dignity, loved to make an entrance on her father's arm at gala events and grace the dais at civic functions; she traveled with him in America and Europe, savoring the attention lavished upon her and the glowing newspaper accounts. And all the while she was learning the tactics and nuances of politics, invaluable training for the life that lay in store for her.

Did Rose's mother feel slighted, or diminished, because her daughter, so like Honey Fitz in her ease in public life, had more than taken her place? Rose witnessed her mother's flashes of resentment at her husband, who called himself a family man but rarely spent an evening at home. Rose was a party to the mayor's separate social life, and it would have been surprising if there had not been irritation—perhaps even subliminal jealousy—on Josephine's part.

Rose realized that she would always be excluded by the Brahmins of Beacon Hill and Pride's Crossing, but she would match them, and more. Being shut out of their society left a scar, but Rose set out to eradicate it by bringing up a new generation that would outshine those who regarded her, and her family, as inferior simply because they were Irish.

To make this possible, there was the matter of marrying a suitable husband, some young man of rising family and good prospects. Before Rose could think seriously about any suitor, it was necessary for her to be presented formally to the sector of society that was open to her. Just after New Year's in 1911, 450 guests arrived at the Fitzgerald mansion in Dorchester for the debut of the mayor's eldest daughter, the social event of the season for the Irish upper crust. More than sixty years later, Rose lovingly relived every detail. The imposing house, with its stained-glass windows, had been turned into "a bower of roses." The debutante was stunning in a gown of embroidered white satin with a short train. She wore no jewelry or ornaments—such a radiant twenty-year-old needed none—"except," she remembered, "a bit of white ribbon in my hair."

The governor, two congressmen, the entire city council, and numerous other dignitaries attended. The event was treated like a municipal celebration, "with yards of photographs and write-ups in the newspapers," Rose, who never shrank from publicity, recounted proudly. Though she would become gracefully at home in embassies, palaces, and the White House, no event was ever

120   FIRST MOTHERS

more exciting than her coming-out party in Dorchester. No Beacon Hill girl had such a debut that year: Rose Fitzgerald was showing them that one could be Irish and elegant.

★ ★ ★

THE point of a debut is to put the eligible daughter on the marriage market. Rose Elizabeth Fitzgerald would have been a catch for any young man in her social circle, except for one thing: she had already fallen in love with Joseph Patrick Kennedy. It was at Old Orchard Beach, a popular vacation spot for Boston's upwardly mobile Irish, that she and Joe first met. Rose, a sparkling seventeen, was swathed in the bathing costume deemed proper in 1906, an ankle-length black dress with sleeves and collar; Joe was eighteen and handsome, with a fine physique. Though Old Orchard Beach was hardly New England's Riviera, to Rose it would always be "magic—it was where I fell in love." And, she said when she was eighty, "I never fell out."

She was thrilled when Joe invited her to a dance at Boston Latin School and dismayed that her father would not allow her to attend. In the first place she was too young, he told her, and in the second place he didn't much like Joe Kennedy. Over the years Honey Fitz and Joe's father, P. J. Kennedy, a ward power, had sometimes been on opposing sides in the complex world of Irish politics. P. J. was another East Boston success story, starting as a saloon keeper, then elected to the Massachusetts House of Representatives and to the state senate, but unlike the flamboyant Honey Fitz, Kennedy preferred to exercise power behind the scenes.

Honey Fitz thought P. J.'s boy was too cocksure, a little loud. Brash, that was the word. With her looks and charm—and his connections—his Rose could do better. "I suppose no father really thinks any man is good enough for his daughter," Rose said in her memoirs. "But my father was a hopeless case, believing that I could take my pick of any beau. He didn't want me pledging my heart prematurely to any young man, however attractive and brilliant he might be." Not that Honey Fitz would have applied those adjectives to Joe.

From the day she met Joe, Rose's heart knew immediately that he was the one for her, but her father's disapproval made an open romance impossible. They would meet surreptitiously in such innocuous venues as the reading rooms of the Boston Public Library, they staged "chance" encounters at the bus stop, and Joe would fill in her dance card with invented names, lest Honey Fitz see the card. Her father, resorting to a strategy comparable to Sara Delano Roosevelt's when her adored son fell in love too young and unsuitably, took Rose to Europe, to Palm Beach in season, and on an official mission to Panama. But if he thought she would forget Joe, he misread his daughter, just as Sara misread her son. Rose would not be budged from marrying her true love, who by then had graduated from Harvard and was on his way to success. At last Honey Fitz gave in—quite certain that the Kennedys were fortunate to be marrying into the Fitzgeralds.

It took three years for her, and Joe, to overcome her father's objections to their marriage—by then she was twenty-four and Joe was twenty-six. On October 7, 1914, they were married by the cardinal in his private chapel, attended only by their families and closest friends, who made up a sizable congregation. Rose was a picture in a splendid concoction of tulle, worn with a remarkably unbecoming veil and carrying a rather cumbersome bouquet. Joe was dressed with the elegance he deemed appropriate for the country's youngest bank president, no matter how small and unsteady his bank might be.

After a honeymoon at the fashionable Greenbrier in West Virginia, the couple moved into the house Joe had bought in Brookline, a suburb of Boston inhabited not by aspiring Irish but by Yankees. They had already furnished the seven-room gray frame house, along with two rooms for servants on the third floor. Rose remembered it as a house that "would have blended perfectly into most of the main streets of America," and still would, with the notable addition of its National Trust marker as the birthplace of the thirty-fifth president of the United States. (Much of their "solid, serviceable, conservative" furniture is in the house today, including the Ivers and Pond grand piano, a wedding gift of two Fitzgerald uncles, and a china tea service from Sir Thomas, like the one used on his yacht.)

The following July Rose gave birth to a son, who was named for his father, and a second son followed less than two years later—an heir and a spare, as the British say. A new generation was in place to take the Fitzgerald-Kennedy political dynasty to the summit of fame and power. Over time, there would be nine bouncing Kennedy babies. Four of them grew up to die violently and a fifth would live a limited life, separated from the family.

★   ★   ★

EARLY in her marriage Rose had grappled, unhappily, with Joe's immersion in moneymaking—and other women—while she was left at home. In 1920, after a long stint of coping with sick children and handling the household responsibilities alone—with servants, yet alone—and eight months pregnant with her fourth child, Rose, neglected and exhausted, snapped. Suddenly this young woman who had loved being at the center of things saw her life replicating her mother's. She went back to her girlhood home—and her father—to think things out. One evening, wearing a black-net gown that concealed her "condition," Rose attended the winter ball, the event of the season for Irish society, and at the end of the evening she went home to Dorchester with her sisters; Joe returned alone to the family home in Brookline.

What could this mean? Divorce was unthinkable for a devout Catholic, and in those days a divorced woman had no place in proper society. After three weeks Honey Fitz stepped in to deliver unvarnished advice for his daughter. Without rubbing it in that she had married Joe Kennedy against his wishes, he reminded her of her duty: "You've made your commitment, Rosie, and you must honor it now. What is past is past; the old days are gone. Your children need you and your husband needs you. You can make things work out. If you need more help in the household, then get it. If you need a bigger house, ask for it. If you need more private time for yourself, make it. There isn't anything you can't do once you set your mind on it. So go now, Rose. Go back where you belong." There was no suggestion that Joe should be reminded to honor his marriage vows; the double standard was an accepted part of married life. Rose knew it was up to her to make things work out, and she returned home to Brookline, to her children.

★   ★   ★

SHE was at his side as he leapt from the power of money to the status of presidential appointments, as the powerful first chairman of the Securities and Exchange Commission, then to the most prestigous diplomatic post, London. In mid-April 1938 His Excellency the Ambassador of the United States of America to the Court of St. James's and Mrs. Kennedy were invited to join the King and Queen for a weekend at Windsor Castle. It was an interlude of royal splendor, of footmen in satin livery and peruke wigs, of white tie and tiaras. In their guest suite, with its sweeping view of Windsor Great Park, Joe turned to his wife and said, "Rose, this is a helluva long way from East Boston."

It was a distance that could be measured neither in miles nor travel time. Only two generations away from the blight-stricken potato fields of Ireland, Rose and Joe were now seated on the right of the monarchs. Such a thing would have been beyond imagining by their grandfathers. In about 1849 Thomas Fitzgerald had left County Wexford, along with a million others fleeing the scythe of famine, bound for North America, their bags packed with little but desperation and a parcel of hope. They sailed west in vessels so unseaworthy, with conditions so grim, disease so rampant, that they were dubbed "coffin ships." Another million stayed behind and perished.

Despite all, throngs of men and women, mostly young, crowded the docks of New Ross for each sailing to America. Among them was another son of Wexford, Patrick Kennedy; though the two young bachelors were not acquainted, their Wexford bloodlines would mingle, two generations later, with historic results. Each married an Irish colleen and set about dreaming big—a steady job and plenty of food on the table, meat once a week, maybe twice. In the squalor of Boston's slums those were optimistic goals, given the blunt reality that they were shut out of the jobs that promised a future.

The Irish ban spurred the sons of both couples to make their own future. Tom Fitzgerald went from farmhand to street peddler to a small grocery and "bottled goods" (whiskey) store in Boston's North End, Paul Revere's old neighborhood, which had given way to waves of Irish immigrants. By 1850 the Irish constituted a quarter of the city's population, outnumbering the Irish in Dublin.

Tom and his wife crowded nine sons into cramped quarters in a nine-family tenement—and buried their only daughter at six months, a victim of the cholera epidemic that swept through the slums. (In those harsh times 60 percent of such children were marked for death before they reached the age of five.) In a remarkable move of upward mobility, the Fitzgeralds sent their son John Francis to Boston Latin School, considered the finest public school in America, and to Harvard Medical School. At his father's death, John left Harvard to help support his six younger brothers; abandoning medicine was in reality no deep disappointment to him, for he had already contracted an incurable case of political fever.

Ebullient, gregarious, blarniferous, he was a natural politician who outstripped his Irish Catholic rivals for more than two decades. At twenty-nine, thanks to a lucky break, he became boss of Boston's North End ward, in a time when big-city political bosses controlled jobs and ballots, dispensing favors and expecting favors in return. From then on, the rails were greased for the young man fondly tagged "Honey Fitz"—sending him to the city council, the Massachusetts senate, and in 1895 the U.S. Congress for three terms (at first he was the only Catholic in the House of Representatives). Then he won the prize that meant most of all: mayor of Boston, the first son of an immigrant to rule the city that was the second home of Irish-Americans. Along the way, he acquired a highly profitable newspaper aimed at the flourishing Irish community. Honey Fitz was becoming a man of substance.

On July 22, 1890, the first of his six children was born, a daughter, a blessing to a man who had grown up amid nine brothers. They christened her Rose, for his mother, and from that day she would always be his favored child, and she, in return, adored him. In time her second son, the one who bore his name, would name his presidential yacht the *Honey Fitz*.

★   ★   ★

JOHN Kennedy, just after he was elected and still unaccustomed to the designation "president-elect," pondered the influences in his early life. "They talk a lot about Dad," he told Hugh Sidey of *Time* magazine. "Mother deserves more credit than she gets. She is the one who was there. She is the one who read to us, who took us to Plymouth Rock and the Old North Church and other historic places. She gave me my interest in history."

In all fairness, Joe did his share of imparting the realities of life and politics to his children, though he was more inclined to practical consideration of current events than intellectual contemplation of the past. Rose was indeed "the one who was there." She encouraged the children to think about great events of history and dream of exciting events in the future, as well as keep up with the present. "And," she pointed out, "we instilled in them that politics is a noble pursuit." Message received—and each of her four sons would, in familial order, aspire to make his mark on history as president of the United States.

Unlike those families who split into factions and wrangle over the spoils of the father's fortune in acrimonious lawsuits, the Kennedys melded into a uniquely close family, challenging and enjoying each other, scarcely needing outsiders except for procreation. In-laws were expected to become Kennedys (Ethel enthusiastically joined, Jackie pursued her own path, and Joan, Ted's first wife, cracked under the strain), and grandchildren were Kennedys regardless of their last name. "The Kennedys," explained the matriarch, "are a self-contained unit."

At the Kennedy dinner table the children were required to keep up with what was going on in the world. "My mother was great on self-improvement," President Kennedy said in 1960. "She always saw to it that we read good books, had good conversations." Joe Kennedy was away for weeks, even months, at a stretch when he was in the movie business in California, but when he was home, his son remembered, dinner talk was "mostly monologues by my father, mostly about personalities—we didn't have opinions in those days." The children were fascinated by his stories about the famous people he lunched with and the exciting places he went. Each of them regularly wrote to him, chatty letters about school and sports, and the busy tycoon answered each individually.

At home he was an involved father with fierce pride in his children, whom he saw as a finer legacy than his vast fortune and the influence it commanded. "He made his children feel that they were the most important things in the world to him," said President Kennedy. "He held up high standards for us, and he was very tough when we failed to meet those standards." His message, in whatever they were doing, was win, win, WIN! Coming in second was not acceptable; to Joe, it mattered not how they played the game but whether they won or lost. Demanding as he was, he was always supportive. Ted Kennedy recalls fondly, "I played sports at

Harvard and at Milton, and I can't think of a game that he didn't come up to watch. Sometimes he was the only parent there." (By the time Teddy was growing up, Joe had abandoned Hollywood.)

For her part, Rose managed the tribe like a business; she was a human computer, entering all relevant data on file cards for each child: shots, fillings, shoe size, godparents, achievements. She inspected them as they set out for school every morning, weighed each child every Saturday night, and, slightly obsessed with teeth, herded them all to the dentist every three months. (For that fortunate practitioner, it was the equivalent of a shoe salon having Imelda Marcos as a customer.) During her time as the ambassador's wife in London, the press made much of her card files as an example of American efficiency. Not at all, she laughed—"it's Kennedy desperation."

To maintain a semblance of order, Rose laid down rules, backed up by her ruler, which she would employ with vigor. "Mother would have made a great featherweight," chuckled her son Ted. "She had a mean right hand." Even as experts in child-rearing came to reject corporal punishment, Rose was not repentant: "After a few times of being whacked on the hand or the britches, the mere mention of the ruler would have a healthy effect on a child's behavior." A believer in swift justice, Rose "took the nearest thing at hand, which was likely to be a wooden coat hanger." She caught on, as they aged, that "some of the boys would stuff a pillowcase into their pants to lessen the anguish." Her guidelines for mothering were simple: "I exercise discipline, as well as love. Provide limits, as well as freedom." Though Joe never spanked his children, he administered tongue-lashings that made a lasting impression.

Under Rose's management, meals were served at a precise time, and woe to him who was late—usually Jack. His lackadaisical habits were a constant challenge to any scheduler, all the way to the White House. But Rose came to recognize that this child was of a different turn of mind from the others, more creative, more detached, funny and insouciant. "He was full of charm and imagination," she reflected. "And surprises—for he did things his own way, and somehow just didn't fit any pattern. Now and then, fairly often, in fact, that distressed me, since I thought I knew what was best. But at the same time I was enchanted and amused." Sara Roosevelt had those same conflicting reactions when dealing with young Franklin, trying to be stern while hiding a smile. Instructed by his mother to pray for a happy death, Jack, age six, countered, "I'd rather wish for two dogs," and later when Rose chided him about his grades, he countered, "If you study too much, you're liable to go crazy." What's a mother to do when she really wants to laugh?

Appreciating the differences among her children, Rose stressed their individual strong points: "I taught them to reach for the outer edge of their endowments." Perhaps because she was the first child herself, she believed strongly in giving the firstborn special authority. "Bring up the oldest ones the way you want them all to go," she would advise young mothers. "If the oldest ones come in and say good night to their parents or say their prayers in the morning, the younger ones think that's probably a good thing to do, and they will do it."

With Joe Sr. away so much, Rose, with his approval, gave the eldest son surrogate authority. Joe Jr., his father's pride and joy, cut from the same bolt, a star athlete, gregarious, confident, a natural leader whom the Kennedy patriarch had marked for the presidency. However, Rose may have invested him with too much control, because young Joe, like his father, savored power. Solicitous and loving to the younger children, he played top sergeant to Jack, taunting him into fury. While Rose saw the two brothers' relationship as "friendly enmity," JFK later told biographer James MacGregor Burns that the only thing that had really bothered him in boyhood was Joe's "pugnacious personality . . . it was a problem." Fights would erupt, angry combat that made the younger children run for cover. Later on, Jack said, "it smoothed out," and they became close friends, but the second son would bring to the White House a character markedly different from his older brother's, a temperament shaped by his mother.

Rose's philosophy of child-rearing ran deeper than rules and index cards. "I tried to tend the roots as well as the stems," she said, "and slowly and carefully plant ideas and concepts of right and wrong, religion and social implications." Her goal was to give them "a sense of responsibility and a sense of security." Underpinning that philosophy was her vibrant faith, a piety that was awe-inspiring, an unshakable belief that would sustain her through the cruelest blows that could befall a mother. "God will not give you a cross heavier than you can bear," she would say. But He put her to the ultimate test.

★   ★   ★

IN 1927 the Kennedys clambered aboard a private railroad car heaped with seven children, luggage, domestic staff, and pets, headed for a new life in New York, leaving behind Boston and its snubs. "Boston," Joe stated, "is no place to bring up Irish Catholic children." He mentioned that, among many predictable slights, his daughters would not be invited to Boston's debutante events, "and I didn't want them to go through what I had to go through when I was growing up." At Boston Latin School, Joe, one of the few Catholic students, was senior class president, and at Harvard he was a star athlete and a member of Hasty Pudding, but he was blackballed by the most prestigious social clubs. His income was high, his golf handicap was low, but when the family summered in Cohasset, the province of Boston's old line, the country club turned him down. A success by any measure other than heritage, Joe was a proud man who would rather seethe over rejection than take pleasure in his achievements, a paradox of thick hide and thin skin—tough in business, touchy in personal encounters, a man who never forgot a snub or an affront. And condescension was a stone hurled at his glass ego.

There was another reason for pulling up stakes and taking on the challenge of New York: "If you want to make money," he said, "go where the money is." At thirty-eight, already a millionaire, he was out to build a fortune that would guarantee status for his children. His enthusiasm for moving to suburban Westchester County was not shared by Rose, who was reluctant to be pulled away from all that was dear and familiar: family, friends, parish church, position in the Irish community.

The young Rose had enjoyed a life of considerable privilege in Boston, yet she was well aware of its limits. Significantly, along with invitations and dance cards, Rose pasted in her scrapbook copies of those hateful NO IRISH NEED APPLY signs and newspaper clippings detailing the discrimination against Irish Americans. She wanted her children and grandchildren, surrounded by wealth and power, to know what their forebears had endured. Her experiences had made her sensitive to all discrimination, an attitude she passed on not only to Jack, who would be the first president to declare that racial discrimination and segregation were wrong, but also to Bobby and Teddy, who would be surrogates for the downtrodden, and to the next generation.

★   ★   ★

ABOUT the time they were settling into a five-acre estate in Westchester, Joe gave his wife a permanent Massachusetts base— a rambling eighteen-room, nine-bathroom, turn-of-the-century house on the south shore of Cape Cod, perched on a rise overlooking Nantucket Sound. The house at Hyannis Port, the centerpiece of what would become the famous Kennedy compound, would always be home in Kennedy hearts. In 1933, during the depths of the Depression, Joe bought a spacious Spanish-style villa in Palm Beach, directly fronting on the ocean. In exclusive Palm

Beach he felt fully accepted—perhaps because Palm Beach was predicated on sunshine and money, and much of its money, like its family trees, was newly minted.

In Hollywood in the midtwenties, Joe, a deep-pockets producer and head of a studio, found total acceptance. Who cared what boat anybody came over on—they were all in love with America and the movies. Money and power opened doors, and the most significant door that opened to Joe Kennedy led to Gloria Swanson's bedroom. Impetuous, demanding, and sensual, she was the most glamorous actress in Hollywood, the kind of woman who had a black marble bathroom with a golden tub, the ultimate Movie Star.

Their long-running affair, while discreet, was a topic of insider gossip, particularly when Rose, Gloria, and Joe traveled together, the three of them, on a crossing to France. Rose seemed not to mind Joe's conspicuous attentions to the star. She chose to see "poor little Gloria" as a troubled movie queen whose French husband had dumped her for actress Constance Bennett. And when Gloria and her little daughter stayed with them at the Kennedy estate in Bronxville, Rose seemed not to notice electricity crackling between her husband and houseguest. She would always explain, even forty-five years later in her memoirs, that Joe was simply helping Gloria with her financial muddle. None are so blind, it is said, as those who do not wish to see.

Was the highly intelligent Rose really so naive? Probably not. But just as she had decided to accommodate Joe's wanderings after her father's injunctions, she would see only what she chose to see, and she saw a husband of many years, a serious father who would never give up his children—or their mother. To preserve her own dignity, Rose had to appear to accept Gloria as a family friend. The alternative was to leave him, which would make the humiliating infidelity a public scandal, and she would not, could not, do that. Her father's lecture still rang in her ears.

In closing her eyes to her husband's blatant infidelity, Rose may have unwittingly allowed his behavior to influence their sons. The Kennedy boys knew the Gloria Swanson stories and were aware of their father's eye for pretty women. (He sometimes took their girlfriends to lunch, and at least one claimed he made a pass at her.) Had Rose refused to tolerate his infidelities, a different example might have been set for the sons, but the price would have been high for her and the outcome uncertain.

Whatever Rose really thought about Gloria and Joe, she, by then the mother of eight children, decided she needed a bit of renovation. Her husband, she knew, "was surrounded daily by some of the most beautiful women in the world, dressed in beautiful clothes. Obviously I couldn't compete in natural beauty, but I could make the most of what I had by keeping my figure trim, my complexion good, my grooming perfect, and by always wearing clothes that were becoming." She refused to be, in her mother's phrase, "a home woman"; she would not stand by passively. Rose would, in her word, *compete*. "With Joe's endorsement, I began spending more time and money on clothes." A lot more. To look glamorous, to do him proud, she turned to Paris, to Chanel, Patou, Molyneux. She made the Best Dressed List's Hall of Fame and learned the tricks for looking her best in "unposed" photographs, chin uptilted to foreshorten the longish nose, eyes brightly focused to show interest, standing at a slight angle to the camera. Her reward was to hear her husband say, "You're a real knockout!" Deeply religious she was, beyond question, but if Rose Kennedy wore a hair shirt, it would be from Dior.

Even in her eighties, she would dress meticulously when Teddy was coming for a visit, anxious to "look good for my son." Only alone at home would she pad about in any old thing, with flesh-colored tape "frownies" plastered strategically on her face to stave off wrinkles.

Some call vanity a sin; in Rose Kennedy, vanity was a strategy, a weapon, to hold her husband, to please her family, to stay alert to life. For any woman over fifty who admired Rose's figure, her style, her élan into old age, her vanity was an inspiration.

★　★　★

A Plea for a Raise by Jack Kennedy

Dedicated to Mr. J. P. Kennedy
Chapter 1

That's how it began, the ten-year-old's entreaty for a larger allowance from his father. With "Mr. J. P. Kennedy" away so much, Rose functioned as chancellor of the exchequer, and her fiscal policy was very tight. Weighing his chances, Jack had decided to go over her head and plead his poverty to his father, who understood bottom lines and cash flow. He explained that he used his forty-cent allowance "for areoplanes [his spelling was, at best, idiosyncratic] and other playthings of childhood but now I am a scout and I put away my childish things. . . . I have to buy canteens, haversacks, blankets, searchliagts [*sic*] ponchos things that will last for years and I can always use it when I can't use a chocolate marshmellow [*sic*] sunday with vanilla ice cream and so I put in my plea for a raise of thirty cents for me to buy scout things and pay my own way more around. Finis." After a summit meeting, his parents authorized the raise, but Rose kept a sharp eye on the inflationary increase. She was determined that her children would not be instant-gratification rich kids.

Her own childhood had been happy, in comfortable-to-affluent circumstances in the small town of Concord. At twelve, she had her own horse and small carriage and her dresses were always the prettiest. Even so, her mother was watchful in her spending, and Rose passed on to her children Josie Fitzgerald's preachment that "one should be careful with money, none should be spent without good and sufficient reason—tangible or intangible—to justify each expense." Rose followed that precept assiduously. An allowance of ten cents a week began at age five, and with it a serious lesson in economics: "We wanted them to know the value of money and the painful consequences of heedless extravagance—as painful, let us say, as not knowing at age five where your next gumdrop is coming from."

Rose was not exactly consistent when it came to spending. She would search out the cheapest apples on Cape Cod, fret about lights left on, sell her out-of-fashion clothes to a thrift shop, but spend thousands on a Paris creation. Perhaps she chalked up those expenses to "intangibles" or "public relations," related to the demanding business of being Mrs. Joseph P. Kennedy. Never begrudging her expenditures on expensive clothes and jewels, Joe added to them with an extravagant present at the birth of each child. By the time Jean, number eight, came along, the gift had escalated to a knock-your-eyes-out diamond bracelet. (A friend joked that she kept on having babies for the baubles.)

Bobby, ever eager to comply with his parents' wishes, decided, when he was about ten, to stretch his meager allowance by selling magazines in affluent Westchester. His enterprise was much like little Dwight Eisenhower's in turn-of-the-century Abilene—up to the point that Bobby figured it would be more efficient to have the family chauffeur drive him on his rounds, and then, in the interest of even greater efficiency, he had the chauffeur make the deliveries himself while he, the young entrepreneur, played at home. Rose was delighted with her son's venture. "The 'paper boy' is practically an American symbol of boyhood spunk and ambition," she asserted, until she found out about the chauffeuring. She was mortified: "There he was, riding around all over Bronxville making his deliveries from a Rolls-Royce! Needless to say I put a stop to this at once." In their approach to topping up their allowance, as in most things, Jack and Bobby were opposites. Jack was the grasshopper, Bobby was the diligent ant; Jack took a cavalier attitude to his mother's rules, Bobby took them to heart.

As Joe Kennedy became a major figure in the business world and in the newspapers, then was appointed by FDR to head the new Securities and Exchange Commission, the parents had to deal with a secret they had chosen not to discuss. "The time came," Rose said gravely, "when we had to tell the children that their father was well to do." There is something surreal about that scene—grappling with how to tell the children that daddy is rich, at a time when other families were searching for a way to tell theirs that papa had lost the farm. Rose, emphasizing that money brought responsibility and the debt of service, preached the gospel of Saint Luke (chapter 12, verse 48): "For unto whomsoever much is given, of him shall be much required." (Rose herself did not know the extent of their wealth until 1957, when she read in *Fortune* magazine that her husband was worth more than a quarter of a billion dollars. She was a bit annoyed: "Why didn't you tell me we had all that money?" Joe, not at all contrite, saw no need to talk about money—the important thing was to make it—and felt it had no place in the instructive dinner table discussions with children.)

There was, however, one very public failure on Joe Kennedy's dazzling résumé—his tenure as U.S. ambassador in London. While Rose was a big hit in London, Joe was not cut out for diplomacy, neither in temperament nor in experience, and 1938 was a sensitive time, when an isolationist America did not grasp the implications of the storm brewing in Europe. Joe's manner was brusque and he had little admiration for the British. They, in turn, resented his view that Hitler could defeat Britain. Nor were his attitude and analysis favorably received by FDR, the Anglophile who had appointed him—a major party contributor—as a tip of the hat to the large and heavily Democratic Irish vote. As tensions between the two hardened into a bitter division, Kennedy, seething at being bypassed and ignored, returned to New York before the 1940 election, confiding to Rose that he planned to support the Republican candidate, Wendell Willkie.

Exercising her well-honed political reasoning, Rose pleaded

with her husband not to break with the president, which would cast him as disloyal, a cardinal sin in her father's political catechism. She reminded Joe that FDR had appointed him—an Irish Catholic—to the preeminent ambassadorship and had sent him as his representative to the coronation of the new pope, Pius XII. Convinced by her arguments, Joe bought radio time to pledge his support for a third Roosevelt term. In a surprising postscript to his announcement, Rose Kennedy came on to assure American mothers that this president would not send their sons to war. In light of what was to happen, her sincerity was a cruel irony. But following FDR's landslide, Kennedy unleashed antiwar tirades that gave the president cause to end his mutinous ambassador's political career.

*   *   *

WHILE Joe sharpened his sons' ambition and pushed his children to win, whether in sports or politics, Rose tended spirit and mind. She made books a priority. Ted Kennedy fondly recalls curling up in her bed to read *Peter Rabbit* together, and as they learned to read for themselves, she assembled an eclectic collection ranging from *Uncle Tom's Cabin* to *The Adventures of Reddy Fox*, from *Days of Ancient Rome* to *Biography of a Grizzly*. Jack, with his inquisitive mind and often confined to bed by illnesses, was her most omnivorous reader, discovering very early the fascination of biographies of famous people and history in action (a thread that runs through the young lives of these modern presidents, from Roosevelt to Clinton).

Rose had been Dorchester High's youngest-ever graduate, a bright fifteen-year-old who was also voted the prettiest girl in the class. A photograph of Mayor Fitzgerald presenting a diploma to his daughter Rose was featured on the front page of the local newspaper that day. She was looking forward to going to Wellesley, a college dedicated to the serious education of young women, stretching the intellect and kindling concern for "small-s" social issues. But only days before she was to leave for Wellesley, her father, under pressure from Boston's archbishop, had second thoughts: what kind of ideas might his innocent young daughter pick up in that climate of independent thinking? Better to send her to the Convent of the Sacred Heart, right there in Boston, an order that shaped the moral and spiritual development of daughters from affluent Catholic families.

Thus, instead of rising to the stringent intellectual challenges of Wellesley, Rose was educated to be the perfect wife and mother. "It was assumed," she commented in her memoirs, "that the girls when they married would be devoting their lives to kinder, kirche and küche and needed to prepare for all the duties implied in that expression. It was further assumed that they would have servants to do all the actual work."

The following year Rose and her younger sister, Agnes, were dispatched to the Convent of the Sacred Heart in Blumenthal, Holland, for a final polish—and to be conveniently absent when an ugly scandal involving their father's administration was brewing, threatening embarrassment for his daughters. In an oppressively strict environment, Rose studied German and French and won a gold medal for her mastery of the piano; even in a convent school that demanded "ladylike" behavior, Rose Fitzgerald was an iron-willed competitor.

Doris Kearns Goodwin, in her yeasty biography *The Fitzgeralds and the Kennedys,* recounted how Rose at ninety was still resentful that she was not allowed to attend Wellesley: "With a bitterness of tone which she did not often allow herself to betray, she said, 'My greatest regret is not having gone to Wellesley College. It is something I have felt a little sad about all my life.' " (Yet she insisted on a Catholic education for the daughters, while accepting Joe's case that the sons go to Harvard to compete against boys already marked as achievers from more varied backgrounds.)

Perhaps seeing her granddaughters actively pursuing their many different careers, without sacrificing marriage and children, made Rose think back to 1907 and the untapped abilities of Rose

Fitzgerald—though she never would have weighed her family and the life that actually happened against any other life, real or imagined. Yet in her later years, asked if she would have liked to have been in politics herself, she mused, "I don't know what would happen if I were starting out today. It's possible that I might be tempted."

★　★　★

JACQUELINE Bouvier Kennedy, who was of a more serene nature than her in-laws—in their midst, yet always a bit apart— once compared the family to "carbonated water, where other families might be flat," an apt description for the clan of effervescent siblings and cousins, all fizzing with action.

When the Hyannis Port summers began in 1928, there were only seven young Kennedys romping about; by the time the matriarch marked her hundredth birthday in 1990 the immediate family had exploded to more than forty, counting spouses, and the fourth generation was arriving on the scene. "Being cousins was an accident," Kathleen Kennedy Townsend, Bobby's eldest child, observed of the congenial throng, "being friends was a choice."

To organize a house run on kinetic energy, Rose ran a veritable Camp Kennedy. She posted activities charts and pinned noteworthy clippings on a bulletin board. Experienced young athletes were engaged to coach the kids in tennis and swimming, how to sail a boat or throw a football. At one point the Kennedys were limited in the number of races they could enter at the yacht club— otherwise, they would win them all. They were expected to hew to their father's maxims: Kennedys do not cry, Kennedys must win. It was Rose who taught them to care about people who couldn't win.

A bear for self-improvement and physical exercise, Rose kept her French and German in working order all her life, swam regularly in cold Cape Cod waters well into her eighties, and vigorously walked the beach. When she played golf, said Bobby's wife,

Ethel, "she would never ride in a golf cart—she carried her own clubs and walked briskly." And she always hit the ball straight, because that's the way it is supposed to go. Straight. Rose would always be a work in progress. At sixty she learned to ski, and when she was eighty-five she had a go at skateboarding, but thought better of making a habit of it. Ted Kennedy declares that she was ice-skating with him until she was ninety.

She was not a cookies-and-cuddles mother and her manner was convent-trained formal, yet Rose was not stuffy. Her daughter Jean Smith, who was the ambassador to Ireland for five eventful years, tells of the time the driver of a tourist bus stopped to ask directions to the Kennedy compound, whereupon Rose jumped on the bus, directed him to her front door, and invited the whole astonished crowd to come in and have a look around. Ted claims that Rose, ever the politician, would inspect the license plates of cars slowly cruising in their Hyannis Port neighborhood, and if they were from a state with a primary election, she would stop and chat with them.

For a special issue of *Life* marking her hundredth birthday, her grandchildren shared memories of their unique Grandma. They remembered her singing "Sweet Adeline," Honey Fitz's old campaign song, and reciting Irish poetry by Yeats. NBC anchorwoman Maria Shriver, the first Kennedy to become a celebrity in a career other than politics, declared, "I use her as a role model . . . an intelligent and beautiful woman who led an interesting life in her own right," along with her "honorable profession" as wife and mother.

John F. Kennedy Jr., who continued her Fitzgerald name, carried the family standard, and would be added to the lengthening list of family tragedies, laughed about his grandmother's unexpected participation at a prep school graduation party at the JFK family's house in the Hyannis compound. It had wound up, as such celebrations often do, with his friends sleeping all over the place. On Sunday morning "I was cleaning up and saw Grandma in her bright red coat walking very determinedly across the lawn. I was horrified. She started tapping the people who were sleeping with her walking stick." Would any of them like to join her at Mass? "She was kind of insisting," John said. "Then she pulled up a chair and started talking to everyone. She wanted to know what they were going to do with their lives and why they were sleeping so late on a beautiful Sunday morning—it was only about 9:30. She managed to rope about six of us into going to church with her, which was about the last thing we expected to be doing!"

Seeing her grandchildren carry on the family's commitment to public service gave Rose special satisfaction. There would be two congressmen, Bobby's Joe in Massachusetts and Teddy's Patrick in Rhode Island; in Maryland Mark Shriver serves in the House of Delegates and Kathleen Kennedy Townsend, wife of an educator and mother of four, and the first female Kennedy to run on her own, is lieutenant governor, positioning to run for governor in 2002. John Jr. had not closed the door on a future in politics, if that future had come. Humanitarian service of other kinds has attracted most of the other Kennedys/Shrivers/Lawfords/Smiths, leading programs for the disadvantaged, the handicapped, the helpless.

As the grandchildren grew up, Rose showed the same individual concern for each that she had shown for her own nine, interested in their education and careers, forthcoming with praise and criticism. All children, she felt, needed a little guidance, a little nudging. "Children," she often declared, "are meant to inherit the earth—and should be fit for it." Whether it was sought or heeded, she gave advice. In a letter to Bobby she confessed to meddling in a son's choice of college courses: "As you can see, I am busy as always, but I'm slowing down a little bit as I read that Socrates gave advice to everybody in Athens and they finally poisoned him."

★   ★   ★

ROSE Kennedy was ninety-five years old, fading but still holding on to the essence of her old self. To an old friend she mused, "If I had grown up in today's world, without hesitation I would have become a politician." Her imagination soared: "Who knows, I might have become the first woman President of the United States."

That was undoubtedly a flight of fancy, completely at odds with her declaration, repeated over the years, "I was never tempted to go into politics." She insisted that it was a greater accomplishment to have been the mother of a president and two senators than hold office herself. But in the years between these comments Rose had seen a new generation of women win a place in politics. Her fantasy about playing a leading role in politics is not surprising for one who had been a campaigner since girlhood, reveling in the attention and applause. "It was great fun," she said of those early days at her father's side, and more than half a century later, "It still is!"

She was out there for each of her sons in turn, campaigning and critiquing—"your speech was too long" or "people don't understand it" or "sit up straight, get a haircut." "Somehow, we didn't mind the criticism," said Eunice, whose husband, Sargent Shriver, the first head of the Peace Corps, ran for vice president in 1972. "She knew how to do it."

There's no reason to think that Rose felt more than a twinge of frustration at being two generations ahead of the rise of women in public life; at her core, Rose Kennedy was mother-teacher, and she filled that role to its outer limits. (Eunice declares that if her mother had been of the next generation, "There is no question that she would have been president of a woman's college.") If her political triumphs were vicarious, they were no less satisfying, and she heartily endorsed the insouciant message inscribed on a silver box Jack gave to Bobby: "When I'm through, how about you?" Even in sorrow, she understood the dynastic impulse Jack had expressed with what proved to be chilling foresight: "I came into politics in my brother Joe's place. If anything happens to me,

Bobby will take my place, and if Bobby goes, we have Teddy coming along." In 1983, after living through the horror of two assassinations, Rose could still say "of course" she would like for one of her grandchildren to run for president, and in step with the times, she specifically included her granddaughters. "The family is politically oriented," she said in a massive understatement.

Sara Roosevelt had been the first mother to dip a toe, ever so discreetly, into her son's campaign; Martha Truman expanded a mother's participation, appearing on the platform at Harry's rallies; and Rose Kennedy went national. In her two sons' presidential campaigns she shook hands from Hyannis Port to California and was the featured attraction at hundreds of events.

Dave Powers, JFK's confidential aide from his first day in politics to the tragic last, called Rose "the greatest pol we had." When Jack made his first bid for Congress in 1946, Rose was a valuable resource. Not only did this Gold Star Mother combine the two best known names in Irish Boston, Powers found that "she had a greater understanding of precinct politics than anyone in our organization."

In her first campaign appearance for her handsome bachelor son, she talked easily about bringing up a family of nine and her experiences hobnobbing with royalty in London—and was given a standing ovation. In that first race Rose was the key figure in a political experiment: a tea in a hotel ballroom, with formal invitations to local ladies to meet the candidate—and the candidate's mother. Kennedy's camp was split, the old pols deriding the idea, arguing that a beer bust would be better, versus Kennedy friends betting on the family magic. At the first tea, Jack, doubtful, his father, skeptical, sister Eunice, anxious, and Rose, smiling and self-possessed in a Paris outfit, waited to receive the guests, if any. They came—fifteen hundred women in their best frocks, their hair coiffed, lined up in a queue that curled down the block.

In 1952, when Jack Kennedy made his presumptuous bid to unseat Senator Henry Cabot Lodge Jr., the teas—thirty-three of them—were a centerpiece of his campaign. Twenty-five years later Rose related with relish how Lodge, conceding defeat, had grumbled, "It was those damn tea parties that beat me." (Such was Boston's social divide that she had never met Lodge.) For Rose, the victory was doubly sweet: in 1916 Lodge's father had defeated her father for the Senate. "We evened the score with the Lodges!" she exulted, and in 1960 she said it again when JFK and Lyndon Johnson defeated the Nixon-Lodge ticket. Politicians have long memories. Still, Rose was foremost a mother, and when Nixon came off badly in the first-ever presidential debate, she said, in the midst of her euphoria, "I felt so sorry for Nixon's mother tonight."

From 1946 until her last hurrah in 1980, her ninetieth year, when Ted made a halfhearted effort to challenge the incumbent Jimmy Carter for the Democratic nomination, Rose displayed an instinct for political subtleties: on the way to a blue-collar rally in Boston's North End, she would wear a simple coat, sensible shoes, and perhaps a head scarf tied under the chin; in the car en route to an upmarket hotel luncheon, she would slip into a mink jacket, good jewelry, and high-heeled pumps. Whatever her mode, she would be perfectly turned out, understanding that these women wanted to see the ambassador's wife, the old mayor's daughter, the senator's mother, and she should look her best for them, in their way. She had learned from Honey Fitz that in the voting booth there is no second class.

Unwittingly, she played into the opposition's hands in the 1972 presidential election. Fearful for her last son, who was set to campaign for the Democratic ticket, she asked President Nixon to authorize Secret Service protection. The Nixon tapes revealed that the president was only too eager to oblige. Specifying agents "we can rely on," Nixon ordered, "Plant one, plant two guys on him. This would be very useful." He authorized a forty-man Secret Service detail to keep Kennedy in sight "around the clock, every place he goes," a point that produced knowing laughter on the tape.

And if Kennedy ran for president in 1976, the information could be valuable, Nixon added. "He doesn't know what he's really getting into." Unaware that she had set in motion a surveillance operation aimed at discrediting her son, Rose sent the president her "deep-felt appreciation and gratitude." She could not have imagined the motives behind the favor.

Rose's lifelong fascination with politics may date back to when she was seven years old. Honey Fitz, then a second-term congressman who nurtured useful friendships in both parties, took Rose and her sister Agnes to the White House to be introduced to President William McKinley, a popular Republican who would be assassinated by a mad gunman. The president gave each a carnation from a vase on his desk and said to Agnes, "This is for the prettiest little girl I have ever seen in the White House." Not a word about pretty Rose.

In retelling the story, Rose insisted, "I thought it was a wonderful thing for him to do." The carnations, maybe, but the competitive Rose, accustomed to being Papa's pet, could hardly have been pleased by the president's tactless omission, and she never forgot the slight. More likely, from that moment Rose determined to be as pretty as she could be. Eighty-five amazing years later, happily surrounded by a score of children and grandchildren at Thanksgiving dinner, Rose confessed that she had always worried that her younger sister had been the prettier.

That love of politics reached its grand fulfillment on a glittering January day in 1961, when her second son, the one who bore her name, the creative one "who somehow just didn't fit any pattern," placed his hand on her Bible and took the oath of office as the thirty-fifth president. A blizzard had paralyzed Washington the evening before, and the city sparkled, pristine white, under a sky as blue as paint. It was a day that those who were there will tell and retell to bored grandchildren, priding themselves on their superhuman effort to get through snow-blocked streets leading to the Capitol, remembering John Fitzgerald Kennedy's voice as it rang out across the world. In his distinctive accent and cadence was the echo of the voice of Rose Fitzgerald, who had done so much to bring him to that place, on that day. And how did she remember it? "It was that cold, cold day and he was delivering his wonderful speech—without his coat on, and you think about how he was always forgetting his sweater. It's *your* son who has won the loyalty and the trust of millions of people. . . . It's almost too emotional to talk about."

★   ★   ★

HER Hyannis Port cook, Nellie, once said to Rose, "You've had the best of everything, haven't you, Mrs. Kennedy?" Rose looked into the distance and replied softly, "I've had the best of everything and I've had the worst of everything. . . . The agony and the ecstasy, that's what my life has been like." How was she able to bear up, to remain cheerful, grateful, unwavering in her faith? "There must be no looking back to what might have been," she said firmly, in an introspective moment. "It is not tears but determination that makes my pain bearable."

People tend to forget how early Rose Kennedy's agonies began. The worst that can happen to any mother, any parent, is the loss of a child. The crown prince of the American Kennedys, Lieutenant Joseph Patrick Kennedy Jr., was lost in September 1944, when his plane exploded over the English Channel in a high-risk volunteer mission against German rocket sites. In the space of a moment the courageous twenty-nine-year old and the ambitious dreams his family had settled on him vanished. Not a trace was ever found of the young pilot endowed with everything but luck.

The year before that tragedy, the family spent days of anguish when Jack, a navy officer, was missing in action in the Solomon Islands after his small patrol boat, PT-109, was rammed by a Japanese destroyer. His near-miraculous survival and heroism—

towing a wounded crew member to safety, the lifejacket cord gripped in his teeth, carving a message for help on a coconut shell—was the stuff of legend, but that would be learned only after much agony and prayer while the family feared the worst.

Kathleen Kennedy, the Kennedys' golden second daughter, caused her mother pain of a different nature. In 1938 the American ambassador's daughter had been debutante of the year in London's last gala social season, and returned to London with the American Red Cross in 1943. She fell in love with William John Robert Cavendish, Marquess of Hartington, in line to be the eleventh Duke of Devonshire, whose ancestral seat, Chatsworth, is more imposing than Buckingham Palace. Rose was distraught that her daughter would marry outside the Catholic church, which meant far more to her than the grand titles Kathleen would acquire. But on May 6, 1944, Kathleen and her marquess, an officer in the elite Coldstream Guards, were wed in a civil ceremony. In September, her "Billy" was killed in action. Three years later, the young widow became involved with another handsome nobleman, Earl (Peter) Fitzwilliam, war hero, rake—and married. Again Rose was heartsick, but the story soon ended. En route to the Riviera in May 1948, both Kathleen and her earl were killed in a private plane crash. Now Kathleen lies buried beside Billy in the ancestral graveyard of the Devonshires, at the stately home that was to have been hers.

For years Rose had struggled through another sorrowful, meticulously kept secret. Rosemary, the eldest of the five Kennedy daughters, was mentally disabled, a condition carrying such stigma in the 1940s that an important family—especially one with political ambitions—would go to great lengths to conceal the fact. In the presidential race of 1960, inquiring journalists were told that Rosemary chose to live "an almost nun-like life" in a convent setting. Not until rumors began to circulate did the family acknowledge that their daughter was retarded.

From infancy Rosemary had seemed slow. As she fell further

behind, Rose was increasingly anxious, but took care to include her in all family activities. In London Rosemary, wearing a Paris gown and an ostrich plume in her hair, was presented, along with Kathleen, to the king and queen at Buckingham Palace, and in summers at Hyannis Port, brother Jack, without being prodded, escorted her to dances at the club, seeing to it that his friends showed her a good time. Though Rose worked with her unceasingly, more and more frequently the once-mild-mannered girl erupted in violence, screaming and flailing at whoever came near. (The family now believes she was epileptic as well as mentally handicapped.)

Without consulting his wife, Joe Kennedy, searching for anything to help his daughter, agreed to a lobotomy, a delicate new procedure that left her in worse condition, though no longer violent. To provide permanent care, Joe built a home for her in a community of nuns in Wisconsin. Only after his stroke in 1961 did Rose learn, horrified, about the lobotomy. "It erased all those years of effort I had put into her," she said sadly. "It was all gone in a matter of minutes." Joe Kennedy has been excoriated for authorizing such an extreme operation, but in 1941 lobotomy was hailed as a breakthrough for deeply disturbed mental patients. He had done what was thought best for his wounded daughter—still, that did not justify not discussing it with Rose, who had been the dedicated, caring parent. Over the years Rosemary visited the family occasionally in both Palm Beach and Hyannis Port, but for many years after Rosemary was sent away, Rose did not visit her. Said a family friend: "To Rose, her Rosemary was dead."

With her unforgiving conscience Rose must have felt guilty at hiding her handicapped child like an ugly secret, but she more than made up for it. The Kennedy Foundation became a leading force for research in the field, and Rose influenced her son, as president, to increase funding for programs for the retarded. Rosemary has served as a catalyst for bringing mental retardation into the light and fostering new attitudes toward the handicappeds' role in society. The much-applauded Special Olympics, the worldwide sports competition for the handicapped, was the creation of Eunice Kennedy Shriver, growing from a modest activities program in her suburban Washington swimming pool forty years ago. Most recently, John F. Kennedy Jr.'s will included a large bequest for Reaching Up, a program for the mentally handicapped. A lustrous record of human service was inspired by Rosemary, the Kennedy who couldn't keep up.

"It has been said that time heals all wounds," Rose wrote in her memoirs. "I don't agree. The wounds remain. Time—the mind protecting its sanity—covers them with some scar tissue and the pain lessens, but it is never gone." Meanwhile, "Carry on. Take care of the living—there is a lot of work to be done."

★   ★   ★

WINDING up their day of triumph with a round of five inaugural balls, the Kennedy family took its place in the balcony of the cavernous Washington armory, cheered lustily by the throng celebrating on the floor below. They presented a dazzling array of designer gowns and elegant white tie and tails and the flashing smiles that were the family hallmark. After the quiescent years of the Eisenhower administration, it was a time for youth, vitality, a new spirit in America. Many years later, the scene lingers in the mind, like a favorite page in a yellowing scrapbook, juxtaposed with a similar image of a once-happy royal family on the Buckingham Palace balcony, acknowledging the cheers the day Lady Diana Spencer became the Princess of Wales.

This was the first time America had seen Jacqueline Kennedy at her most breathtaking, a vision in a slender white creation, radiating sleek, unadorned sophistication that announced a new era in White House style, a generational shift from Mamie Eisenhower's bouffant taffeta ball gowns and the rather stiff formal entertaining that had been White House style for decades. There was near disbelief that the beautiful woman on the new president's right could be his seventy-one-year-old mother, glamorous in a gold and silver-beaded lace sheath by Molyneux—the same gown she had worn in 1938 when, as wife of the American ambassador, she was presented to the king and queen of England. In the twenty-three years that had elapsed between these two state events, bookends to her social triumphs, Rose had gained not a single lumpy pound nor spreading inch, statistics she was quick to point out.

After the final ball, Rose and Joe returned to the White House—the *Kennedy* White House, more thrilling to them than Windsor Castle—the first of many nights she would spend in the queen's suite. Best of all were the state dinners when she played hostess for her son in Jackie's frequent absences; the First Lady often traveled on her own, guarding, as Rose had, a private area of her life, distinct from her life with her children and husband. Rose might have stemmed from Boston's North End, but no visiting queen was more regal than the president's mother in the thousand days that Jackie would call Camelot.

To people marveling at her youthful zest and astonishing looks well into old age, Rose freely dispensed advice: "I believe in keeping interested, growing and learning. Sedentary people are apt to have sluggish minds: A sluggish mind is apt to be reflected in flabbiness of body and in a dullness of expression that invites no interest and gets none. Vivacity, intelligence, curiosity, receptivity to ideas, true interest in other people have special relevance as one gets on in years. I try to take my own advice."

In 1960, when the country was beguiled by the colorful Kennedys, the president-elect reminisced about the early years when all those exuberant children were underfoot. Considering the wear and tear caused by such a household, he suggested that his mother "was a little removed, which I think is the only way to survive when you have nine children. I thought she was a very model mother for a big family." He was diplomatically rebutting criticism that Rose was rather cold and distant emotionally, more headmistress than mother—criticism that was never heard from her children. Bobby called her the family's "stabilizing influence" and Ted spoke of her "understanding and support and encourage-ment that made the high family standards livable and perhaps reachable." He observed to his biographer, Adam Clymer, "For all of us Dad was the spark. Mother was the light of our lives. He was our greatest fan. She was our greatest teacher." JFK's daughter, Caroline Kennedy Schlossberg, who named her first child Rose Kennedy, commented in a recent documentary, "People don't real-ize how much she influenced [her family]." Joe Kennedy had praised another side of his wife, after forty-six years: "I have never heard her complain. Never. Not even once. That is a quality that children are quick to see." Her stoicism was put to the test again and again, and still no one heard Rose Kennedy complain.

★   ★   ★

ONLY once did she lose her composure after John Kennedy was killed. In his moving chronicle *The Death of a President*, William Manchester wrote: "Upstairs she embraced Jacqueline Kennedy. . . . Until then she had been composed, but standing by the towering Lincoln bedstead . . . she collapsed in [Sargent] Shriver's arms. 'It just seems so incredible,' she sobbed. 'Jack being struck down at the peak of his career and my husband Joe in a wheelchair. . . . Think of Jackie! I had my nine children. She's so young, and now she doesn't even have a home.' " But taking her place at the mournful ceremonies, the president's mother was stoic, controlled, grieving in admirable dignity.

Jackie once shared a rare moment when Rose spoke of the tragedies in her life: "Her voice began to sort of break, and she had to stop. Then she took my hand and squeezed it and said, 'Nobody's ever going to have to feel sorry for me. Nobody's ever going to feel sorry for me,' and she put her chin up. And I thought, God, what a thoroughbred!"

In 1969 Joe Kennedy, the once-towering patriarch, died, eight long years after a stroke had left him an invalid, a prisoner of his own body, unable to speak intelligibly, angry at his powerlessness.

He had been kept at home, in his own surroundings, with private nurses around the clock and Rose popping in and out, trying to keep him interested, hoping to brighten his day. It was often a depressing, wearying burden, but when he directed his anger at her, she understood his frustration and concealed her own. Their children and grandchildren, always attentive, brought him visible pleasure in their visits. Most of all, Jackie Kennedy's aura of serene glamour soothed and delighted him. Even paralyzed and bedridden, Joe Kennedy could still appreciate a beautiful young woman.

As he slipped away, with his family at his bedside, Rose wept. Death was a release for the shattered patriarch, but she had loved Joe Kennedy, loved the life they had led and the family they had raised. They had shared fifty-five years together, for better and for worse. Just as she would always see what she chose to see, Rose remembered what she chose to remember, and several years after his death she wrote, with fierce sincerity, "Next to Almighty God, I had loved him—do love him—with all my heart, all my soul, all my mind." It was her rebuttal to the rumors that had swirled about their marriage over the years, starting in the long-ago days of Gloria Swanson.

His passing might have freed her to pick up the strands of her old life, but its glamour no longer attracted her. Much of the time she chose solitude in her huge houses at Hyannis Port and Palm Beach, engrossed in the new generation of Kennedys and the merriment they brought. Ted had taken on the role of all her sons—he teased and cosseted her, read to her, shouldered family burdens, stimulated her mind with insider talk of politics. Sometimes the two of them would sing the old Irish songs—"Rosie O'Grady" and, of course, "Sweet Adeline."

Only in her memoirs, published in 1974, did she reveal a glimpse of her fears for him—"he is the only son I have now, and is my joy and support and solace." Responding to political speculation that the senator would run for president in 1976, she said,

"Ted will decide for himself, just as his brothers did. But if anyone wants to know whether I think he should run for president in 1976, my answer is: I hope he will not."

That was the only time Honey Fitz's daughter had ever shied away from a political challenge, but the risk was too great. Why, she agonized, was her family targeted by assassins, by violent threats that went far beyond the bounds of American politics? She had brought up her children to serve their country and help the disadvantaged, altogether commendable goals. Could the hatred be rooted in jealousy? Had the high-flying Kennedys come up in the world too fast, amassed too much wealth, mixed public service with power, lived a bit flamboyantly, made it all seem too easy?

Ted did not run in 1976, to his mother's relief, but in 1980 Rose, then ninety, did what she could to help him in his brief foray to displace President Carter, mostly by sending letters enlisting support for her son. But never, Ted Kennedy assured me unequivocally, never did she ask him not to run.

★   ★   ★

IN her mideighties the indomitable Rose was diminished by a series of strokes—"Age is so maddening," she protested—and on Easter Sunday of 1984 a serious stroke took its toll on her quick mind and reduced her vigorous walks on the beach to wheelchair outings. Birthday celebrations kept on coming. At her ninety-second, she posed for pictures wearing a jaunty big red hat. Noting all the special things her children had arranged for the day, the ever-frugal Rose said, "Who's paying for all this, Teddy?" No need to worry, he chuckled, "I am."

And then she was ninety-five. Boston paid its respects to its home-grown Rose with a one-acre rose garden named in her honor. "I'm one of the most fortunate people in the world," she said in appreciation. "Even though my life has been scarred by tragedy, I have never lost this feeling. God has held us all in his hand."

By the time the family gathered at Hyannis Port in 1990 to celebrate her hundredth birthday, she had been given the last rites of the church five times, but there she was, attending Mass in her dining room, with Ted serving at the improvised altar. Though she was "in and out—the body was closing down," said a family friend, she still knew her children, and for another four and a half years she clung to life and family.

After a fabled life that stretched from President Benjamin Harrison to President William Jefferson Clinton, a tearless funeral of celebration might have been expected, but it was quite the opposite. Her children, now of the grandparent generation themselves, who had been so stalwart in the many earlier family funerals, held themselves together with difficulty.

This was more than the passing of their valiant mother and matriarch of a legendary clan; she was the core of the family, the unbending tent pole that held a sheltering cover above their diverse lives. Though her nine children had dwindled to four, realistically, twenty-nine grandchildren and four dozen great-grandchildren were stamped with the imprint of the legendary Rose, and six of them bear her name. She had been the great supporter of their ambitions, their strength in times of shattering loss. Without her as the magnet at Hyannis Port, their once and always home, what would become of the celebrated unity that made the Kennedys more than the sum of their parts? Who could possibly fill the gap? It was the passing of a storied world, the end of an era that had been the stuff of myth.

Her surviving son, Ted, her solace, paid tribute to "the most beautiful rose of all." At old St. Stephen's Church in North Boston, where she had been baptized and where the story began, in a voice choked with emotion, he repeated his loving litany: "She was ambitious not only for our success but for our souls. From our youth, we remember how, with effortless ease, she could bandage a cut, dry a tear, recite from memory *The Midnight Ride of Paul Revere,* and spot a hole in a sock from a hundred yards away." With her bedrock faith, Rose Fitzgerald Kennedy would have found joy in abandoning her spent body and entering the next life, yet she might have worried about who would do Thanksgiving dinners and who would correct their grammar. Rose Kennedy was on duty as a mother until the day she died. As her second son had said, many years before, "She was the glue that held it all together."

DANCING LESSONS FOR LYNDON

Thou art thy mother's glass, and she in thee
Calls back the lovely April of her prime.
—WILLIAM SHAKESPEARE, "SONNET 3"

*I*N the little Johnson City school, it was a big day for first-graders. The room was filling with parents when Rebekah Baines Johnson, dressed in her Sunday best, walked in in her stately way and took a seat with the other mothers. Her calm demeanor, as she nodded to friends, gave no hint of her anxiety. This was a landmark day, her Lyndon's first public performance: he and Kitty Clyde Ross, the two stars of the first grade, had been chosen to recite at the end-of-school ceremonies. Rebekah beamed as Lyndon, not yet six, youngest in the class but towering over the others, stood straight and proud, just the way she had showed him, took a deep breath, and began to recite.

It's a fair guess that Lyndon had been well rehearsed by his mother, worried that her professional prestige could be tarnished if her son flubbed his debut. College-trained in elocution, Rebekah gave "expression" lessons to the children of aspiring families, for fees the Johnsons sorely needed. Lyndon had to do well, for her. His poem was "I'd Rather Be Mamma's Boy," and Rebekah insisted he had chosen it himself. He made his way confidently through the saccharine final lines: "I'd like to be a bobolink or else a pretty dove / But, Mother dear, I'd rather be your precious love." Rebekah could relax. The future president of the United States, without benefit of cue cards or TelePrompTer, had performed flawlessly.

This achievement was but another triumph for her and the precocious son who had learned the alphabet from the blocks she showed him before he was two. As a three-year-old he was familiar with Mother Goose rhymes and poems from Longfellow and Tennyson (her own favorite), and at four he could spell such words as Grandpa and Dan (a family horse). Rebekah devoted countless hours to honing the mind of her firstborn and never hesitated to tell friends and relatives about his intellectual prowess. Exercising the artful persuasion for which her son would become famous, she convinced the teacher at the nearby one-room school to accept Lyndon, still only four, in her first grade—he was turning up for recess every day anyway. Miss Katie Dietrich agreed, after hearing him read in "his own little way of talking," and would put him on her knee to recite. (More than fifty years later Katie Dietrich would be a special guest at her little pupil's inauguration.)

From the very beginning, Rebekah Baines Johnson knew in her heart that this child was special. Decades later, when Lyndon dominated the Senate as majority leader and was first on any list of prospective Democratic candidates for president, his mother's reaction to the speculation was not coy: "I didn't bring him up to be president, but I'm not surprised. Of course, this is a mother talking, but from the first time I looked into his eyes, none of his accomplishments have surprised me." Even earlier she had declared, with total conviction, to a friend, "I feel certain that someday he will be president." The years to come would suggest that their bond was as strong as the one between FDR and his mother, and equally complex emotionally.

Lyndon's grandfather, Sam Ealy Johnson Sr., shared her conviction, telling his daughter Lucia, quite seriously, that the three-year-old Lyndon "is a mighty fine grandson, smart as you find them. I expect him to be United States Senator by the time he is forty." Lyndon almost met "Big Sam's" deadline: on his fortieth birthday he won the Democratic nomination for the Senate, which in those days in Texas was tantamount to election.

In her middle years Rebekah set down in writing the particulars of Lyndon Johnson's arrival, with details as fresh as yesterday morning. A frustrated writer given to poetic flourishes, she composed the chronicle in the third person, as though she had been a bystander at the event:

> It was daybreak, Thursday, August 27, 1908, on the Sam Johnson farm on the Pedernales River near Stonewall, Gillespie County. In the rambling old farmhouse of the young Sam Johnsons, lamps had burned all night. Now the light came in from the east, bringing a deep stillness, a stillness so profound and so pervasive that it seemed as if the earth itself were listening. And then there came a sharp compelling cry—the most awesome, happiest sound known to human ears—the cry of a newborn baby; the first child of Sam Ealy and Rebekah Johnson was "discovering America."

A beautiful memoir, but to keep the record straight it must be noted that Rebekah rearranged the facts a bit. As she told the story, the esteemed Dr. John Blanton of Buda, Texas, brought her ten-pound wonder into the world. The unvarnished truth, however, makes a much better story. It actually was, to use an old cliché, "a dark and stormy night," marked by rolling thunder and blinding rain. A few yards from the little Johnson farmhouse the Pedernales River, a usually timid stream that was often bone-dry, churned into a wild torrent, surging over its grassy banks and rampaging through low-lying land.

Lyndon's aunt, Mrs. Jessie Hatcher, recounted for posterity that night she would never forget. "Lyndon came in a storm, the

biggest storm we ever had. The doctor was 20 miles away—and
he couldn't get across the river anyhow. The Pedernales was on the
biggest rise it ever had been. My father [Sam Ealy Johnson Sr.]
forded the river on a big fine horse and brought one of the neighbors, this Mrs. Lindig. As a midwife, she brought lots of children
into the world." And so the credit for delivering the future president goes to Mrs. Lindig.

The sister-in-law offered an explanation for the proud mother's
revisionism: "I don't think Rebekah would have ever wanted anybody to say that Lyndon came with a midwife instead of a real
doctor, because she was of that type that, oh, she just wanted
everything just so." This was said not in criticism but admiration.
"Rebekah," she declared, "was a splendid woman." In Rebekah's
view a midwife, no matter how well respected, was not suitable to
deliver her firstborn, and since that was not the way she had
planned it, she simply recorded the arrival the way it was supposed to have been. Her son must have a proper start in the world.

That this infant had been so eagerly anticipated makes it all the
more inexplicable that weeks turned into months with no name
for him other than The Baby. "This sorely irked his mother,"
Rebekah recounted, continuing her detached third person narrative. "So one cold November morning she decided upon a plan."
Her "plan" was a Texas version of *Lysistrata*: she snuggled deeper
into the covers and declared, "Sam, I'm not cooking breakfast
until this baby is named!" A Texas farmer would do without
many things, but not his breakfast. So the two went back and
forth, with suggestion followed by veto. Finally Sam proposed the
name of a lawyer-friend: Linden. Rebekah agreed, provided she
could spell it her way. Lyndon with a *y*, she insisted, "would be far
more euphonious than Linden Johnson." And thus it was that a
president got his name and Sam Johnson got his biscuits.

It might seem that the new parents did not name this son for his
father and grandfather, both Sam Ealy Johnson, to avoid the confusion of multiple Sams in the family, yet when their second son

arrived, he was promptly named Sam Houston Johnson, in honor of the legendary Texas hero. (Rebekah's reverence for him was largely because her grandfather, the Reverend George Washington Baines, claimed credit for bringing the old Indian fighter into the Baptist Church.) The explanation for the second son's name was probably simpler: Rebekah had been set on giving her firstborn son her proud Baines name.

In between the two sons were two daughters. Rebekah was given both her mother's name and her deep love of poetry. (She in turn passed on that love to her son, Philip Bobbitt, an international relations authority, who established the Rebekah Johnson Bobbitt National Prize for Poetry in her honor, a major award administered by the Library of Congress.) Next came Josefa, and after Sam Houston a third daughter, Lucia, completed the family, all within an eight-year span.

By birth order and by nature (the one may have affected the other) Lyndon was the leader. His sisters, as children and as adults, looked up to their protective big brother, proud of him and turning to him for help. Long before her husband's death, Rebekah would ask Lyndon to handle what might have been a father's responsibilities; she had him intercede in sibling problems; she leaned on him as her strong right arm, to the extent that the siblings sometimes resented his authority.

Not surprisingly, the relationship between Lyndon and Sam Houston (he always used the full name) tended to be fractious. Younger by six years, the resentful kid brother was warped by sibling rivalry, jealous of Lyndon's success, yet grudgingly proud. He was a boozer, a blowhard, a failure, yet, his sister-in-law Lady Bird Johnson reflected, "Sam Houston had a lot of charm, he was fun. Everybody liked him—he was more liked than Lyndon. But he was irresponsible, constantly getting into scrapes, most of them he brought on himself." He used his brother cynically and boasted, as an adult, that he was the one his parents loved best. In his mean-spirited book Sam Houston went so far as to suggest that he was the brains behind Lyndon and that "if they [his parents and brother] had been harder on me, I'd have been president." Unlike the physical skirmishes that would erupt when Joe Kennedy Jr. lorded it over Jack, who was less than two years younger, or between "Big Ike" and "Little Ike," the antagonism between the Johnson boys took place in the mind and heart. Rebekah loved them both, and the daughters, of course, but it was her Lyndon who lavished her with attention, gifts, and a place in the sun.

It was fortunate for Rebekah that she had given the Baines name to the son who would be a success. To Rebekah, the Baines name, which dated back to 1741 in colonial North Carolina, stood for something special, a family that prized things intellectual and held to lofty standards, that took pride in the several preachers and one college president on its family tree. The Baineses' style of life, as Rebekah remembered it, was considerably grander than that of the Johnsons. She had married down, in terms of fortune and position, and she would make up for that by raising her children according to Baines values and standards.

But much had to be accomplished between producing the boy and perfecting him, and there would be a stretch of time when the outcome was in jeopardy. From the outset Rebekah, a woman of long-range vision, had a plan. She would mold Lyndon in the image of her father, erudite, dignified, sober—and partial to her; she would strive to prepare this grandson for the public acclaim and financial success that had eluded her father.

To attain this end, she replicated the methods her father had used with her as a child. Proud that "he taught me to read at an incredibly early age," she did the same with her son, reading to him every day. She sensed success when a very young Lyndon would listen to a story and ask, "Is it true, Mama? Did this really happen?" Children's stories were promptly replaced by biographies of great men and stories of great events, all intended to inspire the boy.

Next came the polishing stage. To offset the paucity of culture around Johnson City, when Lyndon was still in elementary school Rebekah arranged for him to take violin lessons with the school principal, Lloyd Brody, in return for elocution lessons for Mrs. Brody. Grumbling that playing the violin was sissy stuff, Lyndon went—he always obeyed his mother—but his mind was on the seasonal rhythm of country boys' games. Marbles in early spring, baseball later, then in the long summers, fishing in the river and swimming in the pond.

So it was under duress that he gave up the real stuff and reported for violin lessons. Since Lyndon had no detectable talent and even less interest, both sides soon gave up. (In response to those who expressed doubts about this brief encounter with the violin, boyhood friends provided convincing details in oral histories for the LBJ Presidential Library. They recalled the teacher, they described the violin, and most of all they remembered how terrible the violinist was.)

Rebekah then tried a new tack: dancing lessons. With girls! Teasing the girls, Lyndon discovered, was more fun than dancing the two-step. The exasperated teacher, Mrs. Stella Gliddon, finally gave him a good whack on the bottom and ejected him from the class. Rebekah was coldly furious that he had dashed her hopes for making him a gentleman, but she would have been pleased at the delayed result. In the Johnson White House years, the president was always first on the dance floor, proficient, even smooth, and clearly enjoying himself. Once at a state dinner in the Philippines' Malacanang Palace, the American president upstaged six other heads of government as he twirled their hostess, First Lady Imelda Marcos, then at the peak of her beauty and power, around the parquet ballroom. Yes, he still teased the girls a bit—but somehow it's different with a president.

★   ★   ★

ONE thing Rebekah neglected to take into account as she mapped her son's future: he came with two strains of familial DNA and the genes were decidedly unalike. It was Joseph Wilson Baines, the sometime newspaper editor, who suggested to his daughter, a budding young freelance journalist, that the bright young fellow who held Baines's former seat in the Texas legislature would make a good interview. His name was Sam Ealy Johnson. When this pretty, self-assured young lady with eyes the color of bluebonnets opened her notebook, a beguiled Sam asked her more questions than he answered, and soon he was asking her out.

"Sam was enchanted to find a girl who really liked politics," boasted Rebekah. A quarter of a century before women won the vote, Rebekah Baines was fascinated by politics. She learned about issues and strategies from her father, a teacher, lawyer, and editor of the *McKinney Advocate,* a weekly newspaper that was so exuberantly biased in favor of the Democrats that the governor made him Texas secretary of state. "I am a Baptist and a Democrat," he declared, and his daughter would be equally faithful to his church and his party. He was elected to the state legislature, and after one term ran for Congress, unsuccessfully, a rebuff Rebekah would always blame on the sheer ignorance of the voters in the district.

Sam Ealy's idea of courting was to take Rebekah to such romantic events as political "speakings" and a Confederate reunion that brought together such aging veterans as Sam Ealy Sr., who had fought with the South for four years and lived to tell the tale of having his horse shot out from under him in battle. As a special treat he took her to Austin to hear William Jennings Bryan, the orator who could win any crowd except on election day. That might not have been every girl's delight, but hearing Bryan address the Texas legislature was a heady experience for a would-be journalist. In August 1907, following a lively and brief courtship—significantly, after the death of her father—Rebekah married the tall, rangy, voluble, rough-hewn Sam. He brought with him a bale of charm and a load of debts.

The young couple set up housekeeping in a shabby cottage on

Sam's father's farm near Stonewall, one hundred acres of thin, stingy farmland and another four hundred of rock-strewn pasture. (A generation later this unforgiving land would be improved and other land added to it to make up the two-thousand-acre spread that the world came to know as the LBJ Ranch.) Between sessions of the legislature, Sam farmed his father's land, growing and trading cotton, buying and trading cattle.

Sam and Rebekah's little three-room "dogtrot" cottage—center corridor with rooms on either side—had no amenities or genteel touches, just the unrelieved plainness typical of rural Texas at the turn of the century. Rebekah Baines, a town girl accustomed to a life softened by a dollop of culture and social graces, was offended by the coarse language and manners she encountered. Given her standards, Sam Ealy's crude ways were a far cry from easy adjustment. Her new life was harsh and lonely, not at all like the one she had shared with her congenial family. "Texas," the state historian Walter Prescott Webb judged, "is heaven for men and dogs, but hell for women and oxen." Men and dogs were able, within limits, to roam freely; women and oxen were obliged, within reason, to live under the yoke.

The Hill Country of mid-Texas is not like the rest of the state. It is a vast disk the size of Rhode Island, a land of chalky, cedar-spiked hills and snaking streams, overarched by a sky as blue as those the radio cowboys sing about. It separates the plains of north Texas from the gullied brush country of the south, and its climate combines the worst of both parts. In winter winds slashed down from the north, finding the smallest chink in a house, overwhelming fireplaces trying to warm a room. In summer smothering heat drained the energy of man and beast, and parched the land to a sullen brown.

For the Texas farmer and cattleman, such extremes were part of the challenge of outdoor life; for the housewife charged with creating an oasis of comfort and caring amid such conditions, the burden shriveled the soul. Yet each spring there came the moment when the Hill Country rewarded endurance with a Persian carpet of wildflowers that fed eyes hungering for beauty and uplifted weary spirits. Freshets of bluebonnets tumbled down the slopes, golden swaths of cutleaf daisies and clumps of purple wine cups painted the meadows, and flights of monarch butterflies on their way home from Mexico stippled the air with their hues. Each year Rebekah could look forward to that glorious moment when her poet's soul would be refreshed.

More testing than nature's abrupt turns was daily life with her new husband. Years later, when writing the family history, Rebekah was remarkably candid about the strains in that union of two such dissimilar individuals: "Normally the first year of marriage is a period of readjustment. In this case, I was confronted not only by the problem of adjustment to a completely opposite personality but also to a strange and new way of life, a way far removed from that I had known in Blanco and Fredericksburg. I shuddered over the chickens and wrestled with a mammoth iron stove. However, I was determined to overcome circumstances instead of letting them overwhelm me. At last I realized that life is real and earnest and not the charming fairy tale of which I had so long dreamed." That was the dread moment when reality set in. She quickly drew a screen over this glimpse into her inner self, adding that "in principles and motives, the real essentials of life, they were as one."

There's a lot of distance between "the real essentials" and day-to-day living. The strains were painful, the fissures were deep. In her self-censored family account Rebekah would not have mentioned the psychological tug-of-war between father and mother over how their son was to be raised. Sam must have scowled at the dancing lessons and Rebekah must have winced at his rough talk and table manners—objections, perhaps not put into words, as each tried to raise Lyndon in his or her image.

Rebekah's nostalgia-burnished memories of her growing-up years in Blanco, a good-size town of some five thousand, depicted

a blissful life in "a well-ordered, peaceful home to which cross
words and angry looks were foreign. I love to think of our home,
a two-story rock house with a fruitful orchard . . . terraced flower
beds, broad walks. . . . Most of all I love to think of the gracious
hospitality of the home, of the love and trust, the fear of God, and
the beautiful ideals that made it a true home." It all sounds a
shade too perfect, a wisteria-entwined paradise. With the selectiv-
ity that Rebekah so often employed, she skimmed over the hard
times in the wake of a four-year drought that devastated her
father's farmlands and left him financially and physically defeated.
Despite all, her memories of her father remained golden: "the
dominant force in my life as well as my adored parent, reverenced
mentor, and most interesting companion." Did she intentionally
contrast her genteel father with unrefined Sam, her perfect, love-
filled girlhood home with the bleak life for which she was totally
unprepared? Had she no intimation of what was facing her when
she married Sam?

Rebekah was understandably proud of her family's history, with
its long line of Baptist ministers. The first Bains patriarch (the family had not yet added the e to the name), a planter and surveyor,
had arrived in the North Carolina colony years before revolution
was even whispered. After the war, in the restless optimism of the
evolving America, each generation of the family pulled up stakes
and moved on—to Mississippi, to Arkansas, to Louisiana. In 1850
they migrated to the new state of Texas, where Rebekah's grandfather, George Washington Baines, achieved distinction as a Baptist
preacher, educator, and editor of the state's first Baptist newspaper.
As president of Baylor College, he successfully steered the institution through the turbulence of the Civil War.

It was her father who molded Rebekah. "He taught me how to
study, to think, and to endure, the principles of mathematics, the
beauty of simple things. I was timid as a child, but my father
would say, 'My little girl can do anything she wants to do.' He
gave the timid child self-confidence." That self-assurance never

failed her—and was passed from father to daughter to son. Her mother, Ruth Huffman Baines, was mentioned rather dutifully in her daughter's life story as "cheerful and energetic, hospitable and friendly, resourceful and ingenious." Left in near penury by the father Rebekah so admired, Ruth became a resourceful widow—she turned her domestic skills into cash by boarding college students. Not until she was a beleaguered young wife herself did daughter Rebekah come to appreciate her mother—before then it had been hard for her to see beyond the father she idolized.

★   ★   ★

THE clan Rebekah married into left Georgia in 1846 to push west to Texas—a very different breed of Texans. Young Sam Ealy Johnson, Lyndon's grandfather, and his brother Tom were both in their early twenties when they got into the unforgiving business of raising cattle, one in which many were wiped out economically or physically—or both. They mastered the bone-breaking, man-making Chisholm Trail, and in 1870 became the biggest cattle drivers in their region, cited in *Trail Drivers of Texas,* a history of those unique American figures. The reality of the trail was brutal. No western movie could begin to capture the aching weariness of driving cattle, sleeping on the ground in any weather, cooking indigestible food over the fire, guarding against a stampede, and praying you didn't break your neck before you got to trail's end, the rail junction at Abilene. It was a town that had all the amenities a cowboy dreamed of, including a bath, a bar, clean clothes, available girls, and the relatively lawful peacefulness that Marshal Hickok imposed on it, making it an attractive "trail's end" for God-fearing folk like the Eisenhowers.

After a time Sam settled down and took a wife, tall, dark-haired, magnolia-skinned Eliza Bunton, Lyndon's grandmother, who proved a gritty match for her husband. Just as Sam figured in *Trail Drivers of Texas,* Eliza was given a place in *Heroines of the*

Highlands of Southwest Texas, Leading Women Among the Pioneers. She was alone—her husband was away with a posse searching for Comanche attackers—when she heard hoofbeats coming closer, the sound of Indians nearing the cabin. She snatched up her two baby boys and climbed down into a root cellar, carefully pulling a rug over the opening. As Comanches ransacked the house, she nursed and calmed her infant and stuffed a cloth in the toddler's mouth to muffle any cries. She stayed hidden all day, until Sam came home, and then she ministered to the men who had been wounded in the skirmish with the Indians. This fearless couple, in their quieter years, read their Bible every day, and Sam Sr., the tough old trail driver, was never heard to use foul language.

There was yet another Johnson hero of mythical dimensions—his great-great-grandfather, the president said, fought at the Alamo. But historians declared that this Johnson ancestor sprang to life in Korea in 1966, born of a moment of LBJ invention when he visited American soldiers serving there. Just as his mother embellished the story of his birth, he figured it could have happened—were it not for the fact that all of those actual heroes of the Alamo were killed.

This turbulent mix of genes was always at war within Lyndon, the Baines boy who had hoped to be the education president, the Johnson boy who wouldn't withdraw from Vietnam. "Cousin" Oriole Bailey, Lyndon's ever-quotable great-aunt, offered her shrewd view of the two tribes: "The Baineses have the brains and the Johnsons have the guts. The Baineses are intelligent but they couldn't put things over; the Johnsons can put things over." Rebekah, in Oriole's opinion, "was one of the smartest women I've ever known—Lyndon got his brains from her." But even Sam's most venomous detractors conceded that Sam the legislator was "straight as a shingle," and in the horse-trading Texas capital, Sam Ealy Johnson was known as a man who could not be bought.

Lyndon grew up in the middle of this clash of familial cultures. Rebekah never let up on her message that "the Baineses were always strong for high ideals. They talked about high ideals. We felt that you have to have a great purpose behind what you do or no matter what you do, it won't amount to anything." Sam's sister Jessie much admired Rebekah: "She was of a high type. She never came down to [Sam's] level on a lot of things, 'cause he was country-like. Sam said what he wanted to any time, but Rebekah was always, always dignified, in everything."

In retirement, the ex-president, talking with his biographer, Doris Kearns, drew an even bleaker picture of the differences between his two parents. "My mother soon discovered that my daddy was not a man to discuss higher things. To her mind, his life was vulgar and ignorant. His idea of pleasure was to sit up half the night with his friends, drinking beer, telling stories and playing dominoes. She felt very much alone. The first year of her marriage was the worst year of her life—then I came along and suddenly everything was all right again." The favored son cast himself as the instrument of his mother's happiness, the hero riding to her rescue. A self-serving interpretation—and probably accurate.

Unfortunately, everything was not magically made "all right" by his arrival. A moment in the family story captures the son's protective love and, by implication, tells much about a young mother left to cope alone with heavy chores, two small children, and another on the way. At their isolated cottage dark was falling after a wrenching day. Rebekah, who had no idea of where her husband might be, was drawing water from the pump when she burst into tears brought on by sheer exhaustion and the accumulated disappointments of her marriage. Lyndon—"he was a little bitty boy," as his wife Lady Bird tells it, "no older than four"— found her weeping. Clasping his arms around her knees, he comforted her. "Don't worry, Mama. I'll take care of you." No mother could forget such a tender promise, at such a tender age. He showered her with little treasures—smooth pebbles, wildflowers, and all manner of pretty things he would "buy" for her from the pages of magazines. (The grown-up Lyndon kept these childhood promises generously, shouldering family and financial burdens, and sending her ever-prettier things.) "I knew how much she needed me to take care of her," her told Kearns. "I liked that. It made me feel big and important. It made me believe I could do anything in the whole world."

Rebekah tried to minimize that painful episode, insisting that she was afraid of being alone after dark. It seems likely, however, that her anxiety was compounded by resentment at being left alone again and again, for reasons that were less than pressing, if any reason at all was given. Everybody, it seemed, came ahead of Sam's own family, and her.

Of all the strains between the two, Sam's drinking was the most damaging. Demon Rum was the brother of Satan in the Baineses' teetotaling Baptist household, and alcohol could turn Sam into a foul-talking roughneck. Lyndon's boyhood friend, John Brooks Casparis, recalled Sam Johnson, a legislator, obviously intoxicated when he came to speak at the Johnson City school. "I've seen Sam come just lit up like a country church and have to climb up the steps on his all fours. He's been at our house for breakfast and sobered up." Rebekah, with her overdeveloped sense of propriety, would have suffered acutely in that small town. Indeed, life was no fairy tale.

Still, Sam Ealy's political record in the Texas legislature, compiled over six terms in two different periods between 1904 and 1924, was one of accomplishment. It was Sam who led the fight to protect the German American citizens of Fredericksburg against the paranoia of the First World War and it was Sam who courageously stood up to the Ku Klux Klan—neither action easy, or without consequences, in bigoted rural Texas. Rebekah backed him unflinchingly, even when death threats were stuffed in their mailbox.

The imposing red granite capitol of Texas was a second home for Sam Ealy. He glad-handed by day and drank with his pals by night, and for this pleasurable work he was paid five dollars a day.

His energy lit up the chamber and his political instincts steered him toward the powerful men who ran the place. Among those friends was rising star Sam Rayburn, who would become a Washington legend as Speaker of the House of Representatives—and would mentor Sam's boy Lyndon in the Congress and staunchly support him in the White House.

When his son, the new congressman whose victory in a special election had more than fulfilled the father's dreams, boarded the train for Washington in 1937, Sam Ealy kissed him good-bye. Within a few months his bad heart gave out. At the end he insisted that Lyndon take him from the Austin hospital back home to the farm: "Bring me my britches, Lyndon," Sam said. "I want to go back to where people know when you're sick and care when you die." In her family album Rebekah wrote movingly of "the brilliant mind, the dynamic personality, the great and loving heart of Sam Johnson" and their happy life together. If her revisionism was at work again, well, selective memories have their place in twenty-one years of widowhood.

★   ★   ★

WHEN Lyndon was growing up, the family moved from the farm to Johnson City and back three times as Sam's luck and whims shifted. For Rebekah the moves into town were exhilarating, the retreat back to the farm, depressing. In those days Johnson City (founded by a distant relative with grandiose visions) was not much more than a one-block string of stores and a population of about three hundred. It was indistinguishable from any other mid-Texas hamlet, flat, dun-colored, with unpaved streets dribbling away to nothing and unassuming houses with dirt yards, kerosene lamps, doors that were never locked—and neighbors who knew each other. It's not all that different today, except for the historic marker proclaiming, "Johnson City, Home of Lyndon B. Johnson."

The little white frame house that has stood on Avenue G since

1886 is a tourist attraction now, gleaming with ever-fresh paint, its gingerbread trim as crisp as the day the carpenter finished it, windows framed in aluminum, cramped rooms crowded with furniture upholstered in red velvet unmarred by wear. Beautification has been at work here; the keepers of presidential homes cannot resist improving them. "It is not so much the home in which the President used to live," wrote Hugh Sidey, who has observed nine presidents, "as the home that he wishes he had lived in."

That house was typical of the community, unprepossessing and pinched, an insular place bypassed by the railroad. In this unpromising setting Rebekah carved out a full life for herself. Drawing on her college training in speech, she coached a winning high school debating team, directed plays, organized poetry readings and declamation competitions, and gave elocution lessons. Rebekah was known, respected, fulfilled. Twenty-five years after her death, one of her young debaters, John Casparis, was still in awe of his tutor: "Lyndon's mother was the only lady in Blanco County that had a college degree. Rebekah Baines Johnson was a princess."

In its small-town way Johnson City provided a lively social life for its young people: church suppers, parlor parties, watermelon feasts, and chaperoned dances. Rebekah, bowing to Baptist strictures, did not allow her daughters to attend but, in a blatant double standard, did not seem to mind that Lyndon was a regular at the town soirees. Perhaps she rationalized that Lyndon must be able to fit in, wherever his future led him, and she had no doubt that his path would lead to high places.

The Johnson family fortunes, always meager, waxed and waned, depending on such unpredictable factors as weather, the cotton market, and luck. Sam dabbled in real estate, cattle, and little business ventures—without marked success at any of them. At best it was a no-frills life; it took everything they could scrape together to keep the family of seven afloat. "They were poor," acknowledged Truman Fawcett, who lived next door. "But everyone was poor, according to standards of today. Then you could get by with so much less." His wife, Wilma, elaborated: "We were all poor, but not poverty stricken. Some families had more culture and more education than others." Despite hard times, all five Johnson children managed to go to college. Judgments, Truman Fawcett suggested, should be based "not on rich and poor, but then and now. It was a different kind of life." Yet Lyndon remembered lean times in the late 1920s when his mother "wrote a cousin and borrowed $25 for us to eat off."

In a touching letter to his Grandmother Baines when he was a student at San Marcos College, Lyndon thanked her for two suits she had sent him: "The suits both fit me and I wear them every day. I haven't had any new clothes since I started school and these sure are handy. I am looking forward to the day when I can reciprocate." On the other hand, Sam Houston insisted that theirs was one of the few Johnson City families that could boast both a Hudson car and a bathtub. With such contradictory family memories, it's small wonder that biographers present conflicting views of the Johnsons' circumstances.

★   ★   ★

AS president, LBJ added to the confusion of the family's status by exaggerating his impecunious beginnings, casting himself in the mold of log-cabin presidents of the past. Rebekah, bridling, would chide him: "Why, Lyndon, you know you were born in a much better house, closer to town, that's been torn down." (The "birthplace" is in fact a meticulous—and enhanced—reproduction of the original.) She was hurt when the president talked about the family as "tenant farmers and poor," taking it as criticism of herself and Sam for having failed the children. "She didn't feel any sense of failure," insisted Juanita Roberts, LBJ's longtime secretary and family friend. "I think they must have been upper-middle class in standards and perhaps lower-middle class in means." That succinctly captures Rebekah's quandary: she was a woman with upper-class aspirations trapped in a social and economic class beneath her dignity and her dreams.

The family enjoyed some comfortable times until the disastrous year of 1921. Sam was wiped out. Lyndon, at the sensitive adolescent age of thirteen, suffered the humiliation of sudden impoverishment. Rebekah, with fierce pride and no money to match, would not compromise her standards. Looking back from her special perspective, Lady Bird Johnson observed that the full cost of that debacle went beyond financial ruin: "Sam lost all the family's money the year the bottom fell out of the cotton crop. With it went a great deal of his pride and heart, but Mrs. Johnson would never give up. She was determined that the children not suffer."

Though not an ideal husband by any measure, Sam Ealy was never a mean man, and when things were going well he was a generous man. Rebekah could forgive a lot when Sam came home one day with two presents for her, a Victrola and a newspaper. Not merely that day's edition of Johnson City's *Record-Courier*—he had bought the paper, lock, stock, and printing press. For a year Rebekah, the would-be journalist, reveled in being editor—and everything—of the weekly. After the children, all under eight years old, were tucked in for the night, she would settle down with her yellow-lined tablet and write every word of the week's issue.

Then she branched out, sending accounts of local doings to three big-city Texas papers *(Dallas News, Austin American, San Antonio Express)*. She had, at last, become a journalist. This was a time when journalism was not considered a fitting occupation for ladies—apart from writing up weddings and temperance meetings or giving advice to homemakers and the lovelorn. Only a few bold spirits broke through the barriers, mostly young unmarried women like the intrepid Elizabeth Cochrane—nom de plume "Nellie Bly"—who raced around the world in less than eighty days in 1889, a feat that made headlines that might have inspired a little girl in Texas. Rebekah, a more conventional married lady with children, signed her articles with a gender-neutral byline, "R.B.J."

Even with this new outlet for her talents, her life remained centered on her children and her home, and she was determined to
improve both. Years later Lyndon complained about wearing
hand-me-down clothes, as though that was not the practice
among a great many American families pinched for cash. Aunt
Jessie Hatcher had a clearer memory of better times: "She dressed
them little children—oh, they just looked like little angels, her little children did. We [Sam and herself] came up hamper-scamper,
you know. Anything that Mama put on us, that was all right. But
Rebekah's children were always dressed." Rebekah's granddaughter, Luci Baines Johnson, would explain, "My grandmother was
not just the matriarch. She had a role in the family—'the lady'
role. She was the personification of grace, refinement, education—
of what was *right*."

In her "lady" role, Rebekah was at her best as a gracious hostess. Her sister Josefa remembered a steady flow of guests crowding
the dining table. "Seemed like every meal the table would be full.
There was always somebody here, company and all." She served
them good country cooking—platters of fried chicken, hot biscuits, fresh honey. "And she always had a cloth on the table—she
wouldn't use oilcloth. She liked everything just so." Once again
"just so" described Rebekah, a woman who set her standards high
and refused to lower them.

At the other extreme, Sam was just as likely to arrive home with
friends in tow and not a word of warning. Looking back, Rebekah
could laugh about it: "I never knew who was coming to supper—
a ranch hand or the governor. Well, I've always enjoyed good conversation and I never had trouble finding something to talk about
with either of them." In one oft-told story, the family had just sat
down to an elaborate Thanksgiving dinner, the table set with the
wedding china and a lace cloth, when a family of farmhands
stopped by to give Mr. Sam a cake. Sam invited them all to join the
family dinner, which turned into a pretty raucous affair. Rebekah's
meticulously planned family occasion, so important to her, was

destroyed. She was furious, in her cold, silent way, and after dinner Lyndon found her weeping in her room. She yearned to pre-
side over a gracious home, with the decorum she remembered
from her upbringing. But Joseph Baines was polished gemstone;
Sam Ealy was natural Hill Country rock.

★   ★   ★

FROM her earliest days at her father's knee, Rebekah had a passion for education. It was her touchstone, not just as the key to
success but for the sheer joy of learning. She was a voracious
reader, a star pupil. As a mother, she wanted more than anything
to implant the Baines bookishness in her son, starting his reading
at a precocious age. From childhood she was set on going to college, and though her decision was slightly less astonishing than
that of Martha Truman and Ida Eisenhower decades earlier, the
number of girls in higher education in Texas at the turn of the
century was minute. It is impressive, and significant, that these
nineteenth-century mothers of contemporary presidents placed
such importance on higher education, which scarcely figured in
the thinking of most young women in those days.

Since her grandfather the Reverend George Washington Baines
had been president of Baylor University, a major Baptist institution in Texas, it was natural for Rebekah to enroll there. She took
her second year, in 1902, at the University of Texas in Austin,
then transferred to Baylor Female College in Belton, Texas, which
had "graduated" from its lesser status as Baylor's Female Department. (The college is now known as Mary Hardin–Baylor.) It was
considerably more than a way station for young ladies of high
character waiting for the right man; entrance requirements for its
275 students included science, mathematics, Latin, and French or
German.

In Rebekah's time, the college consisted of one overcrowded
three-story structure set on a hillside where cows and horses
roamed freely. Even in those smelling-salts-and-vapors times, Bay-

lor girls were keen on competitive sports, which they played in the cumbersome outfits deemed proper for young ladies—ankle-length white dresses for tennis, dark long-sleeved blouses with voluminous bloomers for basketball and baseball.

Majoring in English literature, Rebekah developed a passion for poetry that stirred her throughout her life, but it was her classes in elocution, a favorite course for genteel girls, that contributed most to her later life. She gave "expression" lessons, directed plays, and helped Lyndon become a first-rate high-school debating coach. (Yet as president his speeches were less than effective—he needed his mother's coaching.) As she prepared for her senior year in 1904, Rebekah was dealt a serious blow: her father was wiped out by a depression that flattened central Texas like a hailstorm. In her first test of self-reliance and grit, she got a job managing the college bookstore, paid her own way—and achieved her lifelong goal, a college degree.

Today, a portrait of Rebekah Baines as a student graces the foyer of the college's Johnson Hall. It is a beautiful young face, delicately sculpted and intelligent, with wide-set eyes that rivet the viewer with their level gaze. "I am sure of myself," this face declares. The building was dedicated to her in 1968, as the institution's most celebrated alumna—if you count celebrity gained secondhand, through a son in the White House.

Rebekah turned teacher in earnest when her children reached school age. "I put them around the big round table at night and stopped behind each to see they were getting their lessons," she said. "The thing I wanted was to have them with their faces clean, their hair combed, to be dressed neatly, to know their lessons." And to come home to "a good hot meal," of course, but she believed "in putting something in their heads" as much as in their stomachs. She worked hardest on Lyndon, following him to the gate, trailing him to the barn as he saddled his horse to ride to the country school, stuffing his head with facts, facts, facts.

Maybe she pushed too hard. As Lyndon began to separate himself from her to define his own identity, as he outgrew her educational sessions at the breakfast table, he no longer shone like the little boy rewarded for reading by his mother's hugs, no longer the boy whose mother would always let him win their games, even if it meant changing the rules. Despite the proper necktie that Rebekah insisted on, Lyndon the teenager was no different from the other country boys, who were stirred up more by the hormonal rush of adolescence than great ideas of western civilization. "I doubt that Lyndon, while he was in school, ever read a book that wasn't required," chuckled his friend Emmette Redford. "But Lyndon could astound you at the knowledge that he gained out of conversation—the man must have been almost a genius in absorbing information from others." The instincts of a politician, not a scholar, were coming to the forefront. He won his first election: his high-school senior class of six elected him president, thanks to the four-girl voting bloc. It was an early lesson in bloc voting and the potential of newly enfranchised women. The class prophesied that some day Lyndon Johnson would be governor of Texas. They sold him short.

Imagine, then, Rebekah's devastation when Lyndon, not yet sixteen, announced upon graduating that he was permanently through with school; he'd had enough education. Her tears, her entreaties, her devastating frosty silence—nothing got through to him. A few weeks later he and four friends surreptitiously packed their belongings in a cantankerous old T-Model Ford and drove away in the night, headed for California. They slept by the road, burying their meager handful of cash against robbers. California has always been the place Americans strike it rich, at least in their dreams; it took less than two years for Lyndon to face the reality that it wasn't the right place for him. He was going nowhere but back to Johnson City, back home, back to Mother.

But not back to college. He worked with pick and shovel on a road crew by day, caroused at roadhouses by night. His life was taking a downhill course that led straight to the quagmire that had claimed his father. Lady Bird Johnson recounted Rebekah's story of the worst night of many. There had been a brawl, with Lyndon in the thick of it. His drinking buddies ferried him home, hurt, bleeding. "He was a pretty sad looking piece of youth," said Lady Bird. "They loaded him into bed. Rebekah stood at the bedroom door and wept. 'Oh, my firstborn, my firstborn. To think you should come to this.'" It would be impossible to overstate Rebekah's pain, her humiliation. Lyndon at least had the residual grace to be consumed by shame at this degrading episode.

Not long after that ugly night, after yet another miserable shift of working with pick and shovel on the roads on a cold, raw February day, Lyndon came home and stretched out across the bed, exhausted. "Mama," he said wearily. "I've tried it with my hands. If you will help me, I'm ready to try it with my head." Few men can ever record such a clearly defined turning point in their lives. Rebekah wasted not a moment; she raced to the party-line telephone to call the president of Southwest Texas State Teachers College, Dr. Cecil Evans. It was a small institution in San Marcos, forty miles away. Sam Ealy had met Evans when serving in the state legislature, but it was Rebekah who took action and made the call. Dr. Evans promised a campus job for Lyndon, but first there was the question of finding money to pay the forty dollars or so a month he would need for minimal living. The Johnson City Bank turned him down cold—apparently Sam Ealy's boy was not deemed a safe bet. Rebekah reached back to her girlhood in Blanco, contacting a banker who had been a law clerk for her father. Joseph Baines's grandson, he decided, could be trusted with a princely loan of seventy-five dollars.

Rebekah's determination was fierce and sustained. To make sure Lyndon passed his entrance exams, "she came down and stayed all night with me working on plane geometry," he would remember gratefully. "We didn't close our eyes until I had to go to the eight o'clock class. She sat up all night long." He squeaked by, barely. Rebekah would have stopped short of nothing—legal—to assure a college education for her son. The Lord had answered her prayers in awakening Lyndon, and she would do her part. Without that education, Lyndon could go nowhere; she saw him drifting into the shiftless life of his father. "She wanted him to be important," said Ben Crider, one of Lyndon's close friends in Johnson City. Her own dreams for her son would be vindicated in a measure even she could not have imagined at the beginning of his real career.

★　　★　　★

SO in March 1927, Rebekah could at last point with pride to her son the college boy. Lyndon hit the campus of Southwest Texas State like a tornado, quickly figuring out the way to a leadership role at his school. Within days of enrolling as one of 750 students, he pontificated in a piece for the campus newspaper: "What you accomplish in life depends almost completely upon what you make yourself do. . . . Perfect concentration and a great desire will bring a person success in any field of work he chooses." Rebekah couldn't have said it any better herself. Doubtless, she had said it—over and over—to deaf ears.

But after two years of constant hustling, living penniless on the edge and thin as a beanpole, a despondent Lyndon was thinking of dropping out. Ben Crider, who was working in California, received a depressed letter from his old buddy. "He said he needed money to buy some clothes, that it was embarrassing to go to class. He didn't ask for money; he wanted me to get him a job in California. He was going to quit school." Once again, Rebekah went into action. "Her big struggle was to get Lyndon to finish college. She had written me a letter and asked me to help keep Lyndon in college, so I sent him a hundred dollars and told him that would help keep him in school. Lyndon never knew what his mother had asked me to do. He stayed in college." (Later, a grateful Lyndon would use his new political pull to get Ben a good government job in Houston.) It should be noted that Rebekah was equally deter- mined that her other children would go to college, which she

accomplished by persuading Sam Ealy to move the family to San
Marcos for nearly four years. In 1930 he had gotten a steady job
as inspector for the motorbus division of the Texas Railroad Commission, a political appointment; his ties to the farm were finished,
and now the four younger children could attend Southwest Texas
State Teachers College for little cost as day students.

Lyndon lived hand-to-mouth on the San Marcos campus, and
after his sophomore year, in 1928, desperate for money, he took
time off to fill in as principal of a run-down school for Mexican
American children in nearby Cotulla. Teaching was in his blood—
not only from the Baines side of the family but also from his
father, who had been a teacher when he was young. Lyndon's nine
months with those poor children in Cotulla stamped his thinking
and would influence his policies as president. When outlining his
proposed "Great Society" to Congress, LBJ told of that long-ago
experience: "They knew even in their youth the pain of injustice."

He returned to college to complete his degree. Among Lyndon's
most endearing letters as a student was one he wrote during his
senior year, dated December 13, 1929; it took up three pages in
his even, flowing handwriting,

My dear Mother,

The end of another busy day brought me a letter from you. Your
letters always give me more strength, renewed courage and that
bulldog tenacity so essential to the success of any man. There is
no force that exerts the power over me that your letters do. I have
learned to look forward to them so long and now when one is
delayed, a spell of sadness and disappointment is cast over me.

Christmas was in the air, and Lyndon was tenderly nostalgic:

I have been thinking of you all afternoon. As I passed through
town I could see the mothers doing their Xmas shopping. It

made me wish for my mother so much. I thought of the hard times that you always have in seeing that every child is supplied with a gift from Mother. I hope that the years to come will place me in a position where I can relieve you of the hardships that it has fallen your lot to suffer—and I'm going to begin in a small scale right now. The enclosed is very small but *you* can make it go a long way.

It was a check, as much as a college student could manage, the first of many generous checks that would more than fulfill the son's early promise. In that heart-baring letter, Lyndon made only one mention of his father, and that was negative in tone. "I don't guess Daddy has found me a job—so I may not get home [for Christmas]. I'll be thinking of all of you every minute." One can almost hear his homesickness and the assignment of blame. The garlands of praise and love are all for mother—"my mother has no equal." Father is failing him yet again; father would be the reason that her lonesome son might miss Christmas with her; father has not provided her a comfortable life. He closes, not with the casual "love" scratched by most college sons on the occasional letter, but with a tender "I love you so much." He signed it, with odd formality, "Your son, Lyndon Johnson." No mother could ask for a more beautiful Christmas gift.

On the back of the envelope of her letters to Lyndon, Rebekah inscribed a single biblical word: "Mezpah." It carries a fullness of meaning: "The Lord watch between thee and me while we are away one from the other." It was a reminder of the bond between mother and son that would endure for as long as they lived.

★   ★   ★

TALL, good-looking, and glib, Lyndon was a big man on campus and always popular with the girls. The ones at the top of his list he would take to Johnson City to meet Rebekah, partly to show her off, partly to see her reaction to his current favorite. But the young

Lyndon was also developing deeper concerns, especially after his experience in Cotulla. Then Lyndon at last was out of college and into the world. Following Rebekah's example, he taught public speaking and coached the debating team at a Houston high school, acquiring vital skills for a politician. Then came the break that changed his life: Congressman Richard Kleberg, an influential Texas Democrat, hired him as his secretary. He was off to Washington and the fringes of the big time. The politically astute Rebekah was ecstatic.

After a time, on a trip to Austin, Lyndon was introduced to a soft-spoken, even shy girl with sparkling dark eyes and keen intelligence. She was Claudia Taylor, the only child of "Captain" Thomas Taylor, an affluent merchant in Karnack, close to the Louisiana line in piney woods East Texas, a region more Old South than New Texas. Motherless from the time she was five, she was reared as a proper young lady by a maiden aunt and a doting father in a fine house with echoes of long-gone plantation days. In her teen years she was sent to the fashionable St. Mary's School for Girls in Dallas for added polish, then to the University of Texas in Austin, for a degree in journalism. Everybody called her Lady Bird. And Lyndon called her constantly.

It was not at all unusual for Lyndon to bring home a regular parade of friends, both boys and girls, to meet the mother he was so proud of. But had he told Rebekah that this girl—whom he had met only the day before—was special, that this one would be The One? No matter. Rebekah would have known; she would have sensed it in the way he looked at Claudia. More than sixty years later, Lady Bird Johnson, long since emerged from her early cocoon of shyness to become one of the great First Ladies and a Johnny Appleseed of wildflowers, reminisced about her mother-in-law.

"I remember the first time I met her. She saw me as a potential enemy. Yes, it was actually fear I saw on her face. It was not just that I was the first person he was serious about—Lyndon was also the strongest oar on the ship of her life. She would have been thinking, 'What does this mean to all of us?' " When the other children—three sisters and a younger, increasingly irresponsible brother—needed help of any kind, Lyndon was called on to provide. Even as a teenager he had bossed them around, and as an adult he came through for them whenever his mother asked, using his connections to get them government jobs and lending them money when he had none to spare. "Rebekah," Lady Bird observed, "regarded him as a care-giver, someone to lean on. There was a considerable amount of resentment from the others, mixed with pride and affection." The unspoken was obvious: at only twenty-six Lyndon was a greater source of strength in the family than the father, who, paradoxically, as a legislator would go out of his way to help widows, veterans, and the forgotten— but was not much help for his own.

Lady Bird sensed at this initial meeting that she was preparing the groundwork for a long relationship. "I was trying to reassure this gentle—but very strong—woman that she had nothing to fear from me. She felt her son was too impetuous—and I kind of agreed." This girl of twenty-one, who had been indulged and protected, instinctively understood the anxiety in this devoted mother's heart. Lyndon's father, who would have sized up the girl from a very different perspective—her pretty face, slender figure, and solid bank account—was "jovial and relaxed" from the start. Rebekah was "reserved, dignified." Lady Bird's choice of adjectives to describe Rebekah—"gentle . . . strong"—were precisely right, and precisely right to describe herself, the daughter-in-law, the beloved national figure, in 2000.

Lady Bird soon took Lyndon to Karnack to meet her father, who instantly approved the match. Perhaps Tom Taylor remembered, still with a pang of humiliation, suffering through the same situation years before. Lady Bird's future mother, Minnie Patillo, whose father owned the largest plantation around, was riding her new mount across open fields when she was thrown and badly hurt her leg. Young Tom Taylor rushed to her rescue, cared for her—and fell head over heels in love with her. But when he approached Minnie's imperious father to ask for his daughter's hand in marriage, the older man coldly rejected the proposal—what did this poor farm boy Tom Taylor have to offer his beautiful Minnie?

Despite her father's implacable opposition, Minnie and Tom (like Rose Fitzgerald and Joe Kennedy) continued to meet secretly. Tom worked hard and within six years he could buy the biggest house in town. The father who had scoffed could only acquiesce—and so they were wed. Remembering Mr. Patillo's initial doubts about him, Captain Taylor bet on young Lyndon Johnson as a comer. Throughout the years Lady Bird has happily retold her father's vote of approval: "You've brought boys here before, but this time, daughter, you've brought home a man."

Back in Washington in that autumn of 1934 Lyndon pursued his Bird with a barrage of letters, pleading—demanding—that they marry at once. She was learning that what Lyndon Johnson wanted, Lyndon got. Then, without informing his mother, something almost inconceivable, Lyndon whisked Lady Bird off to San Antonio and, on November 17, married her.

How could he do that to Rebekah, to whom he owed so much, who would always occupy a central place in his life and heart? Asked about why this wedding had taken place so abruptly, Lady Bird was pensive for a moment, then said, "I don't think he wanted to argue with one more person." Her younger daughter, Luci Baines Johnson, the family philosopher, offered her insight: "I don't think he could ever have gone against his mother—or wouldn't have wanted to. He didn't want to risk her disapproval. By not asking her, he didn't run that risk." Luci drew on her own

parallel experience, when she—scarcely eighteen—was determined to marry Patrick Nugent. "My mother told me all the reasons it wasn't a good idea, but my father refused to talk to me
about it. He never did tell me 'no'—but I was so bullheaded I
would have done it anyway. I think he was fearful of being
rejected." Ruefully recalling the pain of that failed marriage, Luci
mused, "maybe if he had told me 'no' . . ." She knows better than
that, of course. Years later, when Lady Bird met transplanted
Scotsman Ian Turpin, who wished to marry Luci, she thought
back to that day in Karnack, some fifty years before, when her
father took the measure of Lyndon Johnson and gave his
approval. She repeated his judgment—this time, daughter, you've
brought home a man—and welcomed Ian to the family.

The true dimension of Rebekah's wisdom was demonstrated in
her reaction to being deliberately left out of her son's wedding, a
slight that would have embittered many mothers. Instead of brood-
ing, she wrote the newlyweds a three-page letter of such generous,
loving spirit that it should be required reading for all new moth-
ers-in-law. From the start, this understanding mother reached out
to bring Lady Bird into the warm circle of her affection.

My Precious Children, Thinking of you, loving you, dreaming of
a radiant future for you, I find it difficult to express the depth
and tenderness of my feelings. Often I have felt the utter futility
of words, never more than now when I would wish my boy and
his bride the highest and truest happiness together. That I love
you and that my fondest hopes are centered in you, I do not need
to assure you, my own dear children.

My dear Bird, I earnestly hope that you will love me as I do
you. . . . It would make me very happy to have you for my very
own, to have you turn to me with love and confidence, to let me
mother you as I do my precious boy.

Her touching promise to "mother" the motherless girl was coupled with the "desire and expectation that Lyndon will prove to be as tender, as true, as loyal, as loving and as faithful a husband as he has been a son." It was an outpouring of love to this young woman who had just laid claim to her adored son. (Noticeably missing from Rebekah's letter was any mention of her husband or their life together.) Rebekah had been reassured by Lady Bird's gentle manner, good education, and a background of the kind Rebekah would have wished for her own children. Imagine the difference in Eleanor Roosevelt's life if the fiercely possessive Sara Delano Roosevelt had written such a letter instead of pushing her insecure daughter-in-law into a Cinderella role.

With her thoughtful handling of a potentially uneasy relationship, Rebekah created years of happiness for herself as well as for her son's shy young wife. "Before long," Lady Bird happily recounted, "Mrs. Johnson was one of my best friends. If I had an extra hour in Austin, she is the one I would call." They shared a love of books, poetry, and history. The two of them were role models for the LBJs' older daughter, Lynda Bird, who followed in her mother's footsteps as an accomplished political wife—the active first lady of Virginia when her husband, Charles Robb, was governor and partner to him in the U.S. Senate—and Rebekah's passion for poetry and books still burns brightly in her granddaughter.

On Rebekah's visits to Washington, she and Lady Bird would prowl the old villages of Virginia and Maryland, poring over genealogical records, enjoying graveyards, pouncing on antique pressed glass for Rebekah's collection. "You might love somebody out of a sense of duty—but I just enjoyed her," Lady Bird mused. "We had the good fortune to be friends, which is much better than loving one's in-laws, I think."

★   ★   ★

REBEKAH had always counseled Lyndon to aim high; she had stroked a Texas-size self-confidence. At twenty-eight he leapt fearlessly into a special election for Congress and won—big—to enter Congress as one of its youngest members. Rebekah marked this victory with a letter resounding with the flourish of trumpets.

My Darling Boy,

Beyond "Congratulations Congressman," what can I say to my dear son in the hour of triumphant success? In this as in all the many letters I have written you there is the same theme: I love you; I believe in you; I expect great things of you.

To me your election not alone gratifies my pride as a mother in a splendid and satisfying son and delights me with the realization of the joy you must feel in your success but in a measure it compensates for the heartache and disappointment I experienced as a child when my dear father lost the race you have just won. My confidence in the good judgment of the people was sadly shattered then by their choice of another man. Today my faith is restored.

How happy it would have made my precious noble father to know that the firstborn of his firstborn would achieve the position he desired. It makes me happy to have you carry on the ideals and principles so cherished by that great and good man. I gave you his name. I commend you to his example. You have always justified my expectations, my hopes, my dreams. How dear to me you are you cannot know, my darling boy, my devoted son, my strength and comfort. Take care of yourself, darling. Always remember that I love you and am behind you in all that comes to you.

My dearest love,
Mother

In this outpouring of devotion, pride, and unwavering belief in her son and her father, there is not so much as a passing mention of another politician in Lyndon's background—his ailing father, Sam Ealy Johnson.

Rebekah's presence continued to bolster Lyndon, even from two thousand miles away, with her steady stream of letters, florid in style and with a passion that is, by current standards, somewhat excessive from a psychological point of view. "You have always been my confidant & friend & I turn to you as the flowers to the sun," she declared in one letter; in another she sent "dearest love to my precious son, my strength & comfort." He was her "noble son," her "darling boy"—who lovingly saved her letters and responded in kind. "Nobody else has ever written me such letters as you do," he wrote in his senior year in college. "They have always been and always will be a new inspiration to me. There never was such a wonderful mother as you or one that could say such sweet things to her boy; it makes me feel unworthy, almost, to hear you say the things you do, but it also makes me feel so fortunate and happy. My dearest love to you now and always."

Twenty-six years later, on Mother's Day, 1956, the powerful majority leader of the United States Senate put aside his work to write: "I am so grateful for having you for my mother—a woman of such fine spirit and unlimited devotion. You have been my inspiration, always, and whatever I am or become, the credit for all that is good will be yours." A generous check was enclosed—to buy something pretty for herself—but his words were, as ever, the gift she would most cherish.

When Lyndon was elected to Congress, Rebekah could bask in the certainty that the victory was as much hers as his. With "loving and shoving," as presidential scholar James David Barber felicitously puts it, she had shaped the boy, pulled the youth back from the brink, pushed the young man to higher goals. The granddaughter who bears the family name Rebekah was so proud of, Luci Baines Johnson, asserts, "There is no doubt in my mind that my grandmother's belief in my father stretched him beyond his own dreams." Rebekah, Luci mused, "was a very strong woman with a very defined sense of character and the expectation that her children—and her grandchildren—if they had nothing else in their life, would have education and character."

But for her firstborn there was an additional expectation: he was destined to excel, to be a leader. For a time she had hoped he would be a preacher or a teacher, following the footsteps of generations of Baines men, and Lyndon himself gave it passing thought. But politics, at its highest expression, can teach and preach, and what forum can command such attention as the bully pulpit of the White House? This is not to say that Rebekah disdained less lofty work; she tried to make all of her children understand that "a man could be as honorable doing work with his hands as working with his head." The dignity of work, an admirable gospel. But Lyndon, she would make certain, would "work with his head."

Underpinning her philosophy was her strong Baptist faith. "She didn't push the children to church," said Lady Bird. "Her way was to set a good example." At a camp meeting down by the riverside in Johnson City, Lyndon joined the Church of Christ, possibly carried away by revival zeal, possibly as a rebellion against the rigid tenets of Texas Baptists in that era. As president, he was a regular churchgoer, sampling various denominations and preachers. His close friend Billy Graham, a frequent guest at the White House and at the ranch, saw in Lyndon Johnson "a sincere and deeply felt, if simple, spiritual dimension." Many times the president, entirely unself-conscious, spontaneously moved, dropped to his knees in prayer with Dr. Graham.

As Lyndon climbed the power ladder in Washington, he regularly turned to Rebekah as a fountain of sensible opinion beyond the Beltway. More than once he interrupted a meeting in his Senate office to telephone her—"Let's hear what Mama thinks." "Mama's the kind of person that everyone in the community has always turned to with every problem," said the majority leader. "You can discuss everything with her. She's one of the world's great women." On her Washington visits Lyndon showed her off at private luncheons, where she would enthrall his Senate colleagues. She relished the attention, the reflected status of the son she had encouraged to reach the top.

Rebekah was his touchstone, his link to "real folks." She had been dead six years, yet the president would still sigh, "I wish I had her here now. She could go over the questions before I have a press conference—because she could always think of every one the teacher might ask on a test. She'd be helping me now, telling me what not to say in some of my speeches." And which words to emphasize—and to stand up straight.

Her stamp on the Johnson presidency was indelible. Her commitment to education became his—and resulted in the massive Education Act of 1965. "At the center of my mother's philosophy," he told biographer Doris Kearns, "was the belief that the strong must care for the weak." That philosophy was the core of the Great Society programs, the hallmark of the Johnson presidency. Rebekah coupled that concern with careful husbandry of her limited resources, a linkage reflected in the president's order to turn off lights in unoccupied rooms in the White House, while pressing for massive government expenditures. "You don't have to be wasteful to be enlightened or progressive," he admonished a group of administration officials. "My mother was the most liberal person I think I ever knew. Yet she always had some pin money hidden under the pillow to take care of our needs in time of distress. We must have a war on poverty, but we must also have a war on waste."

Rebekah's dedication to life's finer things shone through in his comments to a group of educators: "I am glad in our time we are talking about how to improve the soul and improve the mind and improve the body and to live and learn and expand." And would Lyndon Johnson have espoused the advancement of women, through appointments and exhortation, had it not been for the powerful example of two strong women, first Rebekah, then Lady

Bird? Perhaps, but it is impossible to separate their influence from the president who would take a stand and take action.

There was, in all honesty, a less admirable aspect of Rebekah's influence. LBJ's celebrated manipulation of people, a political skill he raised to the level of an art form, may be attributed to his constant need to win his mother's approval, to be the top of her totem pole, to be the child she loved best. Rebekah herself could be manipulative, subtly pitting one child against another to win her praise or turning frosty to show her displeasure. "She had a way of making quite a few members of her family feel that she loved them best," observed Luci Johnson. "There was almost a vying for position."

The jockeying between Lyndon and his wayward brother would be lifelong. "He worships you. You are his hero," Rebekah cooed when asking Lyndon to help Sam Houston, putting her manipulative skills to work. Perhaps she was blind to the fact that the younger brother chafed under the older's dominance, that the mother's nurturing had turned divisive. "It was a sad, sad thing," Lady Bird reflected, "that after Lyndon's death Sam Houston seemed noticeably recovered—he didn't have to compete any more."

Yet this duplicitous and wastrel son was raised by the same caring mother as the achieving son. Beyond doubt, the bond between Rebekah and her firstborn was special, something more than the usual mother and son relationship, something more than the tie between this mother and her second son. In the case of Lyndon, Rebekah's unflagging admiration and approval likely contributed to his demand for total compliance from all around him, his suspicion that disagreement signaled disloyalty—an attitude that cost him dearly in his conduct of the Vietnam War.

★   ★   ★

REBEKAH died of cancer of the lymph system on September 12, 1958, but her influence on her firstborn never ceased. Years after her death LBJ said tenderly, "She helped me on everything until the day she died." Now mother and son lie beneath the gnarled live oak trees in the family graveyard, close by the birthplace where their story began.

For thirty-eight years Lyndon Baines Johnson lived a life of public service and accomplished great things, apart from his tragic stumble into the quicksands of Vietnam. It was a record of achievements for which the credit must be shared by Rebekah Baines Johnson. No one did more to shape, by positive example and relentless encouragement, the character of a son than Lyndon's mother, who dedicated her life to encouraging the better angel in her boy.

FIFTY PIES BEFORE BREAKFAST

Youth fades; love droops, the leaves of friendship fall
a mother's secret hope outlives them all.

—OLIVER WENDELL HOLMES

FROM her front-row seat in the VIP box at Chicago's vast convention hall, Hannah Milhous Nixon watched thousands of delegates to the 1960 Republican National Convention cheering, singing, yelling as the band played "California, Here I Come!" For this singular American tribal rite they wore silly straw boaters and brandished placards bearing one name: NIXON! This was the familiar fighter they had chosen as their standard-bearer. The event was more coronation than nomination. Hannah, who was not given to any show of emotion, was openly elated. This excitement was all about her son Richard.

A confident Vice President Richard Nixon had arrived in Chicago with the nomination in his pocket. After eight years in Dwight Eisenhower's shadow, Dick Nixon (only his mother called him Richard) was stepping into the spotlight on his own, taking over the role with gusto, and the world—particularly the Kremlin—was taking his measure. Seeing her son with his arms stretched high in victory, his face transformed by exuberance, Hannah basked in this grandest moment of her seventy-five years.

Of her five sons—Harold, Richard, Donald, Arthur, and Edward—she had a special bond with Richard. He looked like her; he shared her introspective turn of mind, her self-contained personality, the comfort she found in solitude. Paradoxically, he had reached the top in a vocation in which such traits are not regarded as helpful. Since his was not an outgoing, gregarious personality, in a political setting he often seemed awkward, uncomfortable. Like his mother, he was most at ease in small groups and serious discussions.

It had taken Hannah a long time to become accustomed to being mother of the vice president. Still a bit incredulous, she would proudly show visitors mementos of the two inaugural ceremonies, stroking the square of the red carpet he had stood on when he took the oath of office. At both events he had placed his left hand on not one but two of her Bibles, well-thumbed by several generations of the Milhous family. One special memento he had saved for himself. On the evening of the first inauguration, at a family dinner before the gala balls, Hannah had quietly given him a slip of paper on which she had written in her even, meticulous script:

To Richard—You have gone far and we are proud of you always—I know that you will keep your relationship with your maker as it should be for after all that, as you must know, is the most important thing in this life.

> With love, Mother

The new vice president, deeply touched, had tucked it in his wallet and kept it as his permanent touchstone. Eight years later it was still there on the night in 1960 when his party bestowed on him its highest prize, the nomination that could set Richard Nixon firmly on his path into history. It was a great honor to be vice president, of course, but Hannah was aware that vice presidents fade into little more than quiz show questions. A few weeks before going to Chicago, she had watched the Democratic convention on television, studying, sizing up the young senator from Massachusetts who would be Richard's opponent. She had little doubt that the voters would see her son as more experienced and capable, more than just a handsome face. But immediately after the first debate, in which Nixon had looked terrible, a worried Hannah called his secretary, Rose Woods, to ask anxiously, "Is Richard ill?" With bad makeup and an anxious expression, Richard Nixon presented a poor contrast to his handsome, confident opponent.

On the night of the nomination, she savored the irony when Richard's persistent rival, New York Governor Nelson Rockefeller, stepped to the podium, not as a candidate—Nixon had bulldozed him aside before the first primary—but to introduce "the next president of the United States." If only Frank Nixon, cantankerous husband and overbearing father who had harbored a populist animosity toward the big rich—in particular Standard Oil, the source of the Rockefeller billions—could have lived to see a Rockefeller humbled by a Nixon.

The two adversaries represented the extremes of the Republican spectrum—Rockefeller, the worldly Eastern liberal born to riches beyond counting, and Nixon, the conservative Californian from Everyman beginnings. But they had two things in common—both were close to their mothers and not wholly comfortable with their fathers, one a small-town grocer and the other a synonym for rich and powerful. By his own account, it was his mother, Abby Aldrich Rockefeller, the daughter of a distinguished senator and the founder of the Museum of Modern Art in New York, who implanted in her son Nelson a lifelong passion for art and dedication to public service.

In the Nixon family, Frank was the political talker, argumentative and disputatious; Hannah kept her opinions to herself. It was months after the election of 1916 (California gave women the vote

in 1911) before she dared confess to Frank that she had voted for
the Democrat, Woodrow Wilson. ("Frank went pale," she recalled,
and Nixon would later claim, though he was not yet four when
this took place, "I can even remember my father berating my
mother" for that apostasy.) Hannah's reticence about politics was
such that when an interviewer asked her and Rose Kennedy what
kind of presidents their sons would be, Rose, the politician's
daughter, delivered an enthusiastic sound bite for her son Jack—"I
think he would make a wonderful president!"—while Hannah
would only say cautiously, "I think he would make a good president if God is on his side."

Richard had gone through so much to climb to this place in the
spotlight. In his first race for Congress, in 1946, he was castigated
by the press for conducting a smear campaign, tagging his opponent, by innuendo, as a Red sympathizer. In 1951 freshman senator Nixon set off a national controversy with his dogged
investigation of Alger Hiss, a State Department consultant with
input into FDR and Truman foreign policy, who he suspected was
a Soviet spy. Hiss was prominent in the eastern establishment,
which rose to his defense, launching a counterattack on Nixon.
Gaunt with worry, his Senate career on the line, Nixon went to see
his mother, who was then living near Washington. "Why don't
you drop the case?" Hannah pleaded, solicitous of his health as
well as his career. "No one else thinks Hiss is guilty." But her son
refused to back down. "I've got to stick it out until I prove
whether I'm right or not." With that, Hannah, always one who
listened to her conscience, was with him: "Don't give in then. Do
what you think is right." He won his gamble, and the vice presidential nomination. (Secret Soviet documents that became public
after the collapse of communism supported Nixon's grounds for
making the case against Hiss.)

Again in the following year, 1952, Hannah would be his tower
of strength in the furor that erupted shortly after Dwight Eisenhower picked the young senator as his running mate. The revela-

tion of a "slush fund" provided by California businessmen, who asked no accounting of how it was being spent, hit like a bombshell. Left to dangle in the wind by Eisenhower, Nixon boldly took his case to the people. In a televised appeal he put it to the public: should he stay on the ticket or step aside? Again it was a gamble, again his political future was at stake. As he agonized over every word of his speech, Hannah sent him a telegram: "This is to tell you we are thinking of you and know everything will be fine. Love always, Mother." Tears welled in his eyes as he read her message; once again his mother had stiffened his resolve.

Nor did Hannah stop with the message to her son. Despite her Quaker faith's prohibition against acting out of anger, she was furious that Eisenhower had not dealt with the attacks on her son, but had stood aside, waiting for public reaction. Consulting no one but her own conscience, she dispatched a telegram to Eisenhower:

Dear General:

I am trusting that the absolute truth may come out concerning this attack on Richard. When it does, I am sure you will be guided aright in your decision to place implicit faith in his integrity and honesty. Best of Wishes from one who has known Richard longer than anyone. His mother, Hannah Nixon.

At a rally the next day Eisenhower read Hannah's telegram to the cheering crowd and reaffirmed Nixon as his running mate. Though some might think her innate Quaker reticence would make her reluctant to get involved in campaigning, she quickly disabused them. "All his life I've been his campaigner," she declared, and while she did not make speeches in the Rose Kennedy style, she graced many a head table at women's events and put in countless hours at Nixon headquarters. When John Kennedy won by a sharply disputed paper-thin margin, Nixon's campaign workers, who had gathered for what was to be a victory celebration, wept as their candidate conceded the election after an unbearably long night of seesawing returns. Hannah, however, remained composed. "Everyone else was in tears, but Hannah was calm," recalled Nixon's secretary and family friend Evlyn Dorn. "She just said a prayer when she heard the bad news." Hannah had experienced greater losses, and she was certain that her ambitious son would rise above this defeat and eventually prevail. She would talk to him in the morning, tell him she would be in his corner again.

★   ★   ★

HANNAH Milhous was three weeks away from her twenty-third birthday when she met an unusual young man at a Whittier Friends Meeting social. His name was Frank Nixon and he wasn't a Quaker, which was easy to tell in this company of people given to thoughtful silence, because he never seemed to stop talking. As a Milhous, Hannah belonged to a leading family in Whittier, a genteel southern California town settled in 1887 by and for Quakers. They had named it in honor of their esteemed poet James Greenleaf Whittier. The Milhouses owned one of the biggest homes in town, situated on the main boulevard, and were deeply committed to their faith.

Hannah was a young lady of admirable composure, with a grave face made more arresting by surprisingly dark, straight, heavy eyebrows, and a concave uptilted nose, those same brows and "ski-jump" nose that would be a cartoonist's delight in capturing the thirty-seventh president's face. Her voice was pleasingly deep and mellow, as her son's would be. When she met Frank she had completed two years at Whittier College, and was a good, compliant student majoring in languages. Brought up relatively sheltered, she had never been out with a young man until this newcomer, a twenty-nine-year-old streetcar motorman, asked if he might walk her home. He was a nice-looking young fellow, a natty dresser in high collar, bow tie, and watch fob; he was neither tall nor short; his face was undistinguished, except for a nice straight nose. Not a single feature would be replicated in his famous son's countenance. She accepted—and four months, ten days later, on June 25, 1908, Hannah Milhous, her hair piled high and wearing a long white dress with leg-o'-mutton sleeves, became Mrs. Francis Anthony Nixon.

The large Milhous family dutifully turned out for the ceremony at the Whittier Women's Club, but they had little enthusiasm for this marriage. Frank Nixon was not their sort of fellow. True, it was widely remarked that the Milhous family pride was such that nobody was deemed quite good enough for any of their six daughters. Jessamyn West, a cousin who would become a distinguished author, commented tartly that "*all* the Milhous daughters felt they had married beneath themselves," and the sons-in-law were all too aware of that. Though the family was prosperous and its members well educated, many people found their pride exceeded their actual achievements. Those laurels, however, would come later.

In the early seventeenth century the Milhous family emigrated from Germany (the name had evolved from Melhausen) to England, where they threw in their lot with Oliver Cromwell, the Puritan who brought down the king. For their services in the assault on Ireland, they were given a plot of land in the village of Timahoe in County Kildare—"cow Irish," chuckles the last remaining Nixon son, Edward, who looks and sounds so like Richard. "They were given a piece of land and a cow." At some point the Milhouses became Quakers and were subjected to religious oppression that caused them to make the great leap across the sea, in about 1730, to the Quaker colony William Penn had established in Pennsylvania. As the American frontier pushed west, the Milhouses moved to Ohio, then to Indiana, and finally to California.

Wherever they lived, whatever the generation, life revolved around faith and family. They were at the meetinghouse most of

Sunday and again at midweek. Every meal began with prayers and Bible reading, and each child was called on to recite a verse from memory, a practice Hannah and Frank would continue with their boys. In an environment that tended to narrow a family's horizons, Hannah's grandfather, Joshua, rejected the constraints. A lively and forward-thinking man, he was the first in the neighborhood to have a carriage, an indoor bathroom, and a well-stocked library, and, though Quakers then frowned on music, Joshua bought an organ for his family. This unusual man inspired his great-granddaughter Jessamyn's best-selling novel, *Friendly Persuasion,* the fictionalized story of the Milhous family, and was the model for its hero. (In the film Gary Cooper played Joshua, Dorothy McGuire his wife.) Franklin, their son and Hannah's father, was sent to college in Cincinnati and for a year or so was a teacher; the Milhouses were yet to produce their Pulitzer Prize winner and a president, but they were a cut above the folks next door.

The family tree blossomed with dynamic women: Hannah's grandmother was an active preacher and, following in her high-buttoned footsteps, her daughter, Elizabeth Price Milhous, traversed southern Indiana, organizing new Quaker meetings, spreading the faith wherever she found responsive folk. Elizabeth, in her sixties, continued to evangelize when the family moved to California, preaching almost to the end of her life, at ninety-six. Hannah's mother, Almira, though not born a Milhous, was a teacher who rode her horse to school every day. Almost thirty when she married widower Franklin Milhous, she bore him six daughters and one son, and brought up his two children as well, eventually becoming the commanding matriarch at the center of a family that had ramified into dozens of kinfolk.

While Hannah, quiet and reflective, had neither the evangelistic zeal of her preacher-grandmothers nor the autocratic assertiveness of her mother, she was far from passive. She would take extraordinary steps to try to save the life of her eldest son, Harold; she

would work sixteen-hour days to make the family business profitable; she would bolster her famous son when he was under
attack. She spared nothing in defense of her family, a trait Richard
Nixon appreciated from his earliest years.

Almira and her grandson Richard developed a special affinity:
she recognized in him a fine mind and grit, and he admired her tolerance and keen interest in the world beyond her own. "She set
the standards for the whole family," he said of her. "Honesty, hard
work, do your best at all times—humanitarian ideals. She was
always taking care of every tramp that came along the road, just
like my mother, too. She had strong feelings about pacifism and
very strong feelings on civil liberties. She probably affected me in
that respect." To have an influence on a future president on such
meaningful matters would have given Almira great satisfaction.
The Quaker way to influence others is by living lives that embody
the values they hold dear, particularly tolerance and nonviolence.

In 1897 Franklin Milhous, who followed his father in his successful nurseries and orchards, mixing religious interests and
shrewd business sense, uprooted the family, transplanting them
from Indiana to the new town of Whittier. Extensive irrigation
was transforming southern California's arid land into productive
acreage for ranches and orange groves, a golden opportunity for a
man with a professional green thumb. Hannah was twelve when
they moved, literally lock, stock, and barrel, in a logistical exercise
comparable to that of the Eisenhowers' move to Kansas. They
came with a freight car loaded with lumber, doors and windows,
hardware and feather mattresses, orchard equipment, horses and
cows—and the organ. With such foresight he arrived ready to
build a house, imposing though unadorned by Victorian flourishes, and he wasted no time in opening a tree nursery and setting
out an orange grove.

The Milhous household was one of calm and comity, an
enclosed world of its own, in which the family used "the plain
speech," addressing each other as "thee" and "thou." And there

was never any shouting. Though the family was prosperous, instead of servants Almira had daughters. Hannah and her sister Jane were designated the cooking team. "We had lots of company," Jane recalled when she was an old lady. "People would come out from the east and some would stay just weeks at a time. It was a nice stopping place and it didn't cost as much as a hotel." In a burst of candor she added, "In fact, they just sponged on us a lot." The demands at home were such that Hannah took a year off from Whittier College, with every intention of returning to get her degree. Then she met Frank Nixon.

★   ★   ★

AFTER Hannah said she would marry him, Frank Nixon, who had been a strong Methodist, became one of the Society of Friends and for the rest of his life never left the fold. Even so, this did not make him significantly more acceptable to the Milhous family. Frank had gone to school for only a few months, entering and leaving at fourth-grade level, something that would cause concern in a family that placed a premium on education. And there were other reasons for their reluctance to embrace this prospective son-in-law.

In his emotional farewell to his White House staff as he resigned, President Nixon called his father "a great man." That was undeniably a minority view. Dozens of oral histories, those enlightening reminiscences of relatives and early friends who knew Richard Nixon's father, produce a virtually unanimous negative portrait of the explosive Frank Nixon. The more polite sources described him as "loud"; the less inhibited remembered his "yelling," at home, in his store, at his sons, and sometimes, though not often, at his wife.

Looking back at his parents, Richard observed, "Two more temperamentally different people could hardly be imagined. My father had an Irish quickness both to anger and to mirth. It was his temper that impressed me most as a small child. He had tempestuous arguments with my brothers Harold and Don, and their shouting could be heard all through the neighborhood."

That recollection would lead one to think Richard was the youngest son, watching older brothers brazenly spar with their father, when in fact he was nearly two years older than Don. With the small gap in their ages, Richard and Don would seem a natural team, but in fact Don was closer to Harold, more than five years older. Both were more boisterous and full of fun than their serious brother, whom they dubbed "Gloomy Gus."

Harold was good-looking, more like his father in features and lighter coloring. Photographs suggest a debonair flair, showing him with a scarf flipped over his shoulder, or in the casual pose of a Hollywood actor. Cousin Jessamyn West found him "dashing and bold." Don, who tended toward chunky, was a great favorite among the young people, always at the center of the action and, to his mother's dismay, often driving too fast. Both Harold and Don would stand up to the bombastic father, "like two bulls fighting," said an amazed cousin. Don would say of his president-brother, "Dick was always reserved. He was the studious one of the bunch, always doing more reading while the rest of us were out having more fun."

There was no question which of his "temperamentally different" parents the president identified with. The whole family recognized the affinity between Dick and Hannah. "He had more of mother's traits than the rest of us," brother Don reflected. Like Hannah, he was meticulous: "Dick always planned things, he didn't do things accidentally." Like other, though very different, presidents, Richard Nixon's middle name of Milhous was as much a symbol of the maternal line's influence on his disposition as that of Delano was for Franklin Roosevelt and Wilson for Ronald Reagan.

Showing an early preference for Hannah's ways, Dick took note of how she dealt with his volcanic father: "I tried to follow my mother's example of not crossing him when he was in a bad mood." His prudence spared him the ruler and the strap, Frank's weapons of choice, yet he couldn't convince his brothers that "the only way to deal with him was to abide by the rules he laid down." They were like their father—they relished the combat.

Still, there was kindness beneath the bluster. Edward, who was twenty-one years younger than his oldest brother, Harold, recalls his father with warm understanding. "He was up-front, open-hearted, generous—but very opinionated. He was the leader of the family, no question about that. But my mother was the peace-maker." It was impetuous reaction—"my father would spout off before thinking about it"—balanced against thoughtful considera-tion, heat against cool. "My father loved a funny story and he loved to argue, and he was not often beaten. It was up to mother to make peace between him and my brothers."

Hannah would observe pointedly—after Frank's death—"My father and mother never talked loud, never yelled orders. I tried not to yell at my children. It does something to a child." Could it have been that in admiring his mother's efforts to conciliate war-ring parties Richard Nixon saw peacemaking as a worthy pursuit?

For a woman who longed for harmony and "peace at the cen-ter," the strident clashes between sons and father must have been stressful. Living with Frank required infinite patience and subtle manipulation, which can take a toll on a wife. It was perhaps this inner struggle that triggered her occasional bouts of depression. A number of times she returned to her family in Whittier for extended stays, usually explaining that the boys were getting on her nerves. During one such stay, ten-year-old Richard wrote her the much-analyzed letter addressed to "My Dear Master" and signed, "Your good dog, Richard." His little tale of woe began, "The two dogs you left with me are very bad to me," apparently referring to his brothers. He told of a scuffle, and "I lost my tem-per and bit him." After more tribulations, it ended, "I wish you would come home right now." Whatever the symbolism, it was hardly a letter to make Mother feel better.

More likely, Hannah's emotional exhaustion was caused by Frank's temper and constant combat with the boys, but she could never acknowledge such a problem. That would have been humiliating to concede, and would give her family reason to say, "we told you so," further lowering the Milhous clan's opinion of Frank.

Hannah's sister Jane tried to explain, tactfully, the family's reaction to Frank. First, they didn't want her to marry before graduating from college. And, well, "he was a little different type than we were used to. He was quite abrupt in his speech and had very decided opinions and ideas. He was very outspoken, maybe a little loud, and our family was not given to that. He was jolly and . . . nice to the rest of the family, but I always felt that he didn't like to join in the family gatherings." (Milhous family gatherings were major events, drawing sixty-five or so relatives together at Christmastime, eighty at one reunion.)

Jane then delved deeper into the turbulent mind of Frank Nixon. "I don't know whether he was jealous or just what. I could never figure it out. Maybe he didn't want to share Hannah with the rest of us. I think he was quite possessive—but they were not particularly expressive about their devotion at all." While other in-laws might swallow resentment at their second-class status in the Milhous pecking order, Frank's ego couldn't tolerate it. In later years he would sit outside in his car, impatiently honking the horn, when Hannah stopped by to visit her mother.

Bumptious as he was, it is impossible not to feel sympathy for Frank after the childhood he had. The American Nixons were Scots-Irish who arrived from Ireland in 1731, settling in Delaware's Brandywine Valley and starting off well. Two generations later, a Nixon crossed the Delaware with George Washington's troops; another several generations passed and in 1863 George Nixon III—Company B, 73d Ohio Voluntary Infantry Regiment—was mortally wounded on blood-drenched Cemetery Ridge at Gettysburg, where he lies buried. His wife died soon after, leaving eight orphaned children. The eldest, Samuel, taught school (little formal education was required back then to teach at any level), carried mail, farmed, and married Sarah Wadsworth, also a teacher. She bore five children; the middle one, a boy born in 1878, they named Francis Anthony. They were ordinary hard-working Ohio Methodists and a stable family until 1883, when Sarah began to cough up blood, the telltale sign of tuberculosis. In that era there was no cure; tuberculosis brought death.

Then began a harrowing Dickensian family story. In a frantic search for a cure, Samuel sold out and moved the family to a cabin in the West Virginia Appalachians in quest of purer air. But moun-tain walks were no help to Sarah; they only tired her out. Desper-ate, Samuel piled the five children, his mortally ill wife, and their meager possessions into a wagon. For a year they lived in the wagon, trundling through the Carolinas and Georgia, stopping wherever he could find work, and finding no remedy for Sarah's cough. At last she pleaded to go home to Ohio to die.

After her death seven-year-old Frank was turned over to an uncle, with a harsh admonition from his father: "It's root, hog, or die." Four years later the family was reunited with their father and a new stepmother, who could have stepped from a fairy tale—by the Brothers Grimm. She constantly abused them with stick and words. With his brother Ernest, Frank entered the fourth grade in a one-room school; they were outsiders in shabby clothes, wounded by their bitter life and tormented by the other students. Under such conditions it is no surprise that Frank permanently abandoned the schoolroom. (Ernest, however, stayed the course to become a professor of horticulture at Pennsylvania State College.)

When he was about thirteen, Frank left his unhappy home to work as a farmhand, then began years of drifting from place to place, job to job, west to Colorado, back to Ohio, acquiring a cat-alog of skills: glassworker, potter, house painter, telephone line-man, oil field roustabout, carpenter, sheep rancher, potato farmer. It was his job as a trolley motorman, requiring him to stand for long hours in the car's open vestibule, that would change his life. Ohio's bitter winter left him with frostbitten feet. Frank successfully organized motormen to demand legislation for more tolerable conditions. Then, seeking a warm climate, he fetched up in California in 1907, and got a job on the trolley linking Los Angeles and Whittier. Then his trolley struck a car, and Frank was fired.

Fatefully, he found a job on a ranch just outside of Whittier. There, at long last, Francis Anthony Nixon found love, family, and stability. Yes, he yelled and drove everybody crazy with his arguing, but this man had survived brutal times.

★   ★   ★

THE last nail had been hammered into place; the front stoop was painted; the wood stove was ready for Hannah's good cooking. Frank Nixon could look with pride on the house he had built for his wife and three-year-old son. He was a skilled carpenter, and that little bungalow—three rooms downstairs (a fourth would be added later), plus a little workroom for Hannah, and one bedroom in the steep-pitched attic—would be standing, snug and sturdy, most certainly for a hundred years, and would be visited by tens of thousands moved by the modest origin of an American president.

To have a home of her own was bliss for Hannah, never mind that its seven hundred square feet were far less than the spacious houses she had grown up in. Now she could put out their wedding gifts: a cut-glass pitcher, a fine bone china dinner set, silver serving pieces, a clock that would run for at least ninety more years, and a beautiful old quilt, dated 1837, for their metal-frame bed. For more than a year following their marriage, she and Frank had lived with her family. Though her father made Frank the foreman on his thriving orange ranch, it could only have been a touchy situation, given the tensions between the Milhouses and Hannah's husband.

Soon after their first baby, Harold, was born, the young couple moved north to the San Joaquin Valley for a short time to another ranch owned by Franklin Milhous; then, drawn back to the familiar landscape of desert and mountains—and Hannah's loneliness for her family—they bought ten acres of nothing at all in a community that didn't really exist, called Yorba Linda. There Frank planned to develop a lemon grove and put down Nixon roots.

He built their house on a ridge crowning a bare, treeless sweep with an unbroken view of the slumping contours of Old Saddleback to the east. Scattered across the hills were a handful of families. One of the housewives, Cecil Pickering, described Yorba Linda as it was when the Nixons settled there: "It wasn't a town. It was turkey mullein, cactus, rattlesnakes, tumbleweeds and tracks." The concept of a yard and flowers was fantasy—"there was no grass, no nothing except dust." What they dreaded most in Yorba Linda was the Santa Ana wind roaring in from the east, "devil breath," the Indians called it. Mrs. Pickering could feel it as she spoke: "The wind would sway the house this a-way, you'd think you were rolling over. There were no trees to protect us. You could hear the rocks hitting the side of the house when the wind would blow. You could just hardly breathe. . . . If you laid by the east wall when you'd go to bed, the next morning your hair would be white with dust." Native creatures were reluctant to give up their domain; rattlesnakes, lizards, and horned toads lurked among the rocks and coyotes would boldly lope up to the door.

This was the southern California Richard Nixon was born into on January 9, 1913, an eleven-pound boy whose birth left Hannah ill for weeks. In the groves Frank shouted the news—"I've got another boy! I've got another boy!"—and danced a little jig. They named him Richard for the fabled king, Richard the Lion-Hearted, coupled with Hannah's proud Milhous name. In less than two years another son, Donald, was born, and four years later, in 1918, still another son arrived. They had longed for a daughter, and this boy Arthur, would always be treated more gently. Hannah, like Sara Roosevelt and Rebekah Johnson, could not bear to cut his pretty curls until he grew old enough to protest.

In Richard's memory, the world of his early years was idyllic: California weather, endless outdoor space, floating in the irrigation canal that ran only a few yards from the house, so close that Frank once backed the car into it. Evenings brought the family together in the living room, warmed on chill nights by a fire in the brick fireplace. (Perhaps it was this memory that moved him, in the White House, to have fires crackling and the air-conditioning running at the same time.) In that one room the six of them would read, do their lessons, and, with no radio or television to distract, talk to each other, evening after evening. Now and then visiting relatives were somehow accommodated in the tiny cottage.

At mealtimes all squeezed around the oak dining table in the same room, but before any food was served there was the blessing, usually silent, and all were expected to recite a Bible verse. Sometimes the family would gather around the piano and sing popular tunes to Hannah's accompaniment. Life was spartan but secure. Richard would say later that they never thought they were poor, but Hannah remembered it differently: "While we were there, the lemon grove only kept us poor. Many days I had nothing to serve but corn meal." She put the best face on the cornmeal—as she would on her life—and the family "would gobble it up as if it were the most delectable of dishes." While the Nixons might have been no worse off than most of the families trying to eke a living from Yorba Linda's sorry soil, a friend recalled that in those days "she seemed sometimes overwhelmed by the poverty," and would retreat to the comforting arms of family and home in Whittier.

When Richard entered the first grade, Hannah politely but firmly told his teacher, Miss Mary George, "Please call my son Richard, never Dick." Miss George remembered Richard-never-Dick as "a very solemn child [who] rarely ever smiled or laughed," which suggests that his saturnine personality, a darker version of his mother's traits, might have been a product of his genes, as the science of genetics now suggests, and not necessarily the later family tragedies, as is often assumed. (His brother Don, experiencing the same life circumstances, would always be light-hearted.) Hannah had taught him to read before he started school, and he was soon devouring books; like virtually every one of the modern presidents, he was fascinated by history. Years later Miss George still recalled how Richard "absorbed knowledge of any kind . . . like a blotter. . . ."

Donald, lively and more mischievous than Richard, was not keen on school; he was happy to let Richard set the record for reading. "Don's just like a big friendly dog with his tail a-wagging," said cousin Floyd Wildermuth. Edward Nixon, the much-younger brother, a geologist in Washington state, remembers that "Don was loved by everybody. He was unlike any of the family—gregarious, fun-loving, entertaining." It might be that Don was like his father without the anger, without the brutal memories, without the need to dominate. (Edward could not compare Harold, the brother he never knew.)

Home, school, and church were the three pillars of the Nixons' life in Yorba Linda. Frank helped build the Friends meetinghouse, where they would be stalwart members. Meeting-goers would nod approval of Hannah's boys, who always went to meeting wearing ties and little hats, and were very mannerly. Mrs. Pickering knew where the credit lay: "Hannah taught them well."

The poor soil of Frank's land, coupled with his poor management, led to failure. After ten years he sold out. "The land in Yorba Linda," Nixon once wryly commented when he was vice president, "was better for raising kids than for raising lemons." But with foresight that in 1922 was shrewd, Frank foresaw the automobile changing the life of California. Acting on his hunch, he bought an acre of ground on the road between Whittier and the next town, La Habra, and put up a two-pump filling station, the only one on that stretch of highway. On the same plot he built a small house and a garage with a big room above it for the boys.

Richard was nine, Don seven, when the family moved to Whittier, and even at that age they handled small chores while Harold, thirteen, pumped gas. Arthur, then a very young child, was best remembered at that time for sneaking a cigarette at age five, only to have a nosy neighbor catch him smoking and tattle to his mother. Soon Frank expanded, adding groceries to the gas station to become "Nixon's Market." It required hard work from his sons, before and after school, as well as on weekends. But it turned out to be more fruitful than the sour lemon grove.

★ ★ ★

THE town was sleeping when the lights came on in the Nixon kitchen at four o'clock in the morning. Hannah tied on her apron and lit the oven for her daily chore of baking pies and cakes for the store. People said they were "the most delicious pies you ever ate in your life." Before the sun had hoisted itself over the blue-gray mountains, Hannah would have turned out dozens of pies and cakes, maybe fifty before she was finished, and she knew they would all be sold by closing time, for thirty-five cents.

She didn't mind the early hours, especially after Frank put Richard in charge of the produce sold in the family's combined gas station and store as soon as he was old enough to drive. He would be in the kitchen with her before dawn, downing a quick breakfast, then he would climb in the truck and drive fifteen miles to the produce market in Los Angeles, look over the day's vegetables, bargain with the farmers for their best price, and drive the fifteen miles back. While he washed the produce and Hannah rolled her pie crusts, the two of them would enjoy quiet talk, real conversations. After he had become a famous man, she remarked that even as a youngster "Richard always seemed to need me more than the four other sons did. He used to like to have me sit with him when he studied. He doesn't pour out his heart, but he confides in me."

By contrast Harold was out with the crowd, going on dates, talking cars, tinkering with gadgets. Unlike Richard, he was a natural with anything mechanical. In his late high-school years, Harold's priorities began to tell in his grades and his questionable choice of friends. To put a stop to this, his parents sent him east to Mount Herman School for Boys, founded by evangelist Dwight Moody, strong on Bible study and stern discipline. In cold Massachusetts the southern California boy was required to start the day at 5:30 with a cold shower. It was a decision Hannah would painfully regret.

Meanwhile, Don was working on having fun. "Don, oh, he was the one!" chuckled schoolmate Helen Letts years later. "He always had the car that could go the fastest, around this curve and up that hill. He was very outgoing, he had a personality that drew people to him. Dick had to work at making friends."

For Richard those early-morning moments with his mother made all the washing and culling and arranging the vegetables more tolerable. He disliked doing it, but going to college depended on selling the green peppers and avocados and lettuce. Even with his working and saving, he could only afford to attend local Whittier College, a small Quaker school that had high standards but little recognition outside the community. So not until his last tomato was in place could he hurry off to school.

Hannah had a wonderful way with customers, who would line up to have her wait on them, partly because she would dispense thoughtful advice, and partly because customers wanted to avoid the contentious Frank, who argued "just for the sake of arguing" and whose tongue, they said, "could clip a hedge." And the shouting among Don, Harold, and their father grew increasingly strident, jangling the atmosphere at home and at work. It meant that Hannah spent her life walking on eggs.

Edward Nixon recalls that when Frank administered a good wallop and roared, "You've got to learn a lesson," Hannah "would intercede—but she could discipline as well, mostly with a look of disapproval." Dick Nixon later remembered how "she would just sit you down and talk very quietly, and when you got through you had been through an emotional experience. In our family we would always prefer the spanking."

Franklin and his father rode, hunted, and explored the Roosevelts' hundreds of private acres, the life of a country squire that Sara expected her son to pursue. Though she disdained politics, when he chose that path she did all she could to help him succeed. *Courtesy of the Franklin D. Roosevelt Library*

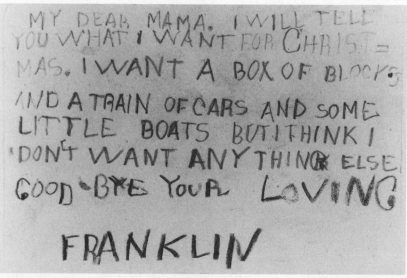

MY DEAR MAMA. I WILL TELL YOU WHAT I WANT FOR CHRIST= MAS. I WANT A BOX OF BLOCKS AND A TRAIN OF CARS AND SOME LITTLE BOATS BUT I THINK I DON'T WANT ANYTHING ELSE. GOOD-BYE YOUR LOVING

FRANKLIN

When he was seven, "your loving Franklin" gave his mother a modest list of things he would like for Christmas—blocks, a train, and little boats. Sara made sure that all were under the tree. *Courtesy of the Franklin D. Roosevelt Library*

Sara was the first mother empowered to cast a ballot for her son for president. On their way to vote for Franklin's historic third term in 1940, Sara and Eleanor gave no hint of their uneasy relationship.
UPI-Bettmann/CORBIS

On December 28, 1881, clear-eyed and confident Martha Ellen Young, in an elegant black wedding dress, started a new life with John Truman. She was lively, astute, and always knew her Harry would amount to something.
Deane Photo, courtesy of the Harry S. Truman Library

At twenty-two Harry was called back to work on the family farm, giving up his job at a Kansas City bank and prospects for a future. Here he is in work clothes with his mother (*left*) and grandmother. The life of hard work and isolation was brightened by the indomitable Martha—and his abiding love for Bess Wallace. *Courtesy of the Harry S. Truman Library*

In 1945 (on her first flight), Martha arrived for a White House visit aboard the presidential plane. She scoffed at the "fuss and feathers" that greeted her and delighted the press with her wit and candor—until she was muzzled. *J. Sherrel Lakey, courtesy of the Harry S. Truman Library*

Ida and David Eisenhower with their sons in 1902: *(left to right)* Dwight, Edgar, Earl, Arthur, Roy; Milton is between his parents. Ida met hard times with unfailing good cheer, trained her boys to cook and clean, and preached the importance of education. *Courtesy of the Dwight D. Eisenhower Library*

On home leave in Kansas in 1938, Ike shed his lieutenant colonel's uniform and relaxed with his mother. His son, sixteen-year-old John, snapped this rare photograph of Ida and her devoted "most troublesome son." *Courtesy of the Dwight D. Eisenhower Library*

As a resolute pacifist, Ida was at first dismayed that Dwight chose a military career, but after victory in Europe in 1945 she joyously shared in her world-famous son's triumphant homecoming. *Courtesy of the Dwight D. Eisenhower Library*

In 1919, Rose Kennedy was the wife of a man on his way to success and the mother of the first three of their nine children—Joe Jr., John *(right)*, and baby Rosemary. *Courtesy of the John F. Kennedy Library*

The political genes of three generations reveled in young Jack's election to Congress in 1946. Rose inherited her love of politics from her father, Boston mayor John "Honey Fitz" Fitzgerald *(left)*, and passed it on to her children. Some called her the best politician in the family. Proud father Joe Kennedy and Rose's mother shared the victory. *Courtesy of the John F. Kennedy Library*

The president led the applause for his mother when she was honored in 1962 for her service to the mentally handicapped. In tribute to daughter Rosemary, the Kennedy Foundation has led in funding research in that field. *Courtesy of the John F. Kennedy Library*

Rebekah Baines as a college student, beautiful and eager to pursue all things uplifting. As her husband fell short of her "noble" father's standards, she poured her hopes into her firstborn son and pushed him toward success. *LBJ Library Collection*

The Senate majority leader might have been pointing toward the White House, for that was his goal, with the joint support of his mother and his wife, Lady Bird. *LBJ Library Collection*

Visiting Austin in 1958, Eleanor Roosevelt paid a call on Rebekah, who idolized the former First Lady for her humanitarian service—and because FDR had given a young, green congressman named Lyndon Johnson a boost up the political ladder. *Bettmann/CORBIS*

In 1895, Hannah Milhous was a serene ten-year-old in a prosperous Quaker family in Indiana. She would grow up to be the peacemaker in an argumentative household and an inspiration to her son Richard. *Courtesy of the Richard Nixon Library*

An exuberant Hannah and Frank hear great news from the 1952 Republican convention in Chicago: Richard had just been chosen to be General Eisenhower's running mate. *Bettmann/CORBIS*

On the platform of the 1960 GOP convention, Hannah, with granddaughters Julie *(left)* and Tricia, and daughter-in-law Pat, shared the jubilation of her son's nomination for president. She stoically accepted his defeat but sadly did not live to cheer his victory eight years later. *Courtesy of the Richard Nixon Library*

To protect her baby son, Dorothy Gardner King fled from her abusive husband, a bold act in 1913. The story had a happy ending for her and the boy who became Jerry Ford. *Courtesy of the Gerald R. Ford Library*

Home from the war, navy hero Jerry Ford pinpoints his South Pacific battles for his parents. His next campaign was in politics, and he never lost a contest until he sought the presidency on his own. *Courtesy of the Gerald R. Ford Library*

With Dorothy nudging him to get married, Jerry won Betty Bloomer and managed to squeeze the wedding into his campaign schedule. Though Betty was not Dorothy's first choice for Jerry, mother and daughter-in-law became devoted friends. *Courtesy of the Gerald R. Ford Library*

The Depression years in rural Georgia offered no frills, as this early snapshot of Lillian Carter with her two older children, Jimmy and Gloria, suggests. But the house was filled with books and boasted the first radio anywhere around Plains, a tennis court, and a pony for Jimmy. *Courtesy of the Jimmy Carter Library*

Graduation day at Annapolis in 1946, and the proud mother and fiancée pin ensign's bars on Jimmy's new U.S. Navy uniform—two strong women with a claim on him. *Courtesy of the Jimmy Carter Library*

Shake every hand, sign his books, give yet another interview, throw out the first ball at a World Series game—there was almost nothing Lillian wouldn't do to help elect Jimmy. *Lawrence Smith photograph, courtesy of the Jimmy Carter National Historic Site*

Frank and Nelle Reagan, circa 1915, with their sons, Neil—"Moon"—always closer to his father, and Ronald *(right)*—"Dutch"—his mother's favorite, the son who performed with her as a boy and shared her optimism and aspirations as a man. *Courtesy of the Ronald Reagan Library*

Reagan with the two women he adored, Nancy and Nelle, on the set of *Tropic Zone* in 1953. Would-be actress Nelle was vicariously fulfilled through Ron's movie career, never dreaming of the great role that lay ahead of him. Nancy happily gave up her own acting career to dedicate her life to her Ronnie. *Courtesy of the Ronald Reagan Library*

From the time he got his first job as a sports announcer, Ron would always watch over his mother. He included her in his Hollywood life and bought his parents the only house they ever owned. *Courtesy of the Lee County (Illinois) Historical Society*

Dorothy Bush, a former junior champion, delivers an overhead smash in a club tournament in 1936. Fiercely competitive in tennis or tiddledywinks, she spurred her children to play hard and win, but only with good sportsmanship. *Courtesy of the George Bush Presidential Library*

Dorothy, in 1950, at the heart of the future Bush political dynasty: Barbara, George holding George W. in Texas boots, and Senator Prescott. Jeb was yet to be born—and the story is not finished. *Courtesy of the George Bush Presidential Library*

In the Oval Office, Dorothy did not hesitate to dispense advice to her son: Don't talk about yourself, don't brag about what you have done. The demands of politics conflicted with her strong sense of propriety. *Courtesy of the George Bush Presidential Library*

It was a night to remember, the brand-new president and his mother at the center of attention—where both always wanted to be—at the Arkansas inaugural ball in 1993. *AP/Wide World Photos*

In a last hurrah, Virginia flew to Las Vegas for the 1993 New Year's Eve concert of her friend and idol, Barbra Streisand. The following week Barbra was among the mourners at Virginia's funeral in Little Rock. *AP/Wide World Photos*

These sharp differences in his parents' personalities and their approach to bringing up children caused "a kind of cleavage in the mind of this lad," observed Dr. Paul Smith, a former president of Whittier College, reflecting on its most famous alumnus. "Two ways of life—one, the compassion of the mother; the other, perhaps, the domination of the father. His mother requested his obedience and got it; his father demanded it and got it." Hannah succinctly summed up her own view of these bipolar influences on Richard: "He had his father's fire and my tact." But Richard must have been torn by these conflicting tugs, and he would have been disturbed by the burden imposed on his mother.

The "cleavage," however, had no ill effect on his performance in school: he was an excellent student, sometimes class president, and won prizes for debate and oratory from the seventh grade on. She "wasn't a pusher or a driver," said her nephew, Hadley Marshburn, "but she was always anxious to see him excel and get to the top in whatever he did."

Unlike most boys his age, Richard took a lively interest in current events, especially when the Teapot Dome oil scandal was rocking the nation during the Harding administration, a scandal that led the twelve-year-old to pronounce in all seriousness, "When I grow up, I'm going to be an honest lawyer so things like that can't happen." Though his exact words differ according to who tells the story, all agree on the essential facts. His brother Don would pinpoint that moment as the genesis of Dick's decision to go into politics. Two years later, Dick, the eighth-grade valedictorian, made that decision crystal clear in his "autobiography": "I would like to study law and enter politics . . . so that I might be of some good to the people."

Never too keen on politics, Hannah was pleased by Richard's desire to do good in the world, and a thought very likely began to take root in her mind, one that she would openly express when he was older: perhaps he should enter the ministry. But, she confessed, "He didn't exactly respond favorably, so I dropped the issue."

If Dick declined to join the crowd at the roller rink or go out on dates, he was not without his own interests. He had inherited a musical bent from his mother's family; he played violin in the school orchestra and clarinet in the band, and by the time he was twelve he displayed what Hannah saw as real talent at the piano.

To develop his talent (like Martha Truman, Hannah probably envisaged her son on a concert stage) she sent him to stay with her sister, Jane Milhous Beeson, a professional music teacher who lived two hundred miles to the north and had two sons his age. Mrs. Beeson reported proudly that in the six months under her tutelage he progressed from simple little pieces to "difficult numbers, some Chopin, and the last piece I taught him, 'Rustle of Spring,' is a difficult number." (That piece would always be one of Richard Nixon's favorites. When Aunt Jane, then seventy-nine, was a White House guest at his inauguration in 1969, she played "Rustle of Spring" for the family on the eagle-legged grand piano in the East Room.)

When school was out in June, Hannah, Frank, and little brother Arthur came to take Richard home, a happy reunion, the last unblemished joy the family would know. In August seven-year-old Arthur suddenly developed headaches and drowsiness. Two weeks later the bright-eyed, curly-haired boy was dead of tubercular encephalitis. Two weeks in August 1925—and the Nixons' world was shattered.

Five years later Dick, a college freshman, still mourning the little brother who was his favorite, wrote a touching memoir of Arthur, based on a small photograph of the little boy in a sailor suit: "This particular boy has unusually beautiful eyes, black eyes which seem to sparkle with hidden fire . . . only his hair is not what we would expect in a portrait of a boy. For instead of neatly combed hair, we see a mass of brown curls which seem never to have known the touch of a comb." He recounted how, two days before he slipped away, Arthur put his arms around his mother and asked to say a prayer—"Now I lay me down to sleep . . ."

Hannah and Frank were devastated by the worst tragedy that can befall a parent, but after the initial desolation mother and father reacted in sharply different ways. Hannah's unshakable faith saw her through her grief, certain within herself that the death was God's will and thus had a purpose. But her sister Olive detected remorse, even guilt, in Hannah for having spent such long hours in the store when she could have been with her young son, a conflict that working mothers, now as then, understand. Frank turned extremely religious—fanatic, some thought—certain that God was punishing him for an unspecified transgression. From that time forward, he closed his service station on Sundays and began testifying frequently in church.

"It was the first sad moment in an otherwise very happy childhood," the former president wrote in his memoirs. "It had a dramatic emotional effect on all of us. We could not understand why one so young . . . could be taken away from us so suddenly. There is no doubt that experiencing such an emotional ordeal at such an early age contributed to my sense of fatalism which, ironically, helped me to prepare for and overcome difficult crises in the years ahead."

*   *   *

TRAGEDY followed tragedy. Less than two years later, in the spring of 1927, Harold came home from his school in the east. He was seventeen years old and had contracted tuberculosis, a killer waiting for the discovery of antibiotics. Hannah would always suspect that the chill of those cold showers at school had made him susceptible to the bacteria. Probably not, but it would layer guilt on a mother.

It was the beginning of a desperate three years for the family, as Hannah gave up everything to try to save her firstborn son—she took him first to a guest cottage, then to sanatoriums. He was not improving and the cost was overwhelming. To pay medical bills, Frank sold a half acre of their Whittier place, then Hannah, ever resourceful, came up with a solution: she rented a house in Prescott, Arizona, four hundred miles east of Whittier, where the warm, dry desert air was thought to be beneficial for TB. To cover the weekly rent of twenty-five dollars, she took in three other "lungers." Many years later, the president spoke of her strength with awe: "It was my mother's finest hour. My mother alone, with no help whatever, took care of them all." She cooked their meals, carried groceries for an uphill mile, did all their laundry, gave them alcohol rubs, handled their deadly infectious sputum cups and bedpans. And she watched them all die.

The physical cost to Hannah was not factored into this solution, nor was the psychological cost to the other two teenaged sons and their father, who were left to shift for themselves, to work in the store, manage their schoolwork, and share the can-opener cooking. A simpler, less destructive solution was available. The county operated an excellent tuberculosis facility a short drive from Whittier, but Frank Nixon would have no part of an institution run on public funds—in his view it was better to split up the family and have Hannah cope with the grim, exhausting burden. "If it hadn't been for the expense of my brother's sickness, we would have been fairly well off, with the store," Nixon said in retrospect. "My father would go to his grave before he would take government help. This attitude gave us pride—it may have been false pride, but we had it."

Every five or six weeks Frank, Dick, and Don would pile into their 1924 Packard for the fourteen-hour (barring mishaps) drive across the desert and unpaved mountain roads to join Hannah and Harold for a weekend. When he was fifteen and sixteen Dick stayed for part of the summers of 1928 and 1929, earning money as a barker at Slippery Gulch, a feature of Prescott's annual carnival. He worked at a booth that was, in fact, a front for an illegal gambling room. Hannah would have been mortified had she known. Was young Richard aware of what he was shilling—or was he like the piano player who didn't know what was going on upstairs? Whatever, the money—a dollar an hour—was good, and it was needed.

It was Christmastime 1929. Hannah was almost forty-five years old and she had a rather awkward bit of news for Frank: she was pregnant. This time Frank was less than ecstatic. The situation that had been difficult for the family now became impossible. Hannah had to leave the hazards of disease and heavy work. Harold was left with another family in Prescott, and Hannah went home to prepare for the daughter she and Frank yearned for. The next June she gave birth to her fifth son, Edward, so much younger—Don, the youngest living brother, was sixteen years old—that he was essentially of a different generation from his brothers.

Harold, after months alone in Prescott, could no longer bear the loneliness and returned to Whittier, to home and mother. By then he had to know he was doomed, but he recklessly ignored the strictures imposed on tuberculosis patients—he rode horseback with abandon, worked in strenuous jobs, chased girls, stayed up late. Perhaps the lively twenty-three-year-old, as outgoing as Dick was closed, had decided that for whatever time he had, he would live life rather than endure it. The rest of the family willingly took the risk of contagion.

The day before Hannah's birthday in March 1933, Harold asked Dick to drive him to town—he'd seen an advertisement for an electric cake mixer that would be a swell present. The next morning he was feeling very weak when Dick left for his classes at Whittier College. About three hours later, a librarian quietly told Dick that he must go home: "I knew, of course, what had happened. I saw the hearse out in front of the house. My mother and father were both crying uncontrollably." Dick was barely twenty years old and he had lost two brothers.

Tuberculosis ran through the Milhous family like radon silently poisoning a home. With no regard for age, it brought death and dreadful illness and fear. At the same time that Harold's young life was seeping away, his cousin Jessamyn West was stricken. "We went around with burning cheeks and hacking coughs," she wrote, but her two years of enforced bed rest allowed her to develop her talent as a writer. There was no such silver lining for Harold, an active, restless young man who had no taste for books. Jessamyn remembered a tormented Frank Nixon lamenting at her bedside, "Why is it that the brightest and the strongest, handsomest and best get taken first?" "I often wondered," she later mused, "whether or not Richard had encountered the same attitude in his own home. The best had been taken, and he, a substitute, a man on the second team, would have to struggle untiringly to take his place."

It is called survivor's guilt, affecting those who live while others around them are killed in combat, die in fires, are crushed in accidents, and, most notably, were slaughtered in concentration camps. Young Dick Nixon was devastated by the deaths of his two brothers, first bright little Arthur, then fun-loving Harold. They were afflicted; he was spared. Psychologists who deal with this syndrome say the survivor is beset by questions: Why did he live, and not the others? Would Harold have hated him for being healthy? Even worse, did his mother ever wish she could exchange him for the bedridden son? Such dark thoughts, no matter that they are baseless, can have a detrimental effect on an adolescent or an uncertain young man, make him solemn, a grind, a loner, that catalog of dour descriptions that clung to Richard Nixon throughout his life.

Joseph Dmohowski, head of the Nixon collection at Whittier College Library, offers the intriguing proposition that tuberculosis indirectly led Nixon to the presidency: the loss of his brothers motivated him to work harder, aim higher, strive more purposefully—to overachieve—to be three sons for his grieving parents.

An insight from Hannah lends support to this theory. "I think it was Arthur's passing," she said, "that first stirred within Richard a determination to help make up for our loss by making us very proud of him. His need to succeed became even stronger."

In his twilight years, biographer Jonathan Aitken revealed, the former president spoke for the first time of the tragic—and shocking—circumstances that he linked directly to his brothers' deaths. His father, who, in his opinionated fashion, decided that pasteurized milk was not good for one's health, bought a cow in the early twenties to supply fresh, "healthy" milk for his growing boys. He saw no need to have the cow tested for tuberculosis, although infected cows were widely known to be a source of the disease. The specter of lung diseases began to settle over the family. Little Arthur died of tubercular meningitis, or possibly encephalitis; Dick was struck by a virulent case of undulant fever, an illness related to infected milk; Don fought off a threatening shadow on one lung; and in the seventh grade, Harold suffered a severe lung hemorrhage, a precursor of the tuberculosis that would take his life.

Thinking back to this needless risk to life, the former president drew a sad, unavoidable conclusion: "He refused to pay any attention to the doctor's warning that the cow ought to be tested for tuberculosis, and for this our family paid a heavy price in the years ahead." And a heavy burden for a surviving son to live with, buried deep within his psyche, throughout the years. He had put this terrible cause and effect together years before, but only in the final canto of his life did he make the chilling accusation—indirect but undeniable—against his father, as if to set the record straight.

★   ★   ★

AT Whittier College, Dick was a BMOC, a leader in student affairs, a champion debater, an able actor in campus dramatics, and a high-ranking scholar, a star in everything except football, where at least he tried hard. In the Friends meetinghouse he had been a young activist, encouraging Hannah's dream of the ministry. (Her suggestion was not that far off the mark. Politics, acting, and preaching have much in common—all involve reaching out to your constituency/audience/congregation, convincing them with your message. Like Lyndon Johnson, Dick chose the political pulpit.)

He hoped to go to law school after he graduated in 1934 but had no money. Duke University came to his rescue with a full-tuition fellowship. Whittier president Walter Dexter had a hand in the good fortune, writing a recommendation that declared, "I believe that he will become one of America's important, if not great, leaders." Dick was off to North Carolina and Duke for three years of total immersion in law books, graduating in June 1937, ready to spread his wings in a bigger world. He went to New York seeking his future with a big law firm, only to meet rejection. He tried the FBI, which wasn't hiring.

So there he was, back home in Whittier with a first-class degree, no prospects, and in a deep funk. He wanted to start fresh, escaping the predetermined status allocated to the Nixons, not an asset, as Dr. Paul Smith pointed out, since "the Frank Nixon family was in no way regarded at the social level of the Milhouses in Whittier." Then Hannah intervened, quietly asking an old friend who was a partner in a Whittier law firm to hire her promising son. Dick dragged his feet in responding to the firm's offer, but when nothing else came his way he joined Wingett and Bewley.

Consider the alternative scenarios: had Richard Nixon been hired by a New York firm or the FBI, his life and the country's political history would have been written differently. In the anonymity of a big Manhattan law office, Nixon, a Californian with no connections, would never have been given a crack at running for Congress in New York. Had he gone with the FBI, politics was ruled out by law. Fate, with a shove from Hannah, had kept him in Whittier.

★　★　★

HANNAH and Frank were asleep in their bed that night in 1940 when Dick came into their room and woke them up, with a reluctant Pat Ryan in tow. He had news that wouldn't wait: he and Pat were engaged! That evening, parked in Pat's Oldsmobile at the very edge of the Dana Point cliff, the Pacific pounding below and the scalloped beach stretching south to San Clemente, they had watched the sun slip into the sea leaving a saffron wake, counted stars as they popped into the darkening sky—and Pat had at last said she would marry him.

For more than two years, he had pursued her relentlessly and was finally victorious. Frank Nixon reacted to the news with delight, but Hannah was noticeably reticent. No woman of Hannah's fastidiousness would have chosen to be presented to her future daughter-in-law in her nightclothes, her hair in a net, her dignity parked somewhere with her bedroom slippers. The awkward reception "broke the romantic spell of the evening," Pat confided to their daughter Julie many years later, and gave Pat "the impression that they were undecided" about her. For Hannah the feeling would have been much deeper. She would be losing Richard, her special son, her strong right arm—Edward was then only a boy of ten—to this beautiful red-haired girl who seemed not at all like the girls she knew.

Pat sensed that Hannah did not know quite what to make of her, but Dick most definitely did. At their first meeting, he informed her that he intended to marry her. Pat's reaction? "I thought he was nuts!" The vivacious Miss Ryan, with flashing brown eyes and a quick smile, was the favorite teacher at Whittier High. Thelma Catherine Ryan—who would later change her clumsy name to the more lilting "Pat"—had survived a hard life, nursing a dying father, being orphaned as a teenager, taking all kinds of jobs, working for a while in a New York hospital, being cast in bit parts in a few movies, putting herself through college, then landing the job teaching commercial subjects in Whittier. Nothing in her life was similar to Hannah's except the common denominator of work—hard work and grit. Moreover, Pat was Methodist, with a beloved aunt who was a Catholic nun, which placed her outside the faith that made up the inner core of Hannah's life.

Even their first brief introduction had been strained. After Dick joined the Whittier law firm, he had met Pat, a new high school teacher, as amateur actors in a Whittier theater group, both performing in *The Dark Tower*. Hannah had seen the play and tended to judge Pat by the sexy, bad-woman role she acted so convincingly. But Hannah's innate kindness and Pat's natural diplomacy soon smoothed over the misgivings, and for the wedding Hannah baked her very best cake, elaborately iced and topped with the traditional miniature bride and groom. The ceremony, on June 21, 1940, in the romantic Mission Inn at Riverside, was very small, with only about twenty close friends and family.

Pat's relationship with Frank was quite relaxed—they teased each other and he enjoyed the light touch that was not part of Hannah's serious nature. The relationship between mother and daughter-in-law was more complex. In her biography of her mother, Julie Nixon Eisenhower wrote, "She admired Hannah's capacity for hard work, her stoicism in the face of the loss of two sons, her gratitude for what life had to offer. And yet, in Hannah Milhous Nixon, Pat Ryan did not find someone with whom she could become close or share intimate thoughts. My mother explained to me once, 'Nana and I were completely different. I admire her for what she was. But, you know, she really wasn't a modern person. We did not have much in common.' " But these two strong women shared one attribute: as mother and as wife, both would sacrifice everything to support and advance the career of Richard Nixon.

Pat Ryan was not Dick's first love. At Whittier College he and Ola Florence Welch, a fellow student, were engaged to be engaged, actually saving money for the ring. But while he was at Duke, Ola decided to marry someone else. No matter; Pat Ryan would be the perfect wife, so perfect she would be criticized for it. The newlyweds were settled into a conventional life in Whittier when the attack on Pearl Harbor changed everything. With Washington gearing up for war, Dick took a job there in January 1942 with the Office of Price Administration—his first real salary—but after six months decided to join the navy. Back in Whittier, Hannah, the peace-loving Quaker, was distressed, but accepted that he was serving his country in crisis. After four years in the navy, Lieutenant Commander Nixon returned to his law practice in Whittier in 1946. There Fate was waiting for him in the form of local Republicans seeking a candidate for Congress—and who better than a smart navy officer home from the war, a crack debater with the ideal political wife. He was scooped up as a fresh face to take on the Democratic incumbent in California's twelfth congressional district. He went at it with his customary diligence and Rottweiler style, and won. It was his first step into history, a big one.

*   *   *

IT is always sad when a president's mother does not live to see her son take office, particularly sad in the instance of Hannah Nixon, who had been the dominant influence on the admirable side of her son's complex persona, who had helped him in his campaigns, been there to support him in his crises, reassured him after his losses, and cheered on his inevitable comeback. Whenever he hit a bumpy stretch of road, she would prod him not to give up. There was the time that he had stood at her bedside as she lay gravely ill and pled, "Now, Mother, don't you give up." As he told it, "Her eyes flashed; she sort of leaned up in the bed and said, 'Don't *you* give up!' " His election in 1968 would have vindicated her defense of her son against what she saw as unfair judgments and slanders by his legion of critics, but she had died, her mind occluded by a series of strokes, just one year too soon.

Reflecting on the influence of his parents, the former president mused, "My father was a scrappy, belligerent fighter. He left me with a respect for learning and hard work, and the will to keep fighting no matter what the odds. My mother loved me completely and selflessly, and her special legacy was a quiet, inner peace, and the determination never to despair." Cousin Jessamyn West gave a writer's more objective evaluation of her aunt: "She was not ordinary; but she did what she did and was what she was through a strength and lovingness which welled up out of her own good heart and because of her own indomitable character."

That selfless love was not expressed in words, nor did it need to be. It was not in Hannah to be demonstrative in her emotions or effusive in praise, not her instinct, not the habit of her times nor the practice of her worship. Years later, the former president would declare, with considerable irritation, to Jonathan Aitken: "No one projected warmth and affection more than my mother did . . . but she never indulged in the present-day custom, which I find nauseating, of hugging and kissing her children. She was unsparing in her praise when we did well but never berated us when we might do badly. She believed only in positive incentives, never negative ones. She could communicate more than others could with a lot of sloppy talk and even more sloppy kissing and hugging. I can never remember her saying to any of us, 'I love you'—she didn't have to!" The bond between mother and son was profound, requiring no reassurance through word or touch, and it would always be the strongest influence on Richard Nixon's life.

For forty years the psychoanalysis of Richard Nixon has been a growth industry in this country. One of the most original theories was offered by Elliot Richardson, who was attorney general when President Nixon fired the top echelon of the Justice Department in the notorious "Saturday night massacre" of the Watergate scandal. He postulated that Nixon's "inner core of insecurity led to his great drive—a secure Nixon almost certainly would never have been president of the United States at all." This view is congruent with Hannah's stated belief that Richard strove ever harder to make up for the deaths of his two brothers.

Howard Phillips, Nixon's director of the Office of Economic Opportunity, argued, "To understand Nixon you have to understand his relationship with his father: the dominant factor in his psyche was rejection by his father, and his love-hate relationship with his father, which was mostly hate, with exaggerated exaltation of his mother." In an analogy between Nixon and Lincoln, Phillips saw "the same hatred between Abraham Lincoln and his father and the exaltation of his stepmother."

Nixon undoubtedly felt annoyed, exasperated, and embarrassed by his blustering, quarrelsome father, but hatred? Perhaps— buried deep, sublimated, as he dutifully tried to be the respectful son. Unlike Lincoln, who refused to go to his father on his deathbed, Nixon left the 1956 Republican convention that was set to renominate him as vice president to rush to Frank's bedside in Whittier Hospital. His father, gravely ill, pled with his son to return to the convention, but Richard insisted on staying with him. Together they watched his triumph over the "dump Nixon" movement. He was back again at Frank's bedside the night he died.

Phillips maintained that "Nixon never worked out his hatred, which was reflected in his approach to public policy." Hours of Watergate tapes that have been made public reveal that Nixon harbored hatred toward many groups, in particular the eastern liberal establishment, Georgetown society, and the media; in addition there were numerous harsh comments about blacks and Jews.

As a parent, Nixon followed Hannah's example, not Frank's, and his two daughters were devoted to him, defending him like tigers in his Watergate travails. Long before he was married, he was like a father to his brother Edward, who was seventeen years younger. "I had three fathers," Edward says good-naturedly of

Frank, Don, and Dick. "I was taking orders from all of them." But his devotion to Dick is clear. He tells how his mother and brother nudged him into reading, which hadn't interested him. "When Dick went off to the South Pacific, he said for every page I read, he would give me a savings stamp—my mother would be the proctor. By the end of the war I had enough savings bonds to buy a secondhand car and help me pay to go to Duke."

Richard followed Hannah's style: motivation, incentives—all Quaker persuasion and no shouting. Edward credits his brother for setting him on his career as a geologist. Not every bachelor of twenty-six would take his nine-year-old kid brother along to buy a new car in Detroit. Dick did in 1939. As they drove across Arizona on the long trip home, Eddie spotted a sign pointing to a meteor crater, which sounded exciting. Dick agreeably turned off the highway and found the crater, a detour not many men would have made. Eddie, transfixed by this result of the collision between Earth and something larger than a mere meteor, was hooked on geology for life, and would always be grateful to a fatherly big brother.

Nixon could be thoughtful, but his complex character, his exaggerated version of his mother's example of restraint, made him unable to exhibit even the most innocuous gesture of affection. He appeared emotionally paralyzed. A California political editor recalled Nixon's visiting Hannah at the family home in Whittier after a long absence: "I just couldn't believe my eyes. When he saw her, he walked over and shook her hand! Here's a guy who hadn't seen his mother in I don't know how long—and all he could do was shake her hand. He couldn't kiss her or hug her. He couldn't show any form of affection." And this was Hannah, the mother he called the inspiration of his life. White House correspondents were similarly jolted when Pat returned to a big South Lawn welcome after a long and trying official trip, to have her husband offer a cordial handshake, rather than even the perfunctory hug most wives would expect.

This was the same man who showered Pat with touching love letters when he was courting her, like the one he wrote, in third person—at one remove from the personal—on the second anniversary of the day they met. It ended, "And when the winds blow and the rains fall and the sun shines through the clouds, as it is now, he still resolves as he did then, that nothing so fine ever happened to him or anyone else as falling in love with Thee—my dearest heart." These expressive yet somewhat oratorical letters were written at a distance; no intimacy was required.

Among the many puzzles surrounding this complex man was the shifting role of religion in his life. His active participation in the Quaker meeting was a fundamental force in his early years, to the point that Hannah could picture him as a minister, yet after leaving Whittier he rarely attended services of any denomination. In the White House, treading close to the line of separation of church and state, Nixon invited well-known ministers from a variety of religions to conduct Sunday worship services that were part religion, part social function—politics with prayer.

In his wrenching farewell to the White House staff on that memorable ninth of August, 1974, a speech of such raw emotion that those present could hardly endure it, the disgraced president's thoughts turned to his parents. (How fortunate that Hannah could not witness that event in the East Room.) In his rambling, excruciating good-bye to everything, he made that curious attempt to somehow make amends to his troublesome father, declaring him "a great man." And then he spoke of Hannah: "My mother was a saint." There was a long pause, painfully long, while he collected himself. He recounted the trials Hannah endured and praised her selfless service to others. He struggled to continue. "She will have no books written about her—but she was a saint." Even his bitterest enemies would not take exception to that.

7

AN ACT OF COURAGE

> The world breaks every one and afterward many are
> strong at the broken places.
>
> —ERNEST HEMINGWAY,
> *A FAREWELL TO ARMS*

*T*HE door of the mansion on a tree-lined Omaha street opened tentatively. Then, quietly, very quietly, a woman slipped out into the steamy July night, holding a tiny blanket-wrapped bundle close to her breast. She climbed into a taxi and told the driver to hurry across the Missouri River, taking her to the safety of Iowa and her waiting parents.

The young woman fleeing into the night was Dorothy Gardner King and the bundle she held so carefully was her sixteen-day-old son. The charming young wife of Leslie Lynch King, the son of a wealthy family, was spiriting their firstborn child, his namesake, out of reach of his violence.

Had Dorothy done what wives did in 1913—swallow hard, concoct excuses for bruises, and stay with an abusive husband—Leslie King Jr. would have been doomed to grow up amid domestic strife and the threat of danger. Only twenty-one years old and married less than eleven months, Dorothy faced the most important decision in her life, one forced on her by an earlier decision she had come to regret. She weighed her action carefully: on one hand there was the argument that a child needs two parents, a boy needs a father, and on the other hand there was the certainty of misery and fear. And if she left, she would be socially ostracized and become the subject of malicious gossip. As days, then weeks, went by with cruelty heaped upon cruelty, she knew she could not spend her life locked in fear and loathing of a spouse who abused her emotionally and physically, in private and in public. But most important to her was her baby—she would not subject her son to such a life. She reached deep into her untested self and found the strength to dare to do it on her own.

Gerald Ford was an adult before he learned about that dramatic event that changed his life, his very identity, and as he spoke of it, eighty-five years later, he marveled at that act of courage: "My mother was a very strong person, from girlhood. The thing she did—to leave Omaha with me, not even a month old, in her arms—she literally escaped. It's hard to comprehend. That was a remarkable action for a mother."

It was especially hard for a young woman of Dorothy Gardner's background. She had grown up in a prominent family in Harvard, Illinois, a small town just south of the Wisconsin line. Her father, Levi Gardner, was a leading citizen, elected mayor of the town when he was only thirty-two, owner of the finest furniture store, and a successful dealer in real estate. Her mother's family tree was rooted in New England from seventeenth-century colonial days. Her grandfather Ayer, like so many other ambitious Americans, had moved west along with the country and had been one of the founders of the town of Harvard.

After growing up in the most pleasant circumstances, Dorothy enrolled in a fashionable small college in Knoxville, Illinois, where she and Marietta King, a personable girl from an affluent Omaha family, became close friends. In the spring of their freshman year Dorothy met Marietta's older brother, Leslie, a wool trader in the west who was tall, blond, handsome, and already thirty. Dorothy was then nineteen, a popular young woman with a sparkling smile and boundless enthusiasm. A photograph taken that year showed her looking sweetly serious in a fluttery white summer frock; her soft brown hair, escaping in neoromantic style from a whimsical little mobcap of dotted swiss, softly framed her face. The portrait, in profile, showed nicely sculpted features—and a strong jaw that hinted of a strong spirit within, the same strong jaw that would be replicated in her son's handsome face.

They fell in love, and after the kind of whirlwind courtship that is the stuff of romance novels, they were eager to marry. Leslie very properly asked her father for her hand, assuring him that he could support Dorothy in the manner she had always known; he mentioned a nice salary as manager of the wool business he owned with his father, a thirty-five-thousand-dollar bank account, and additional income of six thousand dollars a year from various investments. Any father would be reassured by that accounting, especially because it was well known that Leslie's father had amassed a fortune, first operating a stagecoach line between Omaha and Wyoming, then shrewdly capitalizing on the railroads as they thrust westward toward the Pacific. His wealth was estimated in the millions, a solid figure that would impress a prospective father-in-law.

Plans for a diploma gave way to plans for a wedding. Christ Episcopal Church was the setting for Harvard's social event of the season. The maid of honor, Marietta King, was pretty in pink; the bride glowed in white satin "covered with real Breton lace," according to the story on the society page of the local paper. After a large reception, the couple boarded a train and set out on their honeymoon, paid for by the bridegroom's father. It was to be a leisurely trip in their private Pullman room through the spectacular Rockies to the Pacific Northwest and down the California coast before returning to Omaha, where they would live.

Twenty days later Dorothy was cruelly struck by her new husband. It happened in Portland, Oregon, in the Multnomah Hotel. As the former president recounted it, "A man came into the elevator. She didn't know him, but he tipped his hat, just being polite. She acknowledged it with a nod. My father was sure she was flirting with the man—he was obsessed with jealousy." Back in their room Leslie, enraged, called her vile names, slapped her face, and then hit her. It is difficult to imagine a young bride's panic, her disbelief that this could be happening to her.

Somehow they made up, but five days later, in their Pullman compartment, he struck her again, and this time kicked her for added effect. On the following day, in their Los Angeles hotel, he insulted and verbally abused her. When Dorothy and Leslie returned to Omaha they moved into his parents' fourteen-room showplace, but within a week, even under his parents' roof, he struck her again and ordered her out of the house. She raced home to her parents and, humiliated, poured out her terrible story—as she would in legal documents much later. A contrite Leslie soon appeared at the Gardner home, apologized profusely, and promised not only to treat her respectfully but to set up a home of their own.

Dorothy agreed to try it again. That was the response expected of women in 1912, and for decades afterward. Spousal abuse was not acknowledged, and it certainly was not a subject to be talked about openly. Divorce was rare, and in the upper middle class, where moral strictures were most rigid, a divorcée was cast as a social leper. In the lower social ranks, it was widely accepted that a sharp chop across the jaw and the occasional black eye did not mean that a man was not a good husband, and it was probably the wife's fault, anyway.

The ugly reality of the battered wife at every economic level had not yet been brought to light. In the early years of the century, Dorothy's story was that of countless women who lacked her backbone; in her decisive action she was a forerunner of the battered women in the last decades of the century, who would stand on their own feet and say, "No more!" Had public attitudes been different, she might have turned the spotlight on spousal abuse the way her daughter-in-law, Betty Ford, would do so admirably on alcohol abuse and on breast cancer.

But that first time, she went back to Omaha with her husband. The place of their own he had promised turned out to be no rose-covered cottage but a dark basement apartment. Leslie was deeply in debt, which he had concealed from his father as well as his wife.

Their first Christmas she spent alone, miserable, without so much as a token gift from her wealthy in-laws. By then she was pregnant, yet her husband's verbal abuse never let up. In June they moved into his parents' house (the Kings were away), and on July 14, Omaha's hottest day in 1913, Dorothy gave birth to a strapping baby boy. His name, decided in advance, would be Leslie Lynch King Jr.

This happy event did nothing to improve the marriage. The next day, with their newborn son at her side and still weak from her ordeal, Dorothy was yet again roughly harangued. Her doctor stepped in to warn Leslie against his outbursts and insisted on a full-time nurse, more for protection than medical care. Leslie ordered Adele Gardner, who had come to help her daughter, to get out of the house. She refused. Levi Gardner rushed to Omaha, insisting that Dorothy and the child leave as soon as she was physically able. "The sooner, the better," Leslie snapped.

With that settled, Dorothy's father, still deeply concerned, returned home. Within the week Leslie threatened not only his wife and her mother but his infant son—this time with a butcher knife. The terrified nurse called the police. Leslie countered with a court order barring the Gardners from any contact with their daughter. Left alone at the mercy of her husband, Dorothy secretly called a lawyer, whose advice was urgent: get out—immediately. So it was, with the nurse as her coconspirator, that Dorothy escaped into the night with her firstborn.

In order to obtain a divorce, Dorothy and a number of witnesses would spell out the awful details in court records, but she kept the painful details from her son until, as a grown man, he helped her fight for the settlement that was owed her.

★   ★   ★

TWO-year-old "Junie" King, a merry little towhead, was too young to remember, later, the wonderful thing that happened to him in 1915. A big, warm, friendly man frequently called on his mother, who would laugh a lot and look prettier when he was there. She called him Jerry.

After she fled from the life of trauma in Omaha, Dorothy lived temporarily with her sister and her husband; she knew that going back to her small hometown was impossible. Her devoted father backed her up yet again, with a total support few parents could, or would, provide. He wrapped up his business in Harvard and set up a real estate office in Grand Rapids, a city he felt would be good for Dorothy and for his grandson. It was a city known for its industrious, churchgoing citizens (a great number of them brand-new Americans) who took pride in their attractive town with its thriving furniture industry and its good schools. It offered the single mother the chance for a new beginning free of gossip and hope that she could build the safe, warm family life she dreamed of.

The family moved into a substantial house in a good neighborhood, where Dorothy could begin to recover from her nightmare experience and blossom once again into the ebullient young woman she had been. The three adults lavished their affection on the boy they called Junie rather than Leslie, the name that brought back terrible memories of his abusive and menacing father.

At a church social one evening Dorothy met a good-looking young paint salesman, tall—six-foot-one and straight as an oak tree—and pleasant. His name was Gerald Ford, and his upbringing contrasted sharply with her well-to-do background. Jerry Ford had been obliged to leave school after the eighth grade, when his father was killed in an accident, leaving the fourteen-year-old boy to support his mother and three sisters. At twenty-four, he was mature and responsible beyond his years when he met Dorothy, a young mother of twenty-three. Though the mutual attraction was strong, Dorothy was wary of embarking on a marriage until she had a chance to be certain about the young man's character. She soon discovered that Jerry's warmth and good temper were genuine, unlike her former husband's, and there was nothing in his background to indicate that he was anything other than the man he seemed to be, someone a woman with her emotional scars could trust.

As it would always be, her first concern was her son: her new suitor was genial and dependable, but would he be a caring father to her boy? It didn't take long for her to see that the two had formed a natural bond. Now she was sure: this was a man she could love for the rest of her life. "She used a lot better judgment on the second marriage than the first," said President Ford approvingly.

On February 1, 1916, Dorothy and Jerry were married in Grace Episcopal Church, where they had met and which in the years ahead would be at the center of their lives. The young Ford family moved into one half of a two-family house, and overnight Junie King was transformed into Junie Ford—Gerald Rudolph Ford Jr.

It is touching that the new husband immediately gave his full name to his ready-made son, with such total commitment that throughout his early years the boy did not know he had been a different man's "junior." (The Fords didn't bother with legal adoption formalities, which seemed unnecessary in that close-knit family, but when Jerry turned twenty-one he had his name legally changed, to leave no doubt that he was Gerald Ford's son.) No stepfather has ever been more deeply loved and admired than Gerald Ford Sr.: "He was the father I grew up to believe was my father," the president wrote in his memoirs, "the father I loved

and learned from and respected. He was my Dad." Their lifelong
devotion proved that it is love, not blood, that matters.

Dorothy at last was in her element, fulfilled as a natural homemaker with a caring husband, an irresistible son, and her parents
nearby. Three years later a son, Tom, was born; then came two
others, Dick and James, widely spaced. In children's terms, they
were almost different generations, with fourteen years separating
Jerry from James, the youngest. The arrival of his half brothers
made no difference in young Jerry's relationship with his stepfather.

As President Ford fondly recalled, "He treated me as well—if
not better—than the others." Gerald Ford Sr. had every reason to
love this boy as his own; Jerry was the hardworking, achieving
son every father yearns for, a boy to play catch with and take fishing. He more than fulfilled his parents' expectations when he
made Eagle Scout, became a football star, and won a niche in the
University of Michigan's Hall of Fame. Who could not be proud
of a son who put himself through Michigan and Yale Law School
in the bleak years of the Depression? In a string of honors that
ultimately led to the most prestigious of all, Jerry Jr. would bring
luster to the name his stepfather had, without hesitation, shared
with him.

Not long after Tom was born, Jerry Sr. was doing so well in
business that he could buy a spacious house in upmarket East
Grand Rapids and a summer cottage on Lake Michigan; the family took vacations in Florida, in a rather grand touring car with a
windshield that tilted out to catch the breeze. But in the deep
recession of 1921 the paint business floundered, and as Easy Street
reached a dead end, the mortgage on their home was foreclosed.
Dorothy and Jerry Sr. took the reverses in stride, rented a house,
and never allowed their sons to feel insecure.

Given two strong, competent, and loving parents, the four Ford
sons' boyhood was uncomplicated and happy, and since there was

a Ford of just about every age, their home on Union Avenue became the gathering place for the neighborhood boys. (Girls were less than welcome.) The center of Jerry's world from the time he was seven until he was in high school was a large, squarish frame house with a welcoming front porch, a house that defies architectural category other than solid, unassuming American, typical of the comfortable middle-class dwellings that anchored thousands of neighborhoods across the country in the first quarter of the century.

For Jerry and his pals the best part was the barnlike garage with an upper story reachable only by ladder, the perfect place for secret club meetings and minor mischief. But his father was sufficiently agile—not to mention suspicious—to climb the rungs now and then to check up. More than once he interrupted a youthful penny-ante poker game and read them the riot act.

Not that his parents were stuffy—they didn't smoke or drink but were convivial folks. "They loved a good party," the president said, but in matters of propriety they kept a tight rein on their sons. Great emphasis was placed on honesty, and churchgoing was a fundamental part of their life, starting with their christenings in Grace Church, where Jerry Sr. was a vestryman and Dorothy was deeply involved in church activities. Every Sunday the handsome Ford family, the four boys scrubbed and slicked, took their places in a pew near the front. Firmly implanted in the boys' minds were three inviolable rules: Tell the truth, work hard, and don't you dare be late for dinner.

Though bringing up a houseful of boys was not an easy job, as any mother would attest, Dorothy ran a tight ship, handling it all with natural confidence and good humor. "My mother was the catalyst that brought everything together," said the president. "She organized it so that everybody knew what had to be done. We all had our chores. We were *mandated* to help. We had to clean up after a meal, clear the table, wash the dishes." And they did it without squawking. As the oldest and biggest, Jerry drew the unenviable task of coping with the coal-fired furnace, getting the fire going in the early morning, banking the coals at night, emptying the never-ending load of ashes.

From early childhood Jerry showed the traits of a natural leader. His parents were comfortable giving him the responsibility of baby-sitting for young James and keeping an eye on the other two. And on their part, the younger boys were happy to have Jerry as a stand-in for the parents—they idolized their big brother. Thinking back on that experience, President Ford was quite sure that taking care of his brothers was not a bad preparation for his later role as a leader in Congress, with all its political siblings clamoring for attention.

At the hub of it all was Dorothy. "She was loving, strong—a person of great resilience," her son remembered fondly. "I think she got got that from her father—he was a strong, successful man. My grandmother was nice, but my mother seemed to follow in his footsteps." Like the other presidents' mothers, Dorothy was a "papa's girl." From concerned father to strong daughter to victorious son, it was a pattern of achievement.

From Sara Roosevelt to Virginia Cassidy Clinton, each of these women (with the exception of Nelle Reagan, whose father left home, and Ida Eisenhower, whose widowed father placed her in the care of her grandparents) had a special bond with her father, whose attention, conversation, and instruction encouraged his daughter to expand her mind and her sense of independence, which built a self-assurance that was not associated with femininity when these women were growing up.

That specialness is quite dramatic with two contemporary history-making papa's girls: former British Prime Minister Margaret Thatcher and Secretary of State Madeleine Albright, both of whom were the first women ever to hold those powerful positions. Thatcher gave her father all the credit for the legendary self-assurance that powered her political career, and Albright, from girlhood, wanted to follow in the footsteps of her father, a professional diplomat.

And both have sisters who have pursued quiet and conventional lives.

★   ★   ★

PRESIDENT Ford's face twists into an "ow!" as he demonstrates how his mother used to give his ear a sharp little twist when he misbehaved. It had been more than seventy years since she had administered the little punishment that made him acutely aware of her displeasure, but even today he remembers, very precisely, that it smarted—and that she made her point. For more serious transgressions, she dispatched him to his room, which he minded more than the ear tweak. Once, disciplining him for some infraction of the rules, she had him memorize the whole of Rudyard Kipling's "If," which was good for the mind though unlikely to make a Kipling lover out of the miscreant. Dorothy, the disciplinarian, was sometimes stern, always fair. "Despite all the discipline," President Ford fondly recalled, "I never once doubted her love. She was the most selfless person I have ever known."

There was one aspect of her firstborn that posed a real challenge for her. "I had a bad temper," President Ford sheepishly volunteered. "I must have inherited it from my real father. I used to get angry very easily—my mother would give me the devil! She'd tell me how unattractive I was, she would teach me a lesson by ridiculing me. She taught me that I had to control it." When he had cooled off and settled down a bit, she would talk to him seriously, as Ida Eisenhower did to her volatile Dwight. The lesson Dorothy drilled into him was not moralistic but practical: "You shouldn't let anger overcome your good judgment." Over time, she got through to him. "The results were very important to me," Ford said. "I did learn to control a bad temper."

Most important were the values instilled by Dorothy. She had grown up in an established small-town family and, after her recovery from her disastrous first marriage, the early years with Jerry Sr. had promised more of the good life. But when hard times forced the Fords from their fashionable neighborhood to one where the collars were bluer than the blood, she gracefully accepted what could have been an embittering setback and quickly formed friendships with her neighbors. When Jerry was ready to enter high school, it was his mother who steered him in a decision that was to have lasting impact on his life. Because the Ford home was on a block bordering on three school districts, Jerry explained, "I could go to Central High, the elite school, or Ottawa, which attracted the nouveau riche, or South High, where the students were most mixed." By "mixed" Ford meant that South had more students who came from Polish and Italian immigrant families, or were black and poor, than were in the other schools.

Choose South, his mother advised, to gain from the broader experience, "to learn more about living." That mixture plus the football team clinched it for sports-loving Jerry. In retrospect, the president concluded, "that decision—the ethnic mixture, the athletic program—was a major factor in my whole character development." It was his route to college and law school; it opened his mind to the talents of "others"—outsiders—a valuable asset in a career of public service. The congressman who wanted to open wide the doors of his party was reflecting South High, and his mother's persuasion.

Most significant in the overtly racist 1920s, Dorothy fostered Jerry's close friendship with the one young black in the neighborhood, Byrd Garel, a chauffeur's son, who would never forget how Jerry and his mother made him feel at home with them. "I think I was the first colored person Jerry Ford was ever exposed to," he told Jerold terHorst, Ford's first White House press secretary. "Me and Jerry would sit in the kitchen and his mother would give us molasses cookies and milk. For what it was like in those days, he was a rich boy—but a regular guy." The two walked to South High together, skated on frozen ponds, and played the usual kid games. In those times another woman of Dorothy's background might have discouraged, even forbidden, that close friendship, for

reasons of status as well as race. She did not, to the lasting benefit
of her son. Jerry Ford would always be "a regular guy."

★   ★   ★

BEHIND the counter at Bill Skougis's diner, Jerry Ford was flip-
ping hamburgers for the lunch crowd, one of the several jobs he
balanced with schoolwork and football to buy a few clothes, have
a little spending money, and save for college. Skougis paid him
two dollars a week plus free lunches, not a bad deal. This noon-
time Jerry noticed a man—no one he recognized—lingering by the
candy case for fifteen or twenty minutes, staring intently at him.
When lunch orders slacked off, the stranger came over and, out of
the blue, said to him, "Are you Jerry Ford?" Yes. "I'm Leslie King,
your father. Can I take you to lunch?"

For a high-school boy snugly secure in his own Ford family, to
have this interloper arrive from nowhere, knocking his world
askew, was a total shock. Dorothy had told him, when he was
twelve or so, the barebones facts that she had been married before
and he was the son of that unhappy marriage. Jerry had been
remarkably undisturbed by this fundamental revision of his life
story: he must have been relieved that his happy life with parents
and brothers he loved had not been preempted by that unknown
father. By his own account, there had been no real discussion of
the matter and he had put it behind him as irrelevant, a reaction a
psychologist might see as denial. Now, almost seventeen years
after his mother fled with her baby, this unwelcome ghost stepped
out of the shadows.

Jerry looked him in the eye and said, "I'm working." Leslie per-
sisted. "Ask your boss if you can get off." Skougis nodded his
okay. Bewildered, Jerry went out with this man who seemed to be
his father and met his wife and the young half sister he had never
heard of. They climbed into Leslie's new Lincoln, just off the
Detroit showroom floor, which he was driving back to his ranch in
Wyoming. The boy took it all in: the gleaming, latest-model lux-

ury car and talk about thousands of acres in cattle country—while
he had to work two, three jobs to keep himself in high school.
King tossed out a proposition: "Wouldn't you like to come out
and live with us in Wyoming?" Obviously, he had sized up this
sturdy, hardworking teenager and decided he was worth claiming.
Dorothy had made the boy into something special, and now the
delinquent father would like to take him over. It might also have
occurred to Leslie that he was not just reclaiming a son, but also
recruiting an unpaid ranch hand.

It took Jerry exactly no time at all to think it over. He knew who
he was and where he belonged: he was Gerald R. Ford Jr. of
Grand Rapids, South High football star, National Honor Society
student, and determined to go to college. What possible right had
this stranger, his father in no way other than biologically, to suggest that Jerry give up the family he loved, his true father and the
mother this man had brutally abused? After Jerry had turned
down the "invitation," King asked if he could help financially.
"That would be nice," Jerry replied. "My father handed me
twenty-five dollars," Ford wrote in his memoirs. "Now, you buy
yourself something, something you want that you can't afford otherwise?" It was the first support his father had provided and
would have been the last, but for Dorothy's uncharacteristic
reprisal some years later.

That evening, after his younger brothers were in bed, Jerry
struggled to tell his parents about this disturbing encounter: "It
was the most difficult night of my life." Dorothy was incensed.
This was the father who had threatened his infant son with a
butcher knife, the man an Omaha court had found guilty of "extreme cruelty," the deadbeat who had welshed on child-support
payments of twenty-five dollars a month and three thousand dollars in alimony—simply by moving across the state line to
Wyoming. And now he dared to try to take the boy away! King's
wealthy father, Jerry's grandfather—whose fortune by then was
estimated at ten million dollars—had been sending the monthly

check for his irresponsible son, but when the old man died the checks stopped coming. "When I went to bed that night," Ford said, "I broke down and cried." Dorothy was by nature a forgiving woman—except in this matter, which she would pursue relentlessly.

In his senior year at South High, Jerry was named captain of the all-Michigan high-school team; football was his identity and indirectly opened the doors of life to him. "Football," said President Ford, "was my ticket to college . . . and that was the luckiest break I ever had." It was more than luck; Ford gave everything he had when he played sports. (Later, as a college senior, he was picked to play in the traditional East-West game, which brought him an offer from the Green Bay Packers for $110 a game, and a tempting $200 a game from the Detroit Lions. Cherishing a seemingly impossible dream of law school, he turned them down.) At every one of his high school and college games, in the worst of Michigan's weather, Dorothy and Jerry Sr. were in the stands rooting for him. "Athletics, my parents kept saying, built a boy's character," the president recalled. "Sports taught you how to be part of a team, how to win, how to lose and come back to try again." The fact that so many other athletes did not absorb those lessons suggests an additional element, a very determined mother.

★   ★   ★

FOR all their enthusiasm for football, good grades came first in the Ford household. Dorothy rode herd on homework and tried to help with Latin, Jerry's nemesis. The subject he really enjoyed was history, an interest he shared with young Lyndon Johnson, who demanded, "Is it true, Mama? Did this really happen?"; teenaged Dwight Eisenhower, who read about military exploits under the bedcovers; FDR, the college boy who collected naval histories; Harry Truman, who read every history book in the library; and John Kennedy, who devoured tales of heroes. Future presidents, it would seem, have a predilection for reading history—and then going on to make it.

Jerry, with his parents' encouragement, had his heart set on the University of Michigan, a first-rate college with a state institution's added inducement of being inexpensive. As a football star and honor student, Jerry Ford would have had his pick of colleges today, but in 1931 most universities did not offer football scholarships, incredible as that seems in this era of athletes sought and bought. In singularly unfortunate timing, his stepfather and a partner had branched out in their own paint and varnish business just three weeks before the crash of 1929; they were hit hard by the Depression, which all but wiped out Grand Rapids' famous furniture industry. "They really had a very tough time," the president recalled. "I remember that when any of my brothers got sick, they paid off the doctors by giving them paint. That's the way things were in the Depression.

"My parents couldn't give me anything. I got a hundred-dollar scholarship from the high school and that paid a year's tuition. Can you believe that? I got a job at the university hospital, waiting on tables and cleaning up, for forty cents an hour. That bought my food. An aunt and uncle who didn't have any children sent me two-dollars a week, all four years I was at Michigan. That was my spending money—and was it appreciated!" He pledged the DKE fraternity and met room-and-board costs by working in its dining room, then as manager of the house. Showing an instinct for the kind of Republican fiscal policy that he would later espouse, "he put the DKE house back on a paying basis" (as reported in Cannon).

In his junior year, Jerry found himself buried under more bills than any amount of campus jobs could cover. He desperately needed a thousand dollars for his final year, and he would not, could not, give up his—and his mother's—goal of a college degree. He reluctantly wrote to Leslie King, asking for emergency help of one thousand dollars (much less than King's unpaid support) to complete his education. He knew that King had recently inherited fifty thousand dollars, a great deal of money in those days of bankruptcies and breadlines, "but I never got an answer." In desperation, he put his case to a family friend—the same Ralph Conger who also counseled him to choose South High—who helped him with the money, confident that young Jerry was a sound investment.

For the children and grandchildren of the Depression generation, the realities of those years are hard to comprehend. It was a time that indelibly marked the psyches of all but the impervious rich. In normally prosperous Michigan, half of the automobile plants were idle; in the South, where poverty was always visible, textile looms, a major source of jobs, gathered dust. Banks failed, taking with them life savings. With a third of the workforce out of a job, the welfare system in many places broke down. Tramps drifted from place to place, knocking on any kitchen door for something to eat. Like Hannah Nixon in California and Lillian Carter in Georgia, bighearted Dorothy Ford would always give them something. In cities, breadlines and soup kitchens did what they could to ease hunger; men who had always brought home paychecks tried to sell apples, pencils, anything, on the streets. Others, unable to face ruin, leapt from windows or put guns to their heads. Hopelessness spawned strikes and unrest in big cities, causing fear among social scientists that the nation's political structure might collapse. Only the cellophane-wrapped rich—like the Kennedys, the Bushs, and the Roosevelts—escaped the pervasive anxiety, even despair, of those terrible years.

The Great Depression destroyed some Americans and toughened others, producing a generation of resilient men and women who expected little to be handed to them, scrimped in ways large and small, and survived. Then, as they began to regain their footing, they or their children were called upon to fight the most devastating war in history, bloody years that claimed the lives of more than three hundred thousand Americans. Most of them were young, hardly starting to live when they were cut down in farmers' fields and mountain passes in Europe, on beaches and in jungles of the South Pacific, in the skies and seas everywhere. Dorothy Ford met the Depression head-on and came out strong; her son, Lieutenant Commander Gerald R. Ford Jr., coped with the hard times and then fought the war: before he was thirty he had done it all.

★   ★   ★

IN the blackest years of the Depression Jerry struggled to put himself through college. With his Michigan degree, his athletic ability, and his rock-solid character, he landed a job as an assistant coach at Yale; with his unstoppable persistence, he won a place in Yale's prestigious law school, and set the course of his life.

While Jerry was working as never before to make it at Yale, Dorothy continued to smolder at her one-time husband's callousness. "My mother was bitter—and I mean bitter," Ford told biographer James Cannon. "Here was a guy who had inherited quite a lot of money and never paid a dime. With her, it was a matter of principle."

Dorothy renewed her case, this time in federal court in Wyoming, which ordered King to pay her $6,303, plus attorney's fees. He had the gall to claim that he was hard up like everybody else and he offered $1,000, which further fueled her anger. As Cannon reconstructs the acrid episode, King suddenly rediscovered his son, then a coach at Yale, and coaxed him to "talk sense" to his mother. Jerry countered with a compromise of $4,000. When King refused, Dorothy had him arrested and thrown into jail. Out on bail, he made the long trip east to Yale, and the two uneasy father-son-strangers worked out an agreement. Three weeks later King reneged, vowing "to fight this to a finish," blatantly threatening that "the publicity in all the newspapers won't do anybody any good." Almost as quickly, this strange man changed his mind again and settled for $4,000.

Though nearly half of that went to her lawyers, Dorothy exulted that finally, at long last, she had forced justice from the man who had abused her and her son. The whole episode was unlike her, but understandable. As for Jerry, "I was just glad it was over." He never heard from his father again, and Leslie Lynch King Sr. did not live to see his son take his place in history as the thirty-eighth president of the United States—proudly bearing another man's good name.

★ ★ ★

THE unpleasant burden of that wrangle was happily offset by a new element in Jerry's life. It was Phyllis Brown, a dazzling blond eastern sophisticate, bright and outdoorsy, whose favorite indoor sport was flirting—a skill that could trigger Jerry's not-yet-controlled temper in an instant. (Looking back over the years, she would laugh about the night in New York when "Jerry got so mad at me that he put his fist right through the door at 21.") President Ford recalled with the mellow glow that suffuses memories of the first love affair, "For the first time in my life I fell deeply in love." The romance flourished throughout his years at Yale. Phyllis spent a month each summer with the Fords at their Lake Michigan cottage; Dorothy and all of the Fords thought she was wonderful. Then he would join the Browns at their home in Maine. To keep up with Phyllis, Jerry learned to ski, went to the theater in New York, and played bridge, tennis, and golf.

In 1939 one of those unbelievable movie-script stories really happened: a celebrated magazine cover artist crashed into Phyllis on the ski slope, took one look, suggested that she should be a model, and actually set her up with the famous Powers agency. In a twinkling, the college girl morphed into a top model. Jerry spent as many weekends with her in New York as he could manage with the demands of law school and coaching. (Yes, she told James Cannon, she slept with him.) And when *Look* magazine chose her for an on-location feature, she had them use Jerry as her partner.

The good-looking all-American pair was splashed over six pages in the March 12, 1940, issue: "A New York Girl and Her Yale Boy Friend Spend a Hilarious Holiday on Skis." If *Look*'s circulation leapt that week, it was Dorothy buying extra copies.

Both families fully expected the two to marry. After winning his degree from Yale Law School, Jerry turned down job offers from prestigious firms in the East to go back to his roots in Grand Rapids—but Phyllis couldn't bear to give up her glossy New York life. So ended the romance of nearly four years. "The end of our relationship," he wrote, "caused me real anguish."

Back in Grand Rapids, with more optimism than cash, Jerry and an old friend, Phil Buchen (who would later be at his side as White House counsel), opened their own law firm and plunged into local life. "I was a compulsive joiner," Jerry said—small wonder, given the example set by his mother.

Then the event that impacted all lives, the bombing of Pearl Harbor, instantly changed Jerry's: the morning after "a date which will live in infamy," he signed on as a navy officer. His race to action would carve out more than four years of his life and take him within a whisper of death. His ship, the small aircraft carrier *Monterey*, was in the thick of eleven fierce Pacific battles, but it was a monster typhoon that nearly claimed the young lieutenant, hurling him across the flight deck. Miraculously, his feet found the narrow metal rim at the edge of the deck, and somehow in the dark terror, with the raging sea ready to claim him, he caught and clung to a narrow catwalk. As he struggled to hold on to life itself, he recited his mother's favorite passage from Proverbs:

Trust in the Lord with all thine heart; and lean not unto thine own understanding.
In all thy ways acknowledge Him, and He shall direct thy paths.

As ships and men and battles were lost in the Pacific, Dorothy agonized over news reports and kept up her spirits by writing to him regularly, rarely knowing where he might be, never knowing and always fearing the dangers he faced.

At long last peace came and millions of servicemen came back to start living again. Jerry reclaimed his old room with Dorothy and Jerry Sr.—"I'd been away from them for a long time. I was glad to be back home"—and, with Buchen, joined a leading law firm.

His professional life was busy, his community involvement was constant, but his social life was meager. Good-looking, popular, marked as a comer, this eligible young man couldn't seem to find a girl that measured up to Phyllis. Contrary to standard form, Jerry Ford's mother was eager for her eldest son to get married. His three brothers—even James, so much younger—were already married with children. Dorothy had been gently nagging him, as Ford told it in his memoirs: " 'When are you going to start dating again?' Mother asked in her good-natured but very persistent way. 'You're thirty-four years old. When are you going to settle down?' "

★   ★   ★

THE turkey was roasted to perfection; the table was elegantly set with gleaming silver and flowers; the family bowed heads and gave thanks for their blessings. Betty Bloomer Warren, a first-time guest at the Ford home, could be forgiven if she was a little too nervous to be hungry. This man she was beginning to like a lot, a whole lot, had taken her home to meet his parents, and on such a big family holiday.

Worse, she was not a bit sure how they felt about Jerry's interest in her. In 1947 Grand Rapids, a former Martha Graham dancer and sometime New York model who now had a career as a fashion coordinator—and was divorced—did not qualify as the ideal daughter-in-law. "I don't think I was their favorite," Betty wrote in her memoirs. "Mrs. Ford was fond of a young widow Jerry had known since they were both children, and I think she was hoping for them to get together." And Betty suspected, a little, that Jerry was still carrying the torch for his glamorous first love.

Her anxiety was misplaced—big-hearted Dorothy Ford could never have been less than hospitable, especially when she was meeting the pretty young woman her Jerry was taking out. Betty immediately felt comfortable with the Fords. "Jerry's stepfather stood straight as an arrow, and I had the impression that he lived his life that way. His mother was a handsome woman, with tremendous charisma. I could see why my mother had told me everyone liked Dorothy Ford. She seemed such a strong woman, confident and positive, just the right person to raise these four Ford boys. Everyone was courteous and respectful and I thought, 'What a wonderful family Jerry has.' " And she never revised that first impression.

Betty was spared the tensions Eleanor Roosevelt endured with Franklin's possessive mother, Lady Bird Johnson's walk-on-eggs tact to reassure Rebekah, and Mamie Eisenhower's game effort to adjust to Ida's plain ways. For Dorothy, gaining the final daughter-in-law was an unmixed joy; now Jerry could settle down. Eventually, mutual friends got him together with Betty and after a hectic fourteen-month courtship—while he waged his first campaign for Congress throughout 1948—they were wed. A family photograph on the day of the wedding shows Dorothy in her maturity, surrounded by her four sons and her husband. Jerry was his mother's son, all right. "His features," a family friend remarked, "were the image of his mother's." Though heavier than before, Dorothy was still stylish in a dark silk dress and a hat with a plume that swept dramatically down over her left ear.

The wedding would be Betty's first sample of what lay ahead for her as a congressman's wife. At the elegant rehearsal dinner given by the Fords, Dorothy must have wished she could twist her son's ear as Betty, the bride, the center of attention, sat beside an empty chair, while Jerry made campaign stops right up until the dessert course. At the ceremony in Grace Church, Dorothy, remembering her own wedding there, would have twisted Jerry's ear again when she saw him race in at the last minute wearing shoes gummy with mud from the campaign trail. That was not the way her well-brought-up firstborn should appear at his own wedding. But voters saw no mud on his shoes or his character when they went to the polls. The young challenger upended a ten-year incumbent in the Republican primary, then won the House seat by 60.5 percent. He was on his way.

The wedding also marked the beginning of a devoted friendship between Dorothy and Betty. Locked in that most problematic of relationships—mother- and daughter-in-law—they proved that two strong women could forge a strong bond. "They got along beautifully," declared the man they shared. "They were super, super friends." The arrival of grandchildren further sealed their relationship. Grandma Ford's visits to Jerry and Betty's home in Alexandria, across the Potomac from Washington, were happily anticipated by the whole family. The grandchildren loved her and she loved doing for them, always ready to listen to the three boys, reliving the days of bringing up four sons, and especially delighting in Susan, the daughter she never had. Thanks to Grandma's nimble fingers, Susan's dolls were the best dressed in the neighborhood, and the real Shirley Temple might have wished for a Christmas stocking as pretty as the one Grandma Ford made for Susan's beloved Shirley Temple doll. Dorothy's doing for others obviously had no limits.

★   ★   ★

THE notion of going into politics had come to Jerry quite naturally: it had probably been germinating from the time he was voted Grand Rapids' most popular high-school senior and was awarded a trip to Washington. Seeing Congress in action had made a big impression on him, just as shaking President Kennedy's hand would have on young Bill Clinton. Then, between semesters at Yale in the summer of 1940, Jerry got his first taste of presidential politics. In pursuit of Phyllis Brown in Manhattan, he signed on to work in the campaign headquarters of the Republican candidate, Wendell Willkie, and went to the GOP convention in Philadelphia as part of the Willkie staff. With the early lessons learned in that campaign, he would recall, "I got some really ground-level experience in how a presidential campaign was organized." Did he harbor the thought that he himself might some day be the presidential candidate? If so, only subliminally. "More important, I learned about myself. I liked politics, everything about it."

★   ★   ★

EVERY time he dived into politics, he came up with the prize, until the last, ultimate test, the race to win the presidency on his own. As he climbed the political ladder in Washington, the impact of his mother's values would be crucial to his success. In the House of Representatives, Jerry Ford's equanimity and evenhandedness brought him the post of Republican leader, and in 1973, when Vice President Spiro T. Agnew was forced to resign amid charges of corruption, Ford's reputation for scrupulous integrity led Richard Nixon to select him—in a historic first—to be vice president. One historic first led to another, when Ford, the unelected vice president, replaced an American president who resigned his office. Then, in 1976, the lingering disgrace of Watergate and the nation's explosive reaction to Ford's pardon of Nixon may have tipped the balance to Jimmy Carter in that paper-thin election.

Eight years earlier, Ford had made a decision that would profoundly affect his future, a decision he revealed in his memoirs, *A Time to Heal*. At the 1968 Republican convention, Nixon, the shoo-in for the nomination, made a proposal to Ford, the well-established Republican leader who was his old friend from their early days in Congress: "I know that in the past, Jerry, you have thought about being Vice President. Would you take it this year?" Most politicians would have said yes before Nixon had a chance to change his mind, but Ford, projecting recent Republican gains in the House, could see a real chance to become Speaker of the House, a position of far greater power than vice president. He graciously declined, and Nixon picked as his running mate little-known Maryland Governor Spiro T. Agnew.

Had Jerry Ford accepted Nixon's offer in 1968, the Nixon administration—and the nation—would have been spared the humiliation of a vice president who accepted kickbacks from Maryland contractors. Bundles of cash in brown paper bags were delivered to him in his executive office, giving new meaning to the old term "brown-bagging." On the other hand, as Nixon's two-term vice president, Jerry Ford, despite his own probity, would almost certainly have been tarnished by the corrosive atmosphere of Watergate. After six years as a member of the Nixon team, he could not have assumed the presidency as the honest outsider who offered a breath of clean, fresh air. He could not have played the healing role and almost certainly would not have been nominated by his party in 1976, not with Ronald Reagan knocking at the gates. History's judgment of Jerry Ford as the right man for a traumatic transition would have been quite different. As he has always said, "I've been lucky." But luck, Jerry Sr. used to point out, was always accompanied by a lot of hard work—and in a time of national upheaval, an unassailable character.

Though Dorothy exulted in Jerry's unbroken string of victories as a congressman, she did not work actively in his campaigns. That's not to say that she was without strong opinions. "She was very outspoken," said her son. "She had her own views and didn't hesitate to express them. In 1960, when there was an effort to get me on the ticket [with Nixon] in the number two spot, she thought it was a terrible idea—said it would be nothing but headaches for me to be vice president." With a wry smile, the for-

mer president explained, "My mother always disdained politics"—but her political instincts were on the mark.

*   *   *

MORE than any of the eleven mothers of presidents from FDR to
Clinton, Dorothy Gardner Ford embodied the composite portrait
of the typical American wife/mother/homemaker of midcentury
America, and at the same time she was the bridge between the traditional women of another era and the multitasking women of
today. Born in the last decade of the nineteenth century, she was in
the vanguard of two major women's issues of the last decade of
the twentieth century: spousal abuse and deadbeat dads. She
defied convention in an era when such humiliations were hidden,
when wives locked into hollow or abusive marriages would stay,
starch his collars, put a hot supper on the table every night, and
try not to think about the emptiness and unmet dreams. Dorothy
took action when she forced Leslie Lynch King to fulfill his obliga-
tions. She dared to fight for herself and her son. It was an inspiring
act of courage.

She lavished on her four sons and husband all of the attention
and love they needed, but possessed of the modern woman's drive
to do more, to be more, she carved out a life of her own beyond
her home. She was the prototype of a new breed of housewife: the
women who built volunteer service into the equivalent of a career.
This singularly American species of woman would say she never
worked outside the home, yet never stopped working in the community. A new quasi-professional civic volunteer emerged from
the sweeping changes in social mores that followed World War I,
which opened up the role of women and allowed those of the
upper social strata a freedom their mothers had never known.
Modern appliances, the stuff of dreams for housewives and sales-
men, gave homemakers more time of their own, but most of all it
was the proliferation of the automobile—the advent of the ubiqui-

tous family car—that took the stay-at-home mom out into the community. The self-starting car gave rise to the self-starting woman, whose license to drive gave her the license to come and go as she pleased.

Without the gift of time and elbow grease provided by the unpaid, hardworking "just a wife" woman, the Girl Scouts, the YWCA, the local Red Cross—every social and cultural organization in a small city—could not have functioned; fund-raising church bazaars and school festivals could not have been organized. Put down by the cynics as "do-gooders," which is precisely what they were, in the best sense, these women were a special outwardly mobile American breed who did much to raise the level of their family's aspirations and powered noblesse oblige into action in middle-class America.

Dorothy was not only a rolled-up-sleeves volunteer, but also a club woman and church activist. To stretch her mind, a mother like Dorothy could use the school day to pursue hobbies, make new friends, and engage in the American zeal for self-improvement: for mental stimulation there were book clubs and lectures; to explore untapped talents, she could take classes in painting, pottery making, flower arranging; to enhance her home she dug deeply into gardening; and in gourmet cooking classes she mastered fancy recipes that her husband didn't much care for. And she would be a mainstay of her church. Mothers like Dorothy were not static women.

"My mother was a goer, a doer," her admiring son said. "I think I'm the same way—I got it from her. Everybody said my personality was a reflection of hers. And I look quite a bit like her." He was clearly pleased by that. "She was an active member of at least ten organizations. And when I ran for Congress, all of those club friends were beneficial." First on her list was Grace Church: "she was engaged in one church activity after another." For Dorothy, religion was an all-week thing: her faith was deep but not long-faced; she was the friend a neighbor could count on, a person who looked for ways to help others.

Betty Ford was in awe of her mother-in-law's energy. "Grandma Ford was a dervish. She whirled from one thing to another—she had to be busy to be happy. She drove, she swam, she was never home for any meal but breakfast—and Jerry's just like her."

Her quasi-career as a volunteer didn't cut into her family time; every step of the way she was there for her boys, encouraging, nudging them to do their best. "She was a real supporter—not only attending the games, but at home," said Jerry. "If we lost a game, I'd be downhearted. She taught me a good lesson: 'You shouldn't sit around and wring your hands. Think about the next game.' That was applicable even when I lost the presidential race—I didn't sit around and moan and groan. She convinced me there's a new life, a new game." Thirty years after she was gone, her Jerry—the former president—was still listening to his mother's advice.

Dorothy Ford also maintained harmony in her family, which was not easy to do with four boys, by giving her sons what each saw as an equal share of her love and attention. "The four of us always got along well," Jerry recalled warmly. "We had a wonderful relationship." That was no minor achievement, given the fourteen-year age span and the fact that the younger brothers would have known, early in their lives, that Jerry was a half brother. In a home infused with her warm and enthusiastic spirit, she gave them a firm fix on the values that matter and the stuff to withstand hard knocks. Jerry Sr. was her true partner in that family enterprise. "To grow up in family unity is a big asset," President Ford observed. It builds team players and teaches both sharing and competition—all very useful for a successful politician.

There was no evidence of sibling rivalry among the widely spaced four boys. Jerry was their role model, sometimes their hero. "I'm sure that Tom [who was five years younger and is deceased] felt that he was living in my shadow. He played football in high school but was never a star, and at the University of Michigan he never made the squad. I went to Congress in 1948, and six years later he ran for local office, and was elected to the state legislature three times." Thus Tom followed the path set by Jerry but always had to settle for less. In contrast to other presidential brothers, Tom never seemed jealous of his older half brother's greater successes. Did he look up to Jerry as a hero? "Probably," Ford replied with his customary candor, "but we didn't talk about that."

Later, in her empty-nest years, Dorothy was again a forerunner in the lifestyle of her granddaughter's generation of women. She took her out-of-home life one step further and, following in her father's footsteps, entered the woman-friendly real estate business. According to her son, and not surprisingly, she did very well at it.

All of her four sons, she would proudly say, did well, but there was special satisfaction in the national stature of her Jerry, the Republican leader of the House of Representatives. In visits with Jerry's family in Alexandria, she loved to go to the Capitol with Betty to visit the historic House chamber. From her place in the family box in the gallery, Dorothy would see her Junie speaking on the affairs of the nation, a leader embodying those values she had instilled in him.

With a catch in his voice, Jerry Ford said, "One of the saddest things was that my mother wasn't alive when I was vice president and president. It would have been a magnificent experience for her to be in the White House. But she saw me as Republican leader—she could visualize the things that were to come."

On Election Day in 1976, a large mural was unveiled in the Grand Rapids airport, depicting the football star, the navy hero, the successful politician, and the historic president. It would be known, quite logically, as the Gerald R. Ford Mural. The name, said the president, "means so much to me because of the first Gerald R. Ford and his wife, Dorothy, my mother and father. I owe

everything to them and to the training, the love, the leadership"—
tears welled up in his eyes—"and whatever has been done by me
in any way whatsoever, it is because of Jerry Ford, Senior, and
Dorothy Ford."

★   ★   ★

IT was a pleasant Sunday morning in September 1967, and
Dorothy had arrived at Grace Church a bit early. Smiling at old
friends, she settled into her accustomed pew, not as easily as she
once did. She was widowed, alone, and nearing seventy-five. A
long inventory of illnesses had slowed her down a bit—two heart
attacks, a double mastectomy, a splenectomy, diabetes, and
cataracts. That Sunday morning, in her familiar place in the
church that had been a constant in her life, her generous heart
gave out. At the funeral of the grandmother she loved so dearly,
young Susan Ford was doubly saddened: "seeing my father cry
just ripped me apart."

The heart had given out, but not her spirit. When Betty and
Jerry examined her date book, they found that it was completely
filled with activities for the following month. "I want to drop dead
with my boots on," she used to say to Jerry. And so she did.

8

─

MAMA'S IN THE PEACE CORPS

A mother is not a person to lean on but a person to make leaning unnecessary.

—DOROTHY CANFIELD FISHER,
BARTLETT'S FAMILIAR QUOTATIONS

SOMETIMES when I look at all my children," declared Lillian Gordy Carter, "I say to myself, 'Lillian, you should have stayed a virgin.'" That's the kind of statement that makes the nation sit up and listen to the president's mother. And if it sat up late enough it might have caught Miss Lillian holding forth on the *Tonight* show, talking to Johnny Carson about life in a home with an outhouse: "Four-seat," she explained to those whose experience had not encompassed backyard privies. "Two high for adults, two lower for children." Thinking back on her appearances with Carson and interviews with Walter Cronkite and others, her son, the former president who is still a world figure, roared with laughter. "She was hilarious!"

To her friends and to the whole country, she was "Miss Lillian," in that wonderful southern formulation that crosses respect with warmth, dignity with informality. (" 'Mrs. Carter'—that's Jimmy's wife," she explained.) It must be pronounced "Miz Lillian," sliding over the syllable to avoid the image of a buttoned-up history teacher with her hair in a bun; however, spelling it that way seems condescending, like journalists trying to quote a southern accent phonetically.

Miss Lillian could light up a room just by entering. Her cloche of white hair was pulled snugly around the handsome face, so like her son's, but hers, the painter Jamie Wyeth asserted, was "a much stronger face." The finely pleated skin testified to eight decades of steady use, and dominating all were deep-set, Crayola-blue eyes that could twinkle or transfix. (Those same eyes, on an irritated President Carter, became the "icy blues" dreaded by his staff.)

The mother of the thirty-ninth president was like no other First Mother. Full of sass and vinegar, insouciant and irascible, compassionate toward the downtrodden and impatient with the high-and-mighty, Lillian Carter was sui generis. "She was an anomaly in our community, even from her early days," Jimmy Carter said about his mother. "She had many friends, but in many ways she lived in a different world." Revealing Lillian's own little secret, he confided, "She also relished being different."

How does one southwest Georgia girl come to be so different from the rest of them? She was born in 1898 into a seemingly conventional background—respectable Methodist family, relatively well-off—in the little country town of Richland. The Gordys were an old family, originally from Scotland, whose early sons had fought in the Revolution and later ones in the War Between the States. One Gordy was a wagon master or maybe a bugler—family lore tends to wander a bit—in the Battle of Atlanta. The Gordy family of Lillian's childhood was an overflowing household typical of her turn-of-the-century era, thirteen in all—parents, grandmother, and ten children, two of whom were cousins folded into the family after the death of their mother.

All it takes is one strong inspiration to stamp a child's mind, and for Bessie Lillian—"Lilly"—that was provided by her father, James Jackson Gordy. He was tall, imposing, with a dashing mustache, a postmaster/politico known to all as Jim Jack. "I got my liberalism from my father," she declared, proudly claiming the liberal label that is pejorative in Republican enclaves of "the New South." "I'm the most liberal woman in the county, maybe the whole state." Jim Jack treated blacks with kindness and a remarkable degree of equality, sometimes fetching meals from the hotel, which blacks were not allowed to enter, and bringing them back to the post office, where he and local black leaders then ate lunch together in the back room. Such collegiality was unheard of in those times.

Lillian, who worked in the post office when she was sixteen, absorbed her father's attitudes "by osmosis," as her youngest sister, Emily Dolvin, described it. She spoke of "the special bond" between Lillian, a middle child, and their father. Lillian, who was more like him than any of the other Gordy children, often boasted, "I was his favorite." She was unequivocal in her devotion: "He was an understanding father—and strict as the mischief. He was the one person I loved more than anybody in the world."

With her independent spirit, Lillian Gordy was a girl of the new century. She would always be a trailblazer, defiant of old shibboleths, ready to accept new thinking, and she instilled this open-mindedness in all of her children, particularly in the son who would become president. Still in her teens, Lillian set her mind on the nursing profession, a goal she accomplished and practiced through marriage and motherhood; she was the first presidential mother to have a career outside the home.

Jim Jack didn't approve of her decision, but in that one instance, her filial devotion was outweighed by her personal vision. "In those days nurses were not thought of very highly," she explained. "My parents opposed it to the nth degree and I had a sister who swore she was never going to speak to me." Despite their opposition, she was not deterred. For Lillian, nursing was more than a profession; it was a calling, a vocation that she pursued with mind and heart. What she should have been, she ruminated many years later, was a doctor, but in 1915 girls, no matter how bright, were in effect barred from medical schools.

For her nurse's training, she journeyed to Plains, twenty miles away. Beyond its charmless one-block business district, it was a pleasant, "y'all come" kind of town. Substantial white frame houses with fanciful gables and porch rockers were set among glossy green magnolias and whispering pines. Lillian met a nice young go-getter named James Earl Carter, but, she would recall, "I just didn't like his looks." Happily, his courtly manner and easy laugh made a greater impact than his looks, and after a two-year courtship, they were engaged. With the same brand of common sense he showed in buying land and other business ventures, Earl insisted that marriage wait until she finished her training, "so I'd always have it to fall back on," Lillian explained. His levelheaded wisdom, rare in a young man courting the woman he was determined to marry, set the course of her life.

She was twenty-five and he was nearing thirty when they exchanged vows in the Baptist preacher's study, then celebrated with family and friends at Earl's home. "Mr. Carter," the Americus *Times-Recorder* noted in its write-up of the wedding, "is regarded as one of the representative young businessmen of Plains and has hundreds of friends who are congratulating him upon his happiness." It was a good match and the beginning of a genuinely happy marriage that leavened hard work with laughter, a passion for baseball, and a lot of dancing.

The following year, on October 1, 1924, James Earl Carter Jr. arrived (the first American president born in a hospital). "I felt I had just become queen of the universe," she exclaimed to biographer James Wooten. Even a half century later she could still recall—as every woman can—precisely how she felt that day, an awareness that she had a place in the endless skein of humankind. "I'd always been afraid that maybe I couldn't really do what all those women I'd watched do. After I had, it made me proud. I had done it—I had done it! I had had a baby. Lillian Gordy—uh, Lillian Carter—was a mother." Earl Carter was ecstatic; he had a fine boy to carry his name. When the baby was only two Lillian and

Earl, like so many families in that preantibiotics era, almost lost their firstborn to bleeding colitis. It was a frighteningly close call, and everyone agreed that Lillian's around-the-clock nursing care pulled their son through. Twenty months after Jimmy was born Lillian produced a daughter, Gloria, to considerably less fanfare.

With the two children and Earl's expanding farmland, the Carters moved in 1928 to Archery, about four miles away, an unincorporated, unorganized nonplace, just a name and a train stop—if you pulled the red leather flag to signal the engineer in time. The family settled into a white frame bungalow on the farm, a simple house with a low hipped roof and wide front porch, with ample space for the two later children, Ruth, her daddy's darling, born in 1929, and Billy, thirteen years younger than his brother. That was home for Jimmy until, as a serious nineteen-year-old, he left for Annapolis. From the time he was six years old, he had set his mind on entering the United States Naval Academy (mostly because of postcards Lillian's brother, a navy man, had sent from faraway places), and he doggedly held on "while Daddy worked the congressman," until he won the appointment, "a political plum in those days." "Jimmy," Lillian said of her firstborn, "usually gets what he wants." Like his mama.

★   ★   ★

IN other parts of Roaring Twenties America, overextended optimists were playing the stock market with abandon, millionaires were transporting pieces of châteaus from France to gold-plated enclaves on Long Island and in the new mecca called Hollywood, while flappers with rouged knees were dancing the Charleston up a storm and the adventurous were signing up for a trip on an airplane.

In that same period, Jimmy Carter recalled, "my life on the farm more nearly resembled farm life of 2,000 years ago than farm life today" and life within the home was "similar to that of our great-grandparents." In the early years on the farm the Carters had no indoor plumbing, only a fireplace and woodstove for heat, with hot bricks to warm the bed on bitterly cold nights. Not until 1937 would electricity, without a generator, light up their lives. But with an automobile, a hand-cranked telephone, and a radio hooked up to the car battery, they were better off than most of the country folk around them in poor rural south Georgia. And theirs was a happy home.

Lillian and Earl were a fun-loving couple, not at all the worn, benighted stereotype beloved of writers spinning bleak tales about the rural South. They went to dances at the Elks Club, drank a little, and played a little poker, even though Earl was a deacon of Plains Baptist Church. Evenings at home they liked to put records on the Victrola and dance, there in the living room, just the two of them. Earl's favorite was "Sweet Georgia Brown"; Lillian could have danced all night to "I'll See You in My Dreams." Sometimes she just spun and twirled around the room by herself. She had more happiness than she knew what to do with.

"It was a sheltered life," the former president reminisced. "I started out as an isolated farm boy, as a minority member of a predominately black neighborhood—there were two white families and maybe twenty-five or thirty black families. It was a stern life. There wasn't much to do." In fact, Jimmy Carter had plenty of things to do. Saddling up his Shetland pony, Lady, his most wished-for present on his eighth birthday, he could explore the fields and meadows of his father's land, much the way young Franklin Roosevelt rode Debby on his father's Hudson Valley estate. Jimmy and his friends—his closest pal, A. D. Davis, was black—flew homemade kites, spun themselves dizzy in the tire swing, frolicked in the barn, caught catfish in Chocktawhatchee creek, and dog-paddled in a nearby pond. They rambled through the woods, with a soft blanket of pine needles underfoot, guns slung over their shoulders, hunting small game and birds. Jimmy's adventures differed little from those of Harry Truman or Dwight Eisenhower a generation earlier. Earl even built a tennis court—in

Georgia, clay courts are a gift of nature—and while it was less than Grand Slam quality, it was as good as their tennis. Carter thinks of those boyhood years as "isolated but not lonely . . . we felt close to nature, close to the members of our family, and close to God."

What set the Carters apart from other farm families around Plains was their life within the simple house in Archery, a life built around books and that battery-operated radio. The children not only loved to read but had several hundred books to choose from at home, and Jimmy quickly learned that if he was engrossed in a book, his mother let him skip chores. Lillian, who had grown up in a book-reading family with a large library, set the example. "She read day and night, at the breakfast table, the lunch table, the supper table. I do, still," Jimmy Carter reminisced. "She just read books, and so did I. My father didn't—he read the newspaper and that was about it." One can only wonder what the dinner hour was like for Earl, with the rest of his family submerged in their books.

The custom, which the Carters never thought of as antisocial, was continued into the third generation when nine-year-old Amy, seated next to the Mexican foreign minister at a White House state dinner, whipped out her book and read between bites. The president was surprised that Washington was surprised—it seemed perfectly normal to him. Like all of the eleven modern presidential mothers presented, Lillian taught her son to read little books when he was four, and he never slowed down. Books opened the world to him; books were teachers, prized companions, exciting escapes. In Lillian's words, "We'd just have orgies reading." He and his sisters bought books with their spending money, and even Billy, who was not thought of as thirsting for knowledge, kept a stack of books by his bed as an adult.

In the evenings, after homework was done, the family gathered around the radio and listened to favorite programs: *Amos 'n' Andy, Fibber McGee and Molly,* and the Eddie Cantor show. Lillian was always quick to point out that the Carters had the first radio and the first TV anywhere around Plains. "Can you imagine the world without television?" she mused half a lifetime later. Significantly, when she ruminated on the deprivations of rural life in those years, it wasn't bringing in a tin tub for baths, heating the water on the stove, or turning lamp wicks high for reading that struck her in hindsight as the hard life: for Lillian, real deprivation was all those years before the advent of the soap operas that she adored. (Many years later, friends wouldn't dare telephone her while *The Young and the Restless* slogged through their daily saga.)

Days were crowded with school and chores. Like farm boys from time immemorial, Jimmy was saddled with tasks from the time he could say "Yes, sir." He gathered eggs, slopped the hogs, and mopped cotton to control boll weevils, the job he most despised. (It entailed swabbing every bud with arsenic and molasses, which turned him into a walking fly trap.) As a five-year-old, he lugged buckets of cool spring water, with a tin dipper, to field hands sweating under the summer sun. When he was big enough to wield a broom, Lillian had him sweep the yard, which was overlaid with white sand trucked in to cover the Georgia clay. In the rain the clay could become gummy enough to add an inch to a boy's height just by sticking to the soles of his shoes.

When he was about six, Jimmy, with Lillian's help, sallied forth as a very small businessman, boiling peanuts from their farm and hawking them on the one street in downtown Plains. That enterprise, Carter would say, taught him how to tell good people from the bad: the good were the ones who bought his peanuts. By the time he was twelve or so he and his cousin Hugh were selling pork barbecue, one of the culinary glories of the South, and home-cranked ice cream. His mother, encouraging his get-up-and-go, was more than peripherally involved in these enterprises—who else would boil the peanuts? Jimmy, whenever he was needed, helped in his daddy's little store beside the house, stocked with everything sharecroppers and hired hands might want to buy on payday.

Lard and bib overalls, rat traps and Tuberose snuff, home-cured bacon and Octagon soap, Moon Pies and Dr Pepper—you could get whatever you needed from Mr. Earl.

That plain bungalow in Archery provided something of more lasting value than fancy trappings could have brought. "Home for us was a haven and a repository of standards, moral standards," the former president recalled gratefully. Rules were backed up by discipline. "In my family, there was absolute compliance with directives from parents. We never contradicted my father. The only rebel was Gloria—she defied him, and was kept at home. If we did something that warranted Daddy's punishment, we expected to be punished." An example? "If I slapped Gloria." Or the time he fired his BB gun point blank at Gloria's derriere—after she threw a wrench at him. That earned him a hiding he never forgot.

"Daddy had absolute authority over the family, but Mama would try to protect us. She'd say, 'Earl, I've already punished him.' " Earl favored switches from a peach tree—limber and close at hand—while Lillian settled for "a few easy licks with a paddle, or maybe not let us go to a movie. We were eager to accept *her* punishment." More often, she used a touch of ridicule. "She despised pouting," the president recalled. "She'd say to us, 'Why don't you go out in the garden and eat worms?' The vision would often result in laughter, or at least a lessening of tension."

In the Carter household, school was an extension of home: "If a teacher chastised us, our punishment at home was more severe— Daddy might restrict us for thirty days." (The current practice of parents' confronting the teacher, much less suing the school, was unthinkable.)

At the center of their life was Plains Baptist Church; there was Sunday school, preaching and BYPU (Baptist Young People's Union), and the church "proms," well-chaperoned Friday evenings when nice girls could meet suitable boys. Lillian and Earl differed in their approach to religion: he was a church mainstay, a deacon and Sunday school teacher; Lillian, born Methodist, had reservations. "I don't think all church-goers are Christians," she reflected. "I think there's a difference between religion and Christianity."

Her deeply religious son has no doubt that "Mama was obviously a believer. She joined the Baptist Church when she married Daddy—that was the custom then—but she never was part of the Baptist Church. I believe she was a Methodist to the day she died." Her sister Emily Dolvin confided to me that Lillian had been turned off by her first meeting with the church's women's group. "They were enthusiastic about sending missionaries to Africa, but they weren't interested in helping black people right there at home. She never went back." At home it was Earl who heard the children's nightly prayers. Her irregular churchgoing should not suggest any lack of faith: Lillian Carter was moved by a sturdy Christianity that led her to defy pervasive discrimination and to minister to lepers. She set that example for her son, who acted on it when he was in office and, as former president, continues to follow it as he sets an example of personal service to others throughout the world.

★   ★   ★

LILLIAN was not what you'd call a born homemaker. She was a combination of bookworm and tomboy, and neither trait is a prerequisite for running a home and family. What she was, most of all, was a born caregiver. After they moved to Archery, Lillian returned to the nursing she loved, devoting much of her time to the black families who lived nearby. "It was a strictly segregated community," Jimmy Carter explained. "It was an accepted situation, but Mama didn't pay any attention to it. Black neighbors came to Mama—she was in effect the only doctor they had." They trusted her; they depended on her; they were profoundly grateful to Miss Lillian, wife of one of the biggest farmers in the county, who came into their homes and treated them with dignity. They knew all too well the price she would pay on the Plains party lines for crossing that invisible color line.

For Lillian, having complete charge of a case was deeply rewarding—it made her almost a physician. She prudently consulted with the doctors in Plains, who were glad enough for her to handle those desperately poor black patients in their overcrowded shacks, and if the illness was really serious, the president pointed out, "she had the influence to get a doctor to take the case."

When Jimmy and Gloria were no longer babies, Lillian did private duty nursing, with shifts of either twelve or twenty hours a day, for which she was paid four dollars for the "short" day, six dollars for the longer. Much of that little income she used to buy medicine for the black neighbors who had no money.

There were many days when she didn't see her children, entrusting them to the care of a loving black woman who worked for the Carters for many years. Lillian left daily notes for Jimmy and Gloria—motherly reminders of what to do and when to do it—on a table where they couldn't miss them. Remembering this, the president chuckled. "We teased her that she was away so much that we thought the big table in our front room was our mother."

Though the Carters lived in what now seem near-primitive conditions, they were not poor. Their home at Archery was standard for country folk in those days, better than some, and their farm made them virtually self-sufficient. It annoyed Lillian that "Jimmy makes it sound like they should take up a collection for us." (Rebekah Baines Johnson had been similarly irritated that Lyndon did the same, but politicians cling to the notion that their version of "born in a log cabin" strikes a chord with the people.) The Carters weathered the devastating Depression without grave consequences, thanks to Earl's sharp eye for acquisitions—an insurance firm, a dry-cleaning shop, foreclosed houses, farmland—and the peanut warehouse that would grow into a major business. Lillian added income, along with her nursing money, by harvesting the pecans from their grove of the stately trees, a no-risk investment that paid off with pecans in season and yielded dividends of deep shade against the summer heat and beauty year-round.

When she was at home, she didn't spend much time in the kitchen. "I cook only to survive," she told Beth Tartan, a leading southern food editor, in the 1976 campaign, dashing hopes of choice Georgia recipes from the new president's mother. Lillian deliberately downplayed her kitchen skills to preclude an apron-and-cookies image that was most definitely not her, and she made the point that she always had a cook and a baby nurse, who was paid the grand sum of fifty cents a week. While the president agreed that "cooking was not my Mama's favorite pastime," his mouth watered just thinking about her homemade doughnuts— "with the dough slightly sweet, cut, fried and then shaken up in powdered sugar." He added that "my father was the one who really enjoyed trying new recipes." And of course "he was free to choose his own time to do it" and wouldn't think of scrubbing the pots and pans.

Growing up as a tomboy in a household of ten children, and emulating Papa in every way she could, Lillian was ill prepared to be a conventional southern wife. To start with, she had always enjoyed a drink now and again, in moderation. "I know folks all have a tizzy about it, but I like a little nip of bourbon of an evening," she said in her dare-you-to-frown way. "It helps me sleep. I don't much care what they say about it." (In the difficult year following her husband's death, her doctor had told her not to take tranquilizers, and she was only too happy to make the substitution.)

"I've always liked boys better," she declared. "I liked to play baseball. And marbles—I liked that better than dolls." In later life, "I was crazy about basketball"; occasionally she went to Columbus for the wrestling matches, and—being Lillian—became friends with one of the hulks, "Mr. Wrestling No. 2." To the last years of her life, she was content with a fishing pole in her hands. "Mama was an avid fisherman—fisherperson, in modern terms," the president said. "She spent her last years on the shore of a small private lake, and delighted in outfishing any of her children. After catching more fish than I, she would make comments such as 'Jimmy, you may be President, but you're pulling too soon.' Or, a few minutes later, 'Now you're pulling too late.' "

Her favorite indoor game was poker, and for years she and select friends in Plains met to play every week. Such a shrewd practitioner was she that she wiped out members of the White House staff on a campaign trip before they realized that this was not some sweet little old lady holding the cards. In her first visit to Las Vegas, during the 1978 congressional campaign, she was · ecstatic to discover another deliciously sinful pleasure: "I got a taste of blackjack and it has infiltrated my whole being. I would go broke in two days if I lived there!"

This image of a poker-playing, bourbon-nipping sports buff did not define her personal style, however—she favored pants suits in blues chosen to enhance her dramatic eyes and pinks that reflected a glow. Pretty as a young woman, she had matured into a handsome older one. She was at her best around men, teasing and bantering, with a sprightly irreverence that never edged into coarseness. She set a tone and a spirit that made life with Miss Lillian fun and unexpected. And she was never boring. Never.

★   ★   ★

BILLY loved to say, "I got a mama who joined the Peace Corps and went to India when she was sixty-eight. I got one sister who's a holy roller preacher. I got another sister who wears a helmet and rides a motorcycle, and I got a brother who thinks he's going to be President. That makes me the only sane person in the family." It always brought down the house, especially coming from the renowned beer-swigging, easy-cussin', joke-telling, outrageous proprietor of the best-known filling station in Plains, or maybe anywhere. Billy Carter was the quintessential southern good ol' boy.

Clearly, this was not your usual family in Plains, Georgia; they got their unconventionality from Miss Lillian, who was not the usual mother. She prized individualism, as anyone could see, and brought up her four offspring to follow their own drummers, who turned out to be playing very different beats. The drummer who led James Earl Carter Jr.—who insisted on being "Jimmy" even in official documents—was his mother, with her lively interest in public issues and yearning to serve the afflicted, who could be the poor blacks or oppressed dissidents.

Into this family of strong-willed individualists came Rosalynn Smith, a pretty eighteen-year-old, the shy girl next door whose soft voice and wide-set brown eyes made her seem more vulnerable than she would prove to be. It was not an easy entry. It was the age-old pull between the mother who had nurtured the boy, shaped the man, looked to him as a strong rod to lean on in old age, and the fresh young thing who waltzes into his life, captures his heart and soul, and takes him away for a life of their own. And in this instance, the triangle was squared to include the sister-friend who was suddenly excluded.

Long after the deaths of his mother and younger sister, Jimmy Carter shared a remarkably candid insight: "It seemed to me, even in retrospect, that Ruth and Mama were somewhat jealous of Rosalynn. I had always been theirs and then, almost overnight, I became totally infatuated with Rosalynn.

"In our inner family matters, Mama was a dominant presence," he explained. "Even within our own household, Rosalynn's forcefulness and self-assurance evolved over a number of years. Mama was intensely interested in public affairs, international events, politics, and had an encyclopedic knowledge of baseball. The interests of the Smiths were more parochial. It was during the last ten years of her life that Mama's friendship with Rosalynn evolved."

Significantly, that was the period during which Rosalynn grew into an accomplished public figure in her own right, respected for her leadership in the field of mental health as well as her performance as First Lady. Those were the heady ten years in which Miss Lillian and Rosalynn, together, threw themselves into Jimmy's quest for the White House, expanded their own roles as a result of his presidency, and bolstered him in the bitter aftermath of defeat. After a somewhat rocky beginning, these two remarkable women forged the kind of bond that could never have been formed between Sara and Eleanor Roosevelt, enabling both to define themselves within a wide range of public issues.

The Carters had known Rosalynn well. As Ruth's best friend, she was often at the Carter home. Secretly, she and Ruth had been conniving to get Jimmy to ask her for a date, as she later admitted, "trying to get him to notice me." The first date, the night before he was returning to Annapolis for his final year, came about in a casual way, but by the time he took her home it was no longer casual. "We rode in the rumble seat of the car," Rosalynn wrote in her memoirs, dreamily reliving that magic summer evening many years, four children, and several grandchildren later. "The moon was full in the sky, conversation came easy, and I was in love with a real person, not just a photograph. And on the way home, he kissed me! I couldn't believe it happened. I had never let any boy kiss me on the first date." One kiss—maybe a few more—and Jimmy Carter had made a life decision: he would marry Rosalynn Smith.

As soon as he got home, he sat down at the kitchen table and confided this to his mother. She was flabbergasted—the first date, Rosalynn was so young, Jimmy was more than a year away from graduating. The mother in her must have ached at losing her son so soon, but being part psychologist as well as physician, she did not criticize Rosalynn or mention the gap between the two families. Jimmy had made up his mind—and so the following June, in postwar 1946, Lillian was present, all smiles, at the wedding in the Naval Academy chapel.

Rosalynn was all too aware of the Carter family's displeasure at the engagement. "They were not so happy for Jimmy," she acknowledged in her autobiography, *First Lady from Plains*. "Mr.

Earl was very disappointed. He was ambitious for Jimmy, had great plans for him, and being married to an eighteen-year-old girl from Plains was not one of them. I was glad he never showed those feelings to me and was always very nice. Ruth tried to be, but she and Jimmy had a very special relationship, and even though we had plotted to get us together, when it became serious she didn't like it. She was very jealous, and it would be years before we were comfortable together again." Jimmy's mother "was my only champion—Miss Lillian was always for the underdog." Rosalynn pointedly dismissed the tattletale book by cousin Hugh Carter, who said that Miss Lillian "hit the ceiling" in that kitchen-table session with Jimmy.

Rosalynn was an uneasy newcomer in an eclectic family—Billy had exaggerated only slightly in his capsule description of the Carters. Ruth was an evangelist and faith healer; Gloria, a motorcycle nut, rode on the wild side with her husband, a farmer with the shoulder-length gray hair and grizzled beard of an overage biker. "Gloria could have started the Plains chapter of Hell's Angels," Miss Lillian commented, bemused by her maverick daughter. Coveys of bikers sporting leather and earrings, revving their 'cycles, often descended on Gloria's plain farmhouse and stayed maybe a week before roaring off to Daytona Beach. (The untamed Gloria, who died of cancer at fifty-four, would have loved her funeral: an "honor guard" of thirty-seven motorcycles escorted the hearse to the cemetery, and on her tombstone was inscribed, "She rides in Harley heaven.")

Lillian, part-time iconoclast, identified more comfortably with Gloria, who boasted, "I'm country and proud of it," than Ruth, whose spirituality was not her mother's kind of religion. Lillian mustered only faint praise for her younger daughter in an interview that Gloria, in a onetime role, conducted for *Ladies' Home Journal:* "About all I admire about Ruth is that she never gives up. She's not a very good mother, but her family just adores her, so I reckon she's all right." Ruth, who was married to a veterinarian, lived two states away in North Carolina, which was probably just as well.

And then there was Billy. "Wherever there's trouble—that's where Billy is!" his mother drawled, quite happily, during the 1980 Democratic convention in New York. "Billy drinks a little too much beer," she acknowledged without a trace of disapproval. "I have never mentioned it to him. I like Billy just like he is. If he ever tried to change, he'd be the biggest mess that ever was." Never has a political mother spoken publicly about her children with such unrestrained frankness. Billy roared with laughter; Ruth most likely did not.

In a somewhat puzzling disparaging comment about her first-born, Lillian said, "There was nothing special about Jimmy. There was nothing outstanding about Jimmy at all," and would often state, "I love both my boys." Proud as she was of Jimmy, she stuck with the fiction that the son who confounded the experts and became president was little different from the one who hadn't accomplished anything. It was the compassionate balancing that comes from a loving mother—but every mother knows that some children turn out better than others, and Billy was no Jimmy. Still, his mother defended and cherished her late-born son, admitting that whatever his deficiencies, "I just *like* Billy. Like what he does. And he looks after me as much as he can. He must be very much like I am." Lillian, the maverick, shared with her second son a disdain for what stuffy townsfolk might think, a wicked pleasure in shocking the self-righteous, in shaking up the stiff-necked.

Jimmy, on the other hand, never shocked; Jimmy was a concil-iator, not a disturber; he played by the rules and heeded the Bible. Billy, "tired of those damn hypocrites," stopped going to church and in abstemious Plains brandished his familiar brown bottle even with his ham-and-grits breakfast at his "club," the Best Western Motel. Lillian took delight in Billy the hellion. When a friend implied disapproval that the president's mother would attend the launch of Billy's beer enterprise, when he went from just drinking it to making it, Lillian replied tartly, "I went to Jimmy's inauguration, didn't I?"

There was, however, another side of Billy that he kept out of sight, lest it spoil the reprobate redneck image that endeared him to reporters. An avid reader (Faulkner was a favorite) and the caring father of six children, he was concerned about the world food shortage, which peanuts, "the almost perfect food," could help to alleviate. (Despite a modicum of self-interest, his message was correct.) As a businessman he showed considerable acumen, until he allowed his big plans to balloon out of control. Billy was a much more complex individual than he let on in his guise as Clown Prince. The question is why he ostentatiously played the rube, the redneck while his brother was trying to shake loose from southern stereotypes in his national race. And once Jimmy was in office, how could Billy have considered accepting $220,000 to be a consultant—or unregistered foreign agent, depending on viewpoint—for Muammar Qaddafi's terrorist state of Libya? The short answer is that being kid brother to a man who has won a place in history isn't easy, and succumbing to flattery is. Lillian's evenhandedness with the brothers clouded her judgment—she should have taken the paddle to Billy more often.

★   ★   ★

WHEN Lillian was fifty-five, life dealt her a blow that left her adrift of her moorings: Earl Carter had just been elected to the state senate when his robust life was cut short by swift and lethal pancreatic cancer. She was flattened by his death. At a time when Billy, at sixteen, was soon to leave the nest, she had looked forward to uneventful golden years with Earl. For two years after his death Lillian was lost, sunk in a life that had little meaning beyond her family: "I wasn't depressed—I was angry, angry with everybody who had a husband."

Gradually, her natural exuberance began to break through

again, like daffodils coming up through a late snow. Tired of being "bitter and bored," she picked herself up and jumped at the opportunity to become housemother at Kappa Alpha fraternity at Auburn University in Alabama. She arrived in style in her new Cadillac: "Jimmy made me buy it. He didn't want people to think I was poor and hired out." There was instant rapport between 105 rambunctious boys and their surrogate mother. She counseled and coached them, altered their pants, and lent them her car. "I just gave them all my life, and in six months I was a different person." She had saved herself by helping others. After almost eight years, she left to manage a nursing home nearer Plains for a year or so, then resigned, she said, "because the patients were getting to be younger than I was."

During Carter's years as governor of Georgia, Lillian became a personage. She gave hundreds of speeches around Georgia, usually to retired people, and never missed the chance to give them advice born of her experience. "Stay active," she exhorted. "Don't ever sit in a rocking chair! The best thing on earth is to be able to give of yourself—and get 100 per cent back." She talked about coping with the trauma of widowhood; it was a lesson learned and a lesson taught. "Find something to do. Work, work, work!" she urged women left to lonely lives. "Get a job the minute your husband dies and stay busy!" It was sound advice, and Lillian Carter was her own best example, which her son wisely emulated in building a new life after the White House.

"I thought when my husband died that my life was ended," she reflected years later. "But I've had so many opportunities that I never would have had if he had lived. We would have just had a quiet life together—which I would have liked—and Jimmy wouldn't be president now." And in that she saw the hand of God at work.

★   ★   ★

CARTER called his mother, a self-declared Dodgers fanatic, "a walking encyclopedia on baseball." She had enjoyed a long love affair with the Bums: she and her husband used to plan their vacations around big-league schedules, more than once making the pilgrimage to Brooklyn, back in the days of the Real Dodgers. She recalled the time in 1947, when she and Earl were in the stands "the day Jackie Robinson entered the game. I was the only one standing up and applauding in my section. I went back and raved about the Dodgers, and of course the people down home didn't agree with me." It is impossible to overstate, in this new millennium, just how courageous it was in 1947 for a well-born southern lady to make Jackie Robinson her hero in the Ku Klux Klanland of south Georgia. "I do so like being for the underdog," she often declared. She was almost reluctant to see her Dodgers advance from that category—almost. Her great reward, her true Walter Mitty moment, was the day she threw out the first ball in the fourth game of the 1977 World series between the Dodgers and the Yankees. It was surely the first time Dodgers catcher Steve Yaeger and manager Tommy Lasorda were kissed by a white-haired pitcher wearing a sky blue pants suit.

The underdog, the oppressed, the forgotten—their plight was the central theme of her life and would be important in her son's policies. "Mother always was a champion of disadvantaged people," Carter said proudly. "In our area her concern was poor whites and all blacks." Both mother and son lived their beliefs. "Our lives were dominated by unspoken, unwritten, but powerful rules, rules that were almost never challenged. A few people challenged them, not in politics but in the way they lived their lives. My mother was one of those people. My mother knew no color line." Jimmy, as a child, absorbed her rejection of prejudice. "In my town," Lillian stated, "I have stood alone, Jimmy and I, and my daughter Gloria. I have stood there all these years alone."

That moral loneliness contributed to Lillian's tendency to have many acquaintances but few close friends. Living in rural isolation or in a town of six hundred souls, working long hours, she was never one to drop by for coffee or observe other customs of small-town neighborliness—and, voracious reader that she was, she would have preferred her books to idle chatter. As her liberal attitudes deepened, she was even more standoffish. She loved Plains, but "small town people," she observed, "can sometimes be afraid of independent minds." Her mind had always been independent and she had no intention of conforming to the ideas of others.

The former president acknowledged that his mother's acceptance of blacks "shocked some people, sometimes even my father, who was very conventional in his views on race." Carter remembered the day Lillian invited Alvan Johnson, the son of the local black bishop, a young man who had gone north to college in Boston, into her living room—by the front door—for a social conversation. Earl Carter stalked out of the room and for days afterward was stony-faced and cold, unmoved by Lillian's explanation that she had wanted Jimmy to meet "somebody he was not normally going to be exposed to."

Race was a subject that Jimmy, visiting home as a navy officer, learned to avoid if he and his father were to have a comfortable conversation. In his political campaigns the candidate explained that Earl Carter, known as a generous man and respected in both sectors of the community, held the views of his time and place. He died in 1953, a year before the Supreme Court ruled that segregated schools violated the Constitution, a historic ruling that eventually overturned the old social structure of the South, a leap toward equality for black Americans. "Earl was of his time," Lillian said in his defense. "He was not like me, certainly, but he did not stop me from doing what I wanted to do."

In later years, as sexual orientation became a national political issue, Lillian Carter's compass did not waver. Turning eighty, of a different generation, and living in the heart of a fundamentalist stronghold, she supported the cause of gay rights as staunchly as she had stood up for the rights of black Americans: "To me human rights means the chance for everybody to make the most of himself . . . the right to do what he wants."

This southern individualist had an innate sense of the injustice, the poverty, and the inequities that gnawed at the human beings consigned to the bottom of the heap. She was not a firebrand nor a placard-wielding protester against the system. To her the cause was simply people needing help: "I just saw what needed to be done and did it." It could be frustrating. "I felt the blacks had been treated so unfairly. I had seen so many black people and civil rights workers mistreated. And there was nothing I could do about it—not one woman in fifty thousand people." She was one of six—the Carter family and one other member—who voted to open Plains Baptist Church "to all worshippers"—in other words, to admit blacks.

Earlier, in 1964, Lillian stepped forward to do something specific and very visible. LBJ's Civil Rights Act had made him such a pariah in Georgia that in Plains no Democrat would work in his campaign—except Lillian Carter. She welcomed blacks as warmly as whites to LBJ headquarters, and as she left the office at the end of the day she would regularly find her car soaped up, the antenna knotted, ugly messages left on the seat. "Children would throw eggs and grapefruit—not at me—but my car would be a mess." She jutted her jaw, steeled her blue eyes, and refused to be intimidated. Observing her mother-in-law's courage over the years, an admiring Rosalynn stated, "Miss Lillian was never afraid of controversy. I think she thrived on it." Lillian concurred: "I've never been afraid of anything. Except snakes."

★ ★ ★

LILLIAN Carter, stretched out in bed, peered through her toes at her television screen. Johnny Carson, her regular late-night companion, was taking a commercial break and a public service announcement began its pitch: "Is your glass half empty or is it

half full? Join the Peace Corps and help others. Age is no barrier!"
Lillian was sixty-eight years old, but didn't the man say age was
no barrier? "That's when I decided to go," she said ten years later,
still savoring the audacity of her decision. She had been living in
Plains again, retired, and "just miserable, just bored to death. I
love Plains but there was nothing in this world to do in Plains. I
fished and I played bridge. I just got tired of the same old thing."
And then came the Peace Corps message on the *Tonight* show.

"I put in my application—but I just knew the children were not
going to let me go. It was a joke, me going to the Peace Corps,
really ridiculous. But they let me down!" She asked for "a dark
country and a warm climate"—India. She broached the subject
with each of the four children cautiously, expecting a torrent of
objections. If she was bluffing, they called it. "I think you'll enjoy
it," Jimmy said with a touch of envy. Gloria grumbled fondly,
"Who'll go fishing with me?" And Ruth said, "India? I'll come
visit you!" Billy was relieved that she had found something to do
with herself. Nobody said, Don't do it. "So I had to go, you see, to
keep from losing face with my children."

She had no conception of how hard it would be, how lonely,
how frustrating—and how much she would miss her Jimmy's first
campaign for governor. The Peace Corps is generally considered
an experience for young idealists, but older Americans have
proved to be effective volunteers, bringing practical expertise,
patience, and an understanding of human nature that comes with
(or causes) gray hair. As the Peace Corps looked for mature
women, between fifty and sixty, to help Third World women, rural
nurse Lillian Carter offered a much-needed expertise, though her
sixty-eight years were pushing the limit. (One-quarter of all volunteers drop out before their two-year stint is completed, according
to Peace Corps statistics. Among those over sixty-five the dropout
rate rises to more than one-half.) These days the Peace Corps
rarely sends volunteers over forty-five to countries of extreme heat

and harsh living conditions, which precisely describes the setting in which Lillian Carter was posted.

On December 21, 1966, as she unkinked herself after thirteen hours on the New Delhi–Bombay train, she stepped into a life more suited to Mother Teresa of Bombay than to Miss Lillian of Plains. By the time she left, having put in her full two years, she described working in India as "the greatest thing [that] ever happened to me. It took two years out of my life. But it was wonderful, the most meaningful experience of my life. I was so fulfilled." Her sister Emily said of her, "Lilly was an enigma to me. I never exactly understood her. But we learned more about her after India." Lillian's letters revealed that she was motivated by a deeper missionary spirit than her family and friends had ever recognized.

She was assigned to manage the politically controversial family planning program in a clinic outside of Vikhroli, a small city some thirty miles from Bombay. Reading her letters to her family, which Gloria compiled into a book, *Away from Home,* it is hard not to be affected by the physical and emotional struggle Lillian Carter endured. She was homesick, growing ever thinner until her few clothes hung on her, her feet swollen beyond her shoes. Most of all the letters convey the overwhelming misery, the black pit of despair that she witnessed every day, all around her.

Lillian Carter's personal book of revelations ranges from her Christmas wish "for some clean rags and cheddar cheese—I'd rather have a chunk of cheese than diamonds"—to her lament for a woman lying by the side of the road, untended, dying a little every day. "What is to be done?" she cried. Her own commitment was put to the test the first time she treated a girl with leprosy. She had to force herself to touch the child and afterward scrubbed her hands and body until her skin was almost raw—and was ashamed of her own revulsion. With unflinching self-discipline, she overcame her shudders in dealing with this disease dreaded since biblical times. "My Christianity is shown by deeds and not words,"

Miss Lillian once explained. The story of her loving care of the leper girl was proof enough. It moved her son to write a touching poem about her experience.

Day after day the thermometer soared to 110 degrees and the stream of patients never let up. ("We saw 360 patients today, besides the regular injections . . . it would be a pleasure to faint, but I'm too busy.") She found herself worn down by the sheer physical demands of the job and pining for her beloved south Georgia as she tried to cope with the overwhelming misery of rural, Third World India. She wrote home saying, "I'm so home-sick every night. As long as I'm busy, I'm fine, but as soon as I'm idle, I just die." She longed for news—"I don't even know who played in the World Series"—and she grieved at the murder of her beloved Bobby Kennedy. But, continuing to set an example for her family, she was resolved to serve out her full two years. And she understood that much of her personal anguish came from her acute awareness of the poverty all around her, poverty beyond a solution. "Oh, God, I have seen it all," she wept. "I can never again take bread or warmth for granted."

Yet all of these trials deepened and strengthened her faith. "I have always believed in prayer, and in India I needed God more than I ever have. I feel like it was his purpose for me to go to India. I was so frustrated one time, I said, 'God, you got me over here. If you don't let me do something beside family planning, I'm going to have to go home,' because I could see by the side of the road people who needed me and I couldn't do anything for them." Within days, she was made an assistant to the doctor in charge of the clinic—Lillian's conversations with God often produced solutions to her needs. "I am convinced," she wrote home, "that it was really intended for me to come to India. I don't think I could have NOT come!" Those poignant letters had a more lasting effect than she knew: more than thirty years later, her great-grand-son Jason Carter joined the Peace Corps as a volunteer.

On her seventieth birthday, shortly before returning home, Lillian shared with her children the philosophy India had taught her: "I didn't dream that in this remote corner of the world, so far away from the people and material things that I had always considered so necessary, I would discover what Life is really all about. Sharing yourself with others, and accepting their love for you, is the most precious gift of all. If I had one wish for my children, it would be that each of you would dare to do the things and reach for goals in your own lives that have meaning for you as individuals, doing as much as you can for everybody, but not worrying if you don't please everyone." Her son Jimmy would fulfill that wish, reaching for a goal beyond her imagining, and he no longer worries that he didn't please everyone, not when there's still so much to do.

★　★　★

IN the governor's mansion in Atlanta one night in 1973, Jimmy came into his mother's room to talk. Clad only in his shorts, he put a foot on her bed and started to speak. "Take your foot off the bed," she commanded. So he did—he always obeyed Mama—and then he could tell her his news: he was planning to run for president. "President of what?" she asked. Well, you know, the United States. She thought he must be joking, "until I saw that vein in his forehead that throbs when he's excited. So I knew he was serious." Thus began yet another chapter in the multiple lives of Lillian Gordy Carter, a heady three years that catapulted her son—the ultimate underdog—into presidential history and transformed Miss Lillian into a national personality, a new kind of First Mother.

James Earl Carter Jr., the governor of Georgia, was now considering a run for the White House. How did all of this come about? As an Annapolis man, Jimmy's long-range goal had been the navy's top position, chief of naval operations. He had no thought of returning to live in Plains, especially since Rosalynn was dead set against it. Miss Lillian's judgment, "There's nothing to do in

Plains," was not an exaggeration. But Earl Carter's early death changed everything. With Billy so young, what would become of the family's lucrative peanut warehouse? "I have no alternative," Jimmy decided: his duty lay in Plains. He stepped into his father's shoes and gave up a promising career in the navy's nuclear submarine program, after being handpicked by the powerful Admiral Hyman Rickover. Rosalynn, Lillian said, "hated to come back to Plains."

His return to his small-town roots thrust him into community leadership, which was a way station to state politics and in due course to the ultimate prize. Like Richard Nixon, Jimmy Carter came home unwillingly and found his path to glory there. "I think God has a hand in everything we do," Lillian said with deep conviction in 1976. "Jimmy would never be running for president if he had not come back to Plains. This was God's will."

Echoing Sara Roosevelt's words, Lillian stated, "I never brought up my son to be president, but as long as he is in this, I want to do everything I can to help him win." This was not a great sacrifice for a woman who rejoiced, "I've loved politics all my life. I've never run for office, but I've aided and abetted!" At her husband's death, she was asked to take his seat in the state senate, but, grief-stricken, turned it down—and later regretted having passed up this opportunity to become a player herself.

"She was a natural politician," the former president observed. "She actually liked politics more than I do. She got that from her father. My daddy was aloof from local politics until Roosevelt was elected in '32, but then FDR's farm policy turned him off." Jim Jack Gordy was what his grandson called "a nimble politician"— he managed to keep his patronage job as postmaster through four presidents. While he had a passion for politics, he never ran for office, preferring the role of a backroom insider, a man who was virtually clairvoyant in predicting the vote in any local election. In the Gordy household politics was as much a staple as grits.

When Jimmy ventured into politics, first as a state senator, then for higher stakes, Lillian leapt into campaigning with all the vigor of Rose Kennedy; both got their zest for politics from their fathers. Campaigning was great sport for her: she would shake every hand, visit every factory, toss out quotes for every reporter while ignoring a bruised hand, sore feet, and the occasional incident. (In one memorable moment, as she walked under a conveyor belt in a poultry-processing plant, cold chicken innards dropped down and slid inside her dress. Her response would have made a memorable sound bite: "Get those chicken guts out of my clothes!")

"She loved the entire political arena, and had since she was a young woman," the president said. "When projected into the national scene, she relished contending with Johnny Carson, Walter Cronkite and other TV personalities, usually dominating the interview with her humor and almost total lack of restraint." The media ate up her injudicious declarations and down-home southernisms, delivered in the authentic voice of her region: her *r*s were jellied, final *g*s tossed aside like peach pits, one-syllable words drawled into two.

For her campaign minders, any time Lillian stepped up to the podium or sat down for an interview was a Maalox moment. They could never be sure what might slip out. For example: "Jimmy says he'll never tell a lie. Well, I lie all the time. I have to— to balance the family ticket." It made lively copy; no one had seen a candidate's mother like this one before. At Carter's national headquarters in Plains's abandoned depot, an old white frame building spruced up for its new function, Lillian, the candidate's surrogate, would take her place in a rocking chair parked prominently on the porch and hold court. Tourists lined up, cameras at the ready, for their moment with the celebrated Miss Lillian and she made it worth the wait. Among those waiting, her son noticed, was "a stream of would-be vice-presidents."

She drew the line at the physical contact that is the lot of candidates. At the Plains headquarters they put up a sign, as at a zoo:

DON'T TOUCH MISS LILLIAN. "She bruised easily," her son explained. "She didn't like to be embraced—except gently by handsome men." Maybe so. More likely it was because Lillian Carter, a self-assured professional woman who was anything but the prototypical grandma, did not consider herself public property.

★   ★   ★

SHE reveled in the campaign, relished the adrenaline rush, the fuss and photographers, the attention, attention, attention. After the excitement of winning, when the serious work of the presidency was beginning in Washington, Lillian, back home in Plains, fell into an uncharacteristic torpor. It was a syndrome well known to staff and journalists whose lives were subsumed by the presidential race for months. After the stimulation and camaraderie of campaigning, real life can seem flat, and for Lillian victory brought a gaping hole in her life since her granddaughter—"Amy is my heart"—was no longer Grandma's girl but the White House daughter.

As the family worried about their mother's depression, the new president came up with a cure that no doctor could have ordered: he asked her to lead the American delegation to the funeral of the president of India. Her itinerary took her back to Vikhroli, where she was given a maharani's welcome. She was again the center of attention; it was good for what ailed her. The success of her first mission led to others. She met presidents and prime ministers, assorted royalty, one pope, and one emperor (in Upper Volta) and charmed them all. She visited rural villages, colorful markets, and struggling health clinics. But it was in Mali that she made her most unforgettable impact: the mother of the American president visited a hospital for lepers, shaking hands, talking with them as friends. It was real; it was Lillian.

In Rome she had a private audience with Pope Paul VI, which turned out to be his last, a meeting enlivened by his visitor's irrepressible humor. As Lillian told it: " 'Holy Father,' I said to him—

I was concentrating on not calling him 'Your Majesty' or 'Your Excellency' because there had been so many titles on this trip. He knew I was nearly eighty and he was eighty-one, and he told me, 'I am now ready to meet God.' I said, 'When you see God, please mention my name.' And the Pope said, 'and you tell him about me.' " No doubt he did—not even the pope could forget Lillian Carter. She was awestruck by his presence: "I have never felt so close to God as looking at that man's face." Not one to pass up an opportunity, she asked the pope to pray for human rights and rain in Africa. "When I got to Gambia, it began to rain. In Senegal it rained. In Mali it rained. It rained so much I thought I better call the Pope and tell him to lay off."

Would her entreaty for human rights have a negative effect on relations with Russia? an interviewer asked. "Russia? Who's talking about Russia?" snapped Miss Lillian. "I asked the Pope to pray for human rights in my own country—the deep South."

★   ★   ★

SARA Delano Roosevelt tried to improve the food and ambience of the White House; Martha Truman thrilled to its history; Rose Kennedy relished playing hostess there; Dorothy Bush chatted with tourists. But Lillian Carter bluntly stated, "Living in the White House was boring—I never did like it." From the familiar comforts of Plains, she spoke her mind, as always: "It's like a museum. I hate it there. I can't do anything, can't see anything. I don't get to meet anyone. I don't even see Jimmy that often. The one time I did stay there, they ushered me into his office one morning, and he said, 'Good morning, Mama, how are you. . . . : Well, I'll see you later, Mama, I've got a lot to do.' And they took him out." Rosalynn was busy with her own heavy schedule and Amy was in school most of the day. The queen's suite was beautiful, but she felt at loose ends in the White House, not at all active and useful as she had felt on her foreign missions for Jimmy.

Over the years Carter said many times, "My mother is the most

influential woman in my life." He should have made it an unqual-
ified "most influential person," for his public life reflects scarcely
an imprint of his father, whom he respected, loved, and wanted so
much to please. A hard worker, a good shot, faithful to the Baptist
Church, and a keen competitor, those were Earl's mark on his son,
but he left no legacy of intellectual substance. In a televised conversation with Bill Moyers during the 1976 campaign, Jimmy
Carter acknowledged, "I'm much more like my mother than I am
my father." Earl, too, was influenced by Lillian. "It was only
later," Jimmy commented, "that I understood my mother's strong
will and how she expressed her views to my father in private, and
often prevailed."

It can be argued that the Carter presidency was very much a
product of the values inherited from and the example set by Miss
Lillian. Had young Jimmy Carter not grown up imbued with her
compassion for the plight of black Americans, had he not witnessed her courageous disregard for the barriers cynically designed
to relegate blacks to permanent inferiority, had his sincerity on
human rights not rung true for the inheritors of Dr. Martin Luther
King's legacy and convinced white liberals, it would have been
impossible for Jimmy Carter of Georgia to be nominated, much
less elected as the first president from the Old South since the Civil
War. Not in the America of 1976.

Miss Lillian, an apostle of tolerance surrounded by prejudice
and outright oppression, had an impact far beyond her own
household. For example, when she spoke to the chamber of commerce in Tuscaloosa, Alabama—not her natural constituency—
the local newspaper ran a front-page editorial: "The Lord places a
few people like Lillian Carter on this earth to serve as an example
to the rest of us. We are a better people and a better nation
because she has touched us in so many ways."

No president's mother has had greater impact on her son's
thinking on a range of issues than Lillian Carter, beginning in the
formative years when most boys' deepest concerns are the baseball

scores. Racial justice, equality for women, and health care were her concerns just as they emerged as cutting-edge issues. She had no voice in policy decisions, but she had always spoken out on subjects that mattered to her and was dependably ahead of her time. From the outset of his presidency, Jimmy Carter appointed a record number of women to high-level positions, backed the Equal Rights Amendment, and advanced the other concerns his mother had implanted in her son.

Miss Lillian, herself a shining example for senior citizens, exacted a promise from her son to address the problems of the elderly. "I want him to get into that because I'm getting old so fast," the seventy-seven-year-old told the Gray Panthers. "If he doesn't," she vowed to a group of seniors in Brooklyn, "he's gonna catch the devil from his mama!" With Lillian as a model, Carter understood that her generation needed more than health care and pensions; they needed to feel useful, to be respected, and not be penalized by age discrimination. In a direct score for Lillian, his 1976 platform declared, "We must do much more to make the elderly feel wanted and to take advantage of their experience, which is a true national asset." In his recent book *The Virtues of Aging,* President Carter wrote of his mother, "she continued to age but never grew old. At the age of eighty-five she was still full of life, never failing to wake up in the morning with determination to make the new day an adventure."

In his campaigns, Jimmy never missed a chance to mention that his mother joined the Peace Corps at age sixty-eight and was literally a lifesaver for the poor of that Indian town. "It saddens me," he said, "to know that because of age discrimination in the United States, my mother's service to India would have been almost impossible in her own country." She was his prime example when he addressed working women's groups, this woman who started working when she was a teenager and kept going until she was past seventy—a woman who should have been a doctor but was barred by gender. The joint impact of Lillian and Rosalynn was behind his declaration that "discrimination against women now is as severe and crippling to our national strength as was religious or racial discrimination in the past." And he could remind public health workers that he cared about their problems because "I was raised by a nurse."

In hundreds of speeches over the years, Lillian was a confidence-builder and a goader, challenging her audiences to live fuller lives, lacing her message with cutting wit and curbstone wisdom. Her philosophy, in her capsule form, was simple: "I've found that if you can just last through the rough spots, and hold onto the good spots, things will always get better. They always have." Sometimes she put it in a distinctly Lillian fashion: "I have always believed you ought to try like hell to take care of the things you can take care of and just let the rest of it go hang."

★   ★   ★

NOVEMBER 1980. Disaster. Of all the vicissitudes of politics, nothing is so devastating as a president defeated for a second term. Rejection, disbelief, humiliation, anger—all must be swallowed, concealed behind the mask of the good loser throughout the long, unnecessary interim between election and inauguration. (In Britain and other parliamentary governments, they lop off the loser briskly, dispatching the moving vans the next day, which is a more humane demise than almost three months as the rejected and powerless lame-duck president.)

Jimmy was crushed; Rosalynn was desolate. Lillian? Her son, calling to alert his mother to the inevitability of a Reagan victory, was surprised at her reaction. "I said, 'Good!' It wasn't a blow to me—I wanted him out," Lillian insisted. "My whole family had been split wide open from Jimmy being President." That was the matriarch speaking; surely the mother ached for her son in this devastation after a life of triumphs.

After a period of dejection, Jimmy Carter, drawing on resilience inspired by his mother, charted a new path for himself and Rosalynn. Channeling his prestige into the cause of peace and human rights, Citizen Carter has lent his integrity and commitment to vouchsafe free elections and work toward resolving conflicts around the world. (A village near New Delhi in India was renamed Carterpuri, in gratitude to the Carters, who visited there in 1978 and gave the people their first television set.) And, hammer in hand, the former president has helped build housing for those who have none. By deed and word, Jimmy Carter has been the most productive of all former presidents, a paradigm for those who follow.

★   ★   ★

"WHEN I look back over my life, I see the pieces fit," Lillian mused when she was seventy. "It has been a planned life, and I truly believe God had everything to do with it." In the spring of 1981, Jimmy, bowed by defeat and unaware of the new career of fruitful service that lay ahead, had returned to Plains. Then came more bad news—breast cancer was discovered in the hardy Lillian, and the following year it flared up again. She fought it with high morale and true grit. "She knew what was coming," said a friend, "but it did not affect her great zest for life." Then in September 1983, sweet Ruth, the faith healer, succumbed to the cancer she had fought solely with prayer and diet. Losing her angelic daughter was too great a blow, and five weeks later the indomitable Miss Lillian died. Cancer, the killer that stalked the Carter family, would also claim Gloria and Billy long before their allotted three score and ten.

"Miss Lillian," declared Rosalynn, the distinguished daughter-in-law who had long since outshone the Carters' early misgivings, "was always, by sheer force of personality, the strongest link that held together her collection of strong-willed and independent children, in-laws, and grandchildren."

The most touching tribute came from her son, the "ordinary freckle-faced boy" who reached the White House on the wings of her immutable standards and undaunted spirit. In dedicating his volume of poetry, *Always a Reckoning,* he wrote: "To my mother, Lillian, who never would let racial segregation, loss of loved ones, ravages of age, or any other principalities or powers stop her sharing what she had or was with the least of those she knew." Fifteen years after her death, he said with wistful tenderness, "I don't think anyone ever met my mother and then forgot her."

9

THAT'S MY DUTCH!

Mother's love shapes son's sense of belonging to the world.
—WILLIAM WORDSWORTH

RONALD—"Dutch"—Reagan was not yet five when his mother asked if he would like to be in the play she was staging for their church. Nelle Reagan was confident that he could handle the part—and so was he. From the minute he got up on the makeshift stage he performed like a trouper. And he really liked it—the way the people paid attention, the way they applauded at the end. It was a propitious debut for Ronald Reagan, movie actor—second-tier star—and president. His mother had launched him.

If Hollywood were doing his story, we would now see a calendar's pages flicking past, one by one, to show the years slipping by. The scene then shifts to a handsome young Reagan delivering his lines, still flawlessly, on the set at Warner Brothers studio in the place that is more a state of mind than defined real estate—Hollywood. Nelle loved to visit the set; always pleasant, always inconspicuous, she would find a spot where she could focus on only one actor, her "Dutch." He never forgot his lines, never blew a scene, and always hit his marks. Nelle could feel a quiet sense of pride for her role in raising a successful actor who would become the Great Communicator.

Cut now to the great dome of the United States Capitol, followed by an interior shot of a committee room where Ronald Reagan, president of the Screen Actors Guild, is testifying with the clear-eyed conviction of a reformer before a powerful congressional committee. This was the era of the great Hollywood Red Hunt, when writers, actors, directors, and others who had been members of the Communist Party, or sympathizers, were blacklisted. The camera catches a key moment in the political evolution of Ronald Reagan as he leads the fight to purge his union of leftists.

Fade to a long-lens shot of a tough Governor Reagan standing up to sixties protesters at the University of California. Up the volume on patriotic background music as Ronald Wilson Reagan, left hand on his mother's Bible, open to a favorite verse (2 Chronicles 7:14), delivers the most important lines of his life, the presidential oath of office. Freeze frame.

What a scenario.

Had there been a canvas director's chair in the house during Ronald Reagan's early years, it would have been occupied by the mother he called Nelle. When the credits roll on the Reagan life story, Nelle Wilson Reagan should be listed as director, producer, and head of casting. As producer, there's no doubt about Nelle's credit: On February 6, 1911, in the aftermath of a raging blizzard in the little country town of Tampico, Illinois, in a flat above the store where her husband worked, Nelle gave birth, with much difficulty, to her second son. She pronounced him "perfectly wonderful" and named him Ronald, adding her maiden name to round it out.

The proud father, John Edward "Jack" Reagan, thought the ten-pound baby looked like "a little bit of a fat Dutchman" and promptly dubbed him "Dutch." The nickname would stick throughout his son's boyhood and sportscasting career, and was shed only when "Dutch" was deemed not quite right for a movie star. The first Reagan son, Neil, born two and a half years earlier, had been similarly dubbed "Moon" (for Moon Mullins, a popular comic strip character) by his father. Neil, an advertising executive, kept his nickname until his death in 1996. (The nicknames complicate references to Reagan, in that he was "Dutch," then "Ronnie" to his Hollywood friends, and "Ron" in his political years.)

Starting virtually from that snowy day, Nelle cast her second son as the genial, moralistic, optimistic character the American people would always love. It was not playacting; it was the essence of Ronald Reagan from the opening scene to the time the final curtain fell. As his director, Nelle did much more than coach him on his lines for the little morality plays she put on for the church and for his leading roles in high-school productions. She encouraged his desire to be in the public eye, built up his confidence as a leader. And she looked to him, both openly and subconsciously, to fulfill needs of her own, to be the actor within her, to be upstanding and achieve the middle-class respectability that meant so much to her. And she wanted him to be as charming as her husband, without his flaws. Dutch was her special child; he fulfilled her dreams.

Mulling over the traits he inherited from his parents, Ronald Reagan wrote in his autobiography, "I learned from my father the value of hard work and ambition, and maybe a little something about telling a story. From my mother, I learned the value of prayer, how to have dreams and believe I could make them come true." (Nelle's exhortation to follow his dreams underpinned his courage to pursue the presidency; Jack bequeathed him the leavening touch of the raconteur.) He pointed out another basic difference in the influence of his parents: "While my father was filled with dreams of making something of himself, she had a drive to help my brother and me make something of ourselves."

Nancy Reagan, his adored second wife and the only person Ronald Reagan loved as much as his mother, could see much of Nelle's influence on her son. "Ronnie is a great deal like his mother," she observed thoughtfully. "Her optimism—the notion that the glass is always half full—has always influenced him. She believed that there was a reason for anything that happened. If something went wrong, you didn't let it get you down. You may not know why now, but it will become evident at some point and it will make room for something wonderful to happen. This became Ronnie's philosophy too."

Nelle's optimism, Reagan once said, "ran as deep as the cosmos," and he happily laid claim to it. Now, twelve years since Reagan, the "willful optimist," left the White House, and eight years after Alzheimer's forced him to bow out of public life, his party longs to recapture the ebullient, Nelle-inspired confidence that fueled his success as president. Nelle was never to know her son's place in history, but his career was built on her solid values; she could not have foreseen the issues of his time, but her teaching led directly to the success of his presidency.

In her own story, *My Turn,* Nancy Reagan could joke about her husband's optimism during their White House years: "It can be difficult to live with somebody so relentlessly upbeat. There have been times when his optimism made me angry. . . . I longed for him to show at least a *little* anxiety. Ronnie doesn't worry at all. I seem to do the worrying for both of us.

"The place to begin in understanding Ronald Reagan," said his wife, who understands the forces that shaped her husband as no one else could, "is with his past—his roots, his parents, and the way he was raised." And that begins with his mother. Without money, without family background, without stability in her personal life, Nelle Wilson Reagan was every bit as influential in her son's life as Sara Delano Roosevelt, who had been blessed with an abundance of advantages, was in her Franklin's. "Nellie," the old-fashioned name recorded on her birth certificate, started life as an Illinois farm girl, but with her early flair for the theatrical and her studied self-improvement, she soon streamlined the name to "Nell" and dressed it up with an added *e*. And, in a style that was quite unusual for the early years of the century, when her sons came along she had them call her not mother or mama, but Nelle.

She was petite, blue-eyed, with hair the russet color of oak leaves in autumn. She had a knack for maximizing her good features and possessed a backbone of steel, which she needed to cope with the rootless life Jack Reagan imposed on his family. He was a shoe salesman—a "graduate practipedist," according to his correspondence-course diploma—a footless and feckless charmer, a raconteur with a million stories, a talker who was sure the main chance waited in the next town, a husband and father whose friendship with Demon Rum more than once cost him his job. And, beset by bad luck beyond his control, he was one of the millions of American workers who were trampled by the Depression.

It was the kind of life, constrained by a persistent lack of money, that kept possessions to a bare minimum. Time after time, Nelle, a clever seamstress, made and remade curtains for windows that were never the size of those in the last place, and if she started a little flower garden, she realistically planted only annuals. For young Dutch Reagan, this nomadic life meant four schools in his first four years of education; it meant pulling up stakes again just as he was beginning to make friends.

Nancy Reagan attributes a certain remoteness in her husband's personality to this family's transient life in his early years. "It's hard to make close friends or to put down roots when you're always moving," she observed. "And I think this explains why Ronnie became a loner. There's a wall around him." She added, surprisingly, "There are times when even I feel that barrier."

The wonder is that he turned out to be a sunny child who was popular in any town that offered Jack Reagan brighter prospects. But even though the view from the kitchen window changed with unsettling frequency, one element was constant: Nelle. Whatever the town, the house, the apartment, however brief the stay, she made a warm and secure home for her sons. Sitting around the kitchen table with her two boys after supper, Nelle read aloud to them, helped with their lessons, talked about what was going on in their lives, and offered a bit of Bible-based guidance. As one new place followed the last, the peripatetic life actually bound the family closer together in the reassuring familiarity of each other and their mother's unfailing positive thinking.

★   ★   ★

TEN times Nelle packed up the family and moved to a different home, in four small Illinois towns and once to Chicago. They moved five times in the town of Dixon alone, always downward to a smaller place, always rented, not owned. The Reagan family itinerary reads like the timetable of a local Illinois railway: all aboard for Tampico, Chicago, Galesburg, Monmouth, Tampico again, and all out for Dixon, the end of the line. And this timetable doesn't count two towns Nelle and Jack lived in before the boys were born and Springfield, a desperation move for Jack, a place to which, for once, his family did not follow him. Though all of those towns could claim to be "the boyhood home" of President Reagan, Dixon, in northwestern Illinois, would always be his hometown, in the record books and in his heart.

When the Reagans moved to Dixon, population just over eight thousand, in 1920, it proved to be the final stop on their itinerary. At long last they were transients no more. From the time Dutch was almost ten until he graduated from college, Dixon was home, and in his imagination it would always be. Many years later, when he was governor of California, he reflected, "All of us have to have a place to go back to. Dixon is that place for me."

Though he was a Californian from the time he was twenty-six, clinging to his midwestern roots he insisted, "Dixon is part of me," and was unmoved by Los Angeles dazzle. Beliefs that became the bedrock of his political convictions were shaped in that pleasant town, a farming community that was more nineteenth century than twentieth when the Reagans moved there. He always remembered Dixon as a place devoted to a simpler life and old-fashioned values, a place where people knew each other and, after the fashion of his mother, took care of each other. In his view, which was formed in boyhood and never darkened by adult experience that might have revealed the downside of small-town life, Dixon was the real America.

As the family embraced permanence, Nelle became more involved in amateur theatrical productions. Early in her adult life, the farm girl–milliner who married an alcoholic shoe salesman had discovered the magic of being onstage. Playing the leading lady in amateur theater productions and giving her dramatic readings to ladies' groups provided an escape from daily reality and the chance to be Somebody on her own. She was pleased when her second son, from his first moment onstage, followed that same impulse, making up little plays and acting in them as a small boy, then shining in high school and college productions. He was not a natural show-off; though he was a friendly kid, he tended to be shy, more reserved than his exuberant brother, Neil. Yet it was Ron who craved the spotlight.

Perhaps there's a biological explanation for wanting to become a make-believe character, for yearning to hear the applause that tells you that those people out front like you. The most recent advances in genetics research lend some weight to the hypothesis that an individual's genes account for quirks, talents, traits, and probably sexual orientation that distinguish the personality. In *Living with Our Genes,* geneticist Dean Hamer maintains that "you have about as much choice in some aspects of your personality as you do in the shape of your nose or the size of your feet." Ronald Reagan's dominant genes were clearly handed down from his mother.

From the time he was in high school, the young Reagan showed a flair for being out front. In small-town America, then as now, the football field was the place a boy could win glory. Though Dutch was too skinny and nearsighted to do much more than warm the bench, he played with heart, but he stepped out front—literally—

as the drum major strutting at the head of the YMCA band. All eyes would be on him, especially the time when, concentrating on twirling his baton, he marched straight ahead while the rest of the parade turned left. (Some might see that as a political parable.) By the time he was in college, Ronald Reagan had picked up the necessary brawn to shine on the football field, to be cheered for his leadership and true grit—and he relished it.

More important, more admirable than being a star in a ball game, was the hero role he carved out for himself. It would be a defining experience in his life. In the summer of 1926, when he was only fifteen, Dutch Reagan, a strong swimmer, was hired as the first lifeguard at Dixon's Lowell Park beach, a sandy strip on the treacherous Rock River where drownings had shocked the town. For seven summers his role was emblazoned across his chest: "Life Guard." He should have added an exclamation mark, for sixty years later the president's face still lit up as he recounted his days as the protector of the weak or foolish at Lowell Park.

On steamy hot afternoons Nelle put aside her housework to spend a little time at the beach. She was past forty, well into what was considered middle age in those days, and still as slender as a girl. More a shade lover than sun worshipper, she preferred a pretty frock to a swimming costume; she'd rather picnic with friends than brave the river. She basked in the praise lavished on her Dutch, a lithe young man with the well-muscled shoulders of a powerful swimmer, and popular with all ages.

In those seasons as a real-life hero, the young lifeguard pulled a total of seventy-seven endangered swimmers to safety and recorded each rescue by cutting a notch on a log. Some of them insisted that they were not drowning and had not wanted to be pulled out, but he rescued them anyway—and cut another notch. Throughout his life, Reagan would tell how he "saved the lives" of seventy-seven people. Enough were real to make him the town hero, and a glowing front-page article in the Dixon newspaper gave him his first taste of public acclaim. It was the first of thousands of Ronald Reagan stories that were to appear on front pages in every language, in every country, more than half a century later.

Reagan's desire to play heroic roles was gratified when he became a full-fledged actor in Hollywood. He was the white-hat cowboy, the courageous pilot, the intrepid Secret Service agent, the ordinary fellow standing up to the forces of evil—always winning. Nelle approved of his good-guy roles, even when they were only second billing, because they seemed so natural for him. Her son's roles confirmed Nelle's unshakable conviction that by the third act right always triumphed. When Reagan took on those roles in real life, battling communism, first in the film industry and ultimately in geopolitics, it was a case of life imitating art. He cast himself as the unintimidated marshal from the West, protecting the town from evil gunslingers, drawing a line in the sand that the enemy would not be allowed to cross. And in his climactic scene, when he issued his dramatic challenge in Berlin—"Mr. Gorbachev, tear down this wall!"—the West cheered like an audience in a great theater. Indeed, his life *was* great theater. Had she been alive, Nelle would have led a standing ovation, confident that the rightness of his course would lead to the happy ending, which, in fact, it did.

* * *

CROWDS lined the sidewalks of Dixon on a beautiful September day in 1941, waving and shouting, "Hi, Dutch!" as the motorcade crawled along the main street. A smiling Reagan—now billed as "Ronald" for the movies—waved in return and the woman at his side in the convertible, his mother, smiled and waved, too. Reagan's car was not first in the line-up; that honor belonged to Dixon-born Louella Parsons, chief dragon of Hollywood gossip columnists, a movieland power who could make or break a young actor. The celebration had been organized as "Louella Parsons' Day," but everybody in Dixon knew it was really all about their Dutch Reagan, the hometown boy who had starred on their own stage, who had rescued some of the very folks cheering him, whose voice and imagination on Radio WHO in Des Moines had brought sports to life for listeners all over the region, and who was making his mark in Hollywood. Some of the glow that lighted up the faces in the crowd was directed at Nelle—in that small town her works of charity had touched many lives.

They had traveled to Dixon in a special railroad car filled with Hollywood celebrities. Ron had declined to join the others at the club car bar, instead staying with Nelle in his compartment, and it is safe to assume that they were the only two in the group who read the Bible and joined in prayer as the train clacked east from California.

It was serendipity that Ron's wife at the time, actress Jane Wyman, was unable to make the Dixon event; her absence gave Nelle her one moment to bask in the reflected light of her son's celebrity. At the gala premiere of Reagan's new movie, *International Squadron,* the centerpiece of the weekend, Nelle was in her glory. Radiant in a new evening gown, she glided over the red carpet on the arm of her son, at ease amid popping flashbulbs and fans pressing against the ropes. It was a heady moment, a far cry from little Dutch in her church plays and their joint appearances wherever a group invited them. She had showed him the magic of pretending, she had sent him on his way, and now he thrust her into the spotlight, the moment every secret actress longs for, and— more important to her—her son's accomplishments had given her a status that her husband could never provide.

For Ron, there was the odd feeling of sitting in the familiar old theater he knew so well and seeing on the screen not Tom Mix but Ronald Reagan. Nelle had first experienced just such a thrill in 1937, when his first movie, an infinitely forgettable effort called *Love Is on the Air,* in which he was typecast as a sports announcer, opened (and closed) with a special screening in Des Moines, the home of station WHO, his launching pad. Watching the film, Nelle dabbed at her tears with her lace-edged handkerchief.

"That's my boy," she declared proudly to a Des Moines reporter. "That's my Dutch! He's just as natural as can be. He's no Robert Taylor. He's just himself." Nelle had succinctly summed up the special magic of Ronald Reagan as actor and president: he was as natural as can be, he was just himself.

Reagan candidly called that first movie "a turkey," but Warner Brothers saw sufficient promise in the sportscaster-turned-actor to put him under contract. Empowered by his new sense of security, he insisted that Nelle and Jack move to Los Angeles and sent train tickets. Nelle was briefly torn by the thought of leaving Dixon, their home for seventeen years, the only town where she had felt accepted and appreciated. Nelle's friends in the True Blue Bible Class were distressed: How could she leave them and Dixon? Her reason was simple: "He said, 'I need you.'" Probably not literally true, and probably she believed it. Ron knew they needed help—a heart attack had left Jack unable to work—and his explanation would convince his mother that she and Jack could be a boost, not a burden, to their rising-star son.

So in 1938 they packed up their belongings and boarded the train for Los Angeles. Not only had Ron rented an apartment for them near his own, he invented a no-stress job for Jack—handling his fan mail—which soon turned into a real job as the volume of letters burgeoned. When Warner Brothers raised his salary, he surprised his parents with a dream gift—a two-bedroom house, in the popular California Spanish adobe style, complete with furniture and the deed of ownership. In nearly forty years of marriage, it was the first home they had ever owned. For the postman, the address was Phyllis Avenue, West Hollywood, but for Nelle and Jack it was Easy Street.

Though leaving Dixon had been tinged with uncertainties and the sadness of leaving old friends, Nelle's new life in Hollywood quickly eased the pain. But even in this glossy new world, she did not forget her priorities: she immediately transferred her church membership to the Hollywood–Beverly Hills Christian Church and resumed her works of charity. She wrapped five hundred Christmas presents for the poor and made regular rounds to Olive View sanatorium, spreading cheer to tuberculosis patients confined to bed for months, even years. Sometimes she gave her dramatic readings. In weekly visits to the county jail, she took the bright ray of her friendship, giving the inmates a touch of mother, an apple, and a lesson from the Bible. Nelle's good works—activities that would come to be known as volunteerism—made a deep impression on her son. As president, he deplored the mind-set that expected the government to take care of people in need; he advocated personal charity, not welfare. People should give and do for others the way his mother had always done, even when she had little. But Nelles, he would have to acknowledge, are in short supply in today's big-city world.

Quite apart from her Christian labors, life in Hollywood offered diversions that Dixon couldn't match. Now and then, when Ron was making a film, she visited the Warner Brothers set in Burbank, and every time she was waved through the gate the would-be actress felt a thrill. The crews greeted her warmly: Nelle Reagan was a friendly lady who stayed quietly out of the way as she watched her son perform before the cameras, exercising the talent she had first tapped in her little productions back in Illinois. It had brought him—and her—a long way from the deprivations the family had endured.

It is not the usual thing for a high-flying, handsome young bachelor on his way to marquee billing to spend a lot of his time with his parents, but Ronnie did just that. When he wasn't locked into a filming schedule, he dropped by to see them every day, and when he was working he telephoned every day and took them out to dinner every Sunday. Even the most demanding of mothers (which Nelle was not) could not ask for a more attentive son.

It is hard to imagine a closer and more lasting relationship between son and mother than that binding Ron to Nelle throughout their lives. Franklin Roosevelt and the redoubtable Sara, perhaps, but there was an undercurrent of tension; Lyndon Johnson and his adored Rebekah, perhaps, but the distance of half a continent minimized irritations that can strain a relationship. Reagan, in his days as a sportscaster—he was a legend wherever Radio WHO could be tuned in—invited Nelle to Des Moines so often that she became a favorite among his friends, something of a surrogate mother.

Ironically, his daily visits and telephone calls made letters unnecessary, leaving a void in the intimate details that give a fuller picture of the family's life and relationships. Harry Truman, whose devotion to his mother closely parallels that of Ron to Nelle, wrote chatty letters to Martha Young Truman at least three times a week—and she saved them all. Lyndon's many letters to his mother, Rebekah, from college and from Washington, offer a valuable insight into a complicated man. If Ron wrote during those few years away at college and in Des Moines, the letters apparently were not saved—a transient life discourages archival conservation. And because Nelle died years before Reagan entered politics, there are no interviews or feature stories about her, no quotes from her about her son the candidate, and, of course, no autobiography like the engrossing life stories written by five of these mothers of the eleven modern presidents. Nelle is less known, but no less crucial in the shaping of her presidential son.

The Reagan presidential library has virtually no original material from Ronald Reagan's mother, which makes the recent emergence of Lorraine Makler Wagner of Philadelphia significant. In 1943 Lorraine, then thirteen, was captivated by Ronald Reagan in the movie *Brother Rat*. She wrote him an enthusiastic letter asking for a photograph and in return received not only an autographed studio glamor-shot but a personal note. It was the first of 275 letters that Reagan wrote to her through his Hollywood years, his governorship, his White House years, and into retirement. In 1995 she received one last note: in it he told her that he had been diagnosed with Alzheimer's disease.

Lorraine, who not surprisingly had become president of the Ronald Reagan fan club in Philadelphia, was invited to Dixon in 1950 to attend the weeklong "Injun Summer Days"—starring her idol—in Dixon, a replay of Reagan's triumphant visit to his hometown in 1941. She found herself included in Reagan's personal party, sharing head tables with him and the VIPs, riding in a convertible in the parade. Out of this weekend came a lifelong relationship between Lorraine and Nelle Reagan, who appointed herself "Moms Nelle" for all the Reagan fan clubs in the country.

In Dixon, Nelle confided her fondest dream to Lorraine—"I want him to dance with me"—and stuck her with the job of persuading Ron to do it. He didn't feel comfortable with the idea, but finally agreed—"All right, Nellie, come on"—and they danced to the music of Lawrence Welk. Nelle was in heaven. Then Reagan gave Lorraine the task of making sure that his mother's little speech about her illustrious son would be brief. "I don't care how you do it," he instructed Lorraine, "but get her off the stage!" (It takes an actor to know the lure of the spotlight.) Lorraine was able to persuade Nelle, tactfully, and all three remained friends.

On a visit to Los Angeles the following year, Lorraine was met at the airport by "Moms," who invited the Number One Fan, along with four fellow Reagan aficionadas, to lunch at Ron's ranch at Malibu. There Lorraine met Nancy, not yet Mrs. Reagan, who, over the years, also wrote the occasional note. Lorraine's memories of "Moms" are glowing: "Nelle was marvelously articulate, a warm, gracious person, wonderful to listen to." It was clear that Nelle had not "gone Hollywood." "She went on at length about their poor background. She remembered how people would come to the house when she was out—in those days doors were never locked—and leave something on the table, some special dish, never saying who did it. People helping each other. She was still grateful."

For about eight years, until she lost touch with the world, Nelle wrote motherly letters to Lorraine and sent little gifts—a pretty container of powder puffs and a letter opener—still cherished by Lorraine, who is now retired with her husband, Elwood. In her letters Nelle shared confidences and personal philosophy. "All we can do," she wrote, "is thank God for each new day as it comes, and live it as though it were to be our last—'pleasing in God's sight'—and know that if we do, we need not fear when our time comes."

When Jane Wyman, Ron's first wife, divorced him, Nelle, who had always opposed divorce, expressed dismay: "I pray God she will not marry [again], it's such a sin to do [it] this way." In a letter to Lorraine and other fan club presidents in 1953 Nelle thanked them for their loyalty to her son at a time when his career was in the doldrums: "I love you all so dearly and will never forget you for all you did for Ronald and for being so grand to his old Mom."

★   ★   ★

AT Barney's Beanery, the clubby gathering place for Ron and his pals, most of whom were transplanted from Iowa, Nelle and Jack were regular members of the crowd. Carving out a spot for herself in the midst of the young people, Nelle re-created the den mother role she played with Ron's friends back in Des Moines. At Barney's, Ron said, Nelle and Jack "were in and out more than I was, and I think Nelle would have given someone an argument if he pointed out she hadn't really given birth to the whole gang."

The future president's thoughtful concern for his parents was one of the attributes that first attracted a sparkling young actress named Jane Wyman. Like Ronald Reagan, she had grown up in a middle-class midwestern home, though hers lacked the love that flowed from Nelle. "Ronnie had this wonderful relationship with his mother," she said many years ago. "I sensed it. I wanted to have a part of it." Jane, who had chafed under the rigid rules of her "claustrophobic" home life, noted that Nelle had never set such rules for Ronnie—yet he chose to remain close to his mother while she had contemplated running away from home to escape hers.

She felt that she had become a permanent "part of it" when she and Ronnie were married on January 26, 1940, at the Wee Kirk O'Heather Church, followed by a reception given by no less than Louella Parsons. Jane was the perfect Hollywood bride in a shimmering blue dress and mink pillbox hat, with orchids pinned to her mink muff, and Ron was the handsome, perfect Hollywood groom. The new bride and her mother-in-law quickly became warm friends; Jane was the daughter Nelle never had, and Nelle was the mother Jane wished she'd had. To Nelle's delight Jane went to church with her and even taught Sunday school.

Best of all, the marriage brought grandchildren who filled the loneliness created by Jack's death in 1941. First came Maureen and later Michael, who was adopted. Maureen always remembered "Gramsie" with deep affection. She stayed overnight with Gramsie on Saturdays—Nelle's way of spending what is now called "quality time" with her granddaughter and at the same time making sure she would get to Sunday school on time. Gramsie cooked their favorite dishes, and the two of them ate at the kitchen table. Later, snuggled in bed, they tuned in to Nelle's favorite radio programs, *Gangbusters* and *Inner Sanctum*. "Gramsie filled much of the void left by my parents' careers," Maureen recalled fondly in her book about the family. "She was a constant in our young lives . . . a steadying influence in mine."

But eight years after their star-crossed marriage, Jane filed for divorce. Reagan was stunned, unaware that things were going wrong (as he had been years before, when his college sweetheart returned his engagement ring), and Nelle, who had weathered rough seas in her own marriage, was distressed to see the family break up and her son deeply hurt. According to court records of the uncontested divorce, based on "mental cruelty," Jane charged that her husband was more interested in politics than he was in her. And her movie career was going much better than his. In this era of baring all secrets and cashing in on personal contacts, Jane Wyman, now in her mideighties, has steadfastly declined to talk about her marriage to the actor who was to become the fortieth president of the United States. (She did once comment on Ronnie's ability to be all things to all men. "Don't ask him what time it is," she warned, "because he will tell you how to make a watch. And he doesn't know the first thing about how to make a watch.")

Several years later, Nelle's generation-spanning appeal proved itself again, when Ron married Nancy Davis, an aspiring actress with the eyes of a fawn wide set in a face of pixie beauty. Unlike Jane Wyman, Nancy chucked her career to devote all of her time to her husband, which became a career in itself. She liked her mother-in-law, admired her generous spirit and the values she had imparted to her son, and appreciated her personal charm. "Nelle was very easy to get along with," said the former First Lady. "She got along well with our friends, and everyone else, to the point of asking a complete stranger who came to the door to come in and have a cup of coffee!" Nelle saw no need to abandon her small-town hospitality just because she was living in Hollywood. After all, didn't most folks there come from small towns, too?

★　★　★

HOLLYWOOD and Easy Street were beyond Nelle's imagination in the hard years, which were, in fact, most of the years in the Reagan family story. On Saturday mornings, in the lean times when the Reagans were living on the South Side of Chicago, Nelle called seven-year-old Neil into the kitchen and carefully repeated her weekly instructions. "Here's ten cents," she said to him, placing a dime firmly in his small hand. "Tell the butcher we want a soup bone—and some liver. . . ." Neil, who was almost seven, knew just what to say: ". . . and some liver for the cat." The Reagan cat was purebred Cheshire, imaginary from the top of his orange tabby head to the tip of his gray Persian tail, but his imaginary appetite for liver made him a very useful invention for putting food on the table.

Off Neil trotted to the meat market on Cottage Grove Avenue at Sixty-third Street, where the accommodating butcher picked out a good soup bone and wrapped a nice piece of liver for the "cat." Years later, Neil, who became a successful advertising executive, recounted his Saturday missions to Reagan biographer Lou Cannon: "Our big meal on Sunday was always fried liver. We ate on the soup bone all the rest of the week. My mother would put it in the pot and keep on adding potatoes and slices of carrots and more water, and we ate on the soup bone until it was Saturday again." Nelle scrimped in every way she could devise, reusing, remaking, sewing neat patches on threadbare pants, putting cardboard in shoes with soles worn thin, stretching a bit of ground meat with oatmeal, disguised—hopefully—with gravy. But she was never long-faced about her legerdemain. "Monkey" Winchell, a boyhood pal of the Reagan brothers, never forgot how Moon and Dutch always looked nicer than the other kids: "We were poor folks," he told biographer Anne Edwards, "but they were always dressed clean, not raggedy."

The two brothers' memories of their childhood revealed fundamental differences in their personalities, one bathed in sunshine, the other often clouded. With a clarity that still pinched, Neil described a much harsher life than the picture his brother had stored in his rather selective memory. "We were poor," said the older brother. "And I mean poor." Dutch, on the other hand, described his boyhood as "one of those rare Huck Finn–Tom Sawyer idylls," a montage of roaming the hills in summer and sledding down the slopes in winter, "of woods and mysteries and a clear-bottomed creek," of rollicking neighborhood games and the grand Armistice Day celebration when he was seven. He could bury the disappointment that he couldn't join the Boy Scouts—he was a natural to achieve Eagle Scout rank—because his parents could not afford the scout uniform. Dutch had learned from his mother to remember only the good things about life; he saw the blue sky, Neil saw the cloud.

When Ronald Reagan was seventy-seven and winding up his second term as president, once again he reminisced, in the husky voice that added effect to the recollection, about his blissful small-town boyhood in Illinois. He was nostalgic for an idealized America that had long since vanished and, even at its best, excluded entire sectors of the population and allowed many desperate Americans to fall through the cracks. In his swan song at the GOP nominating convention of 1988, the president told the adoring crowd, "I confess there are times when I feel like I'm still little Dutch Reagan, racing my brother down the hill to the swimming hole under the railroad bridge over the Rock River." Nor was this the product of a rearview mirror memory in which objects are seen as rosier than they were, the compensation of a lifetime of success. As a senior at North Dixon High, seventeen-year-old Reagan wrote a poem for the yearbook, an indication of the optimism that would be the key to the man, to his popular appeal, throughout his life: "Life is just one grand sweet song, so start the music."

From childhood on, Reagan profited from his mother's example of positive thinking, of rejecting all that was limiting or disheartening. "Ronnie remains hopeful even in the worst of times," Nancy wrote, partly in admiration, partly in wifely exasperation, in their final year in the White House. "Depressed? He doesn't know the meaning of the word. He is not impervious to events, but he is very resilient." She attributed this turn of mind to Nelle's unshakable oft-stated conviction that "everything happens for a reason" and her faith in people. "Nelle never saw anything evil in another human being, and Ronnie is the same way. Sometimes it infuriates me, but that's how he is."

While the family's marginal existence did not sour Ronnie, his brother, growing up in the same circumstances, became a more sardonic, darker humored man, the son much more like his father than his unfailingly optimistic mother. Pressed hard, Reagan later commented, "We were poor, but we didn't know we were poor." (Truman, Eisenhower, and Nixon all said the same, in virtually the same words. Men who have achieved great success tend to recast unpleasant memories in a positive light.)

There were interludes when life was not bleak for the Reagans. When things were going well at the Fashion Boot Shop, a smart little emporium on Dixon's main street where Jack was a partner, they could enjoy real beef instead of "oatmeal meat" and they even bought a new Chevrolet, their only new car in what Reagan remembered as "a lifetime of buying second hand."

The house that was to become the official boyhood home, at 816 South Hennepin Avenue, was the nicest of their many, a pleasant, two-story frame dwelling built in 1891 and typical of the time, boasting a gable ornamented with patterned shingles and a spacious front porch. Though there were three bedrooms, the two boys shared one room and one bed to give Nelle the third one for her workroom, which was most useful when she took in sewing for a little extra money.

At the back of the property was a structure that had been a barn or stables, which the Reagan boys made into headquarters for such enterprises as raising rabbits, an endeavor that ended when Dutch decided he didn't like the killing part. Nelle willingly relinquished the side yard to rough-and-tumble football games that usually ended in a raucous pile-on that Dutch called the best part of the game. On Saturday mornings, when there were two dimes to spare, they would go to the picture show, cheering Tom Mix and William S. Hart, the good guys, and hissing the bad guys. On the surface their life could have been the model for a *Saturday Evening Post* cover by Norman Rockwell.

But the Reagans, with a now-and-then successful shoe salesman as head of the family, were most often holding on by the fingernails to a cracked rung of the middle-class ladder; while Jack was always a white-collar worker, the collar was frayed. President

Reagan, with his becoming quality of making light of the rough patches of his life, used to joke, "We didn't live on the wrong side of the track, but we lived so close we could hear the whistle real loud."

They clung to their middle-class standards when they could barely put food on the table, and through it all an intrepid Nelle kept up a good front, trusting that the hard times were only a passing rainstorm, the clouds would soon clear away. She never lost faith that "the Lord will provide." While the younger son absorbed without question her undaunted—and unrealistic—optimism, the older son, given to skepticism and well aware of the family's dire straits, recognized wishful thinking when he saw it.

★ ★ ★

NELLE had one dream above all: her sons would go to college. Both Jack and Nelle had left school after the sixth grade to help their families, as many children did in those days. But Nelle had a motto—"To higher, nobler things my mind is bent"—and an inborn love of books, poetry, and, most of all, the dramatic arts. Somehow, Nelle vowed, their boys would take that great American step up to an assured life, to success. She began getting them ready when they were scarcely more than toddlers. Each night the three of them piled up in bed, Dutch and Moon cuddled close to her as she read aloud. The younger son followed the words with his finger and then came the day, when he was five, and "all the funny black marks on paper clicked into place." Ecstatic at the achievement—hers as well as his—Nelle sent him scurrying to demonstrate for his father, who was skeptical until little Dutch triumphantly read "the marks." Nelle, like Rebekah Baines Johnson, had given her son a vital early advantage in life.

In a few years he was reading omnivorously, less captivated by such historic figures as Hannibal and Caesar, who fascinated Eisenhower and Truman, than by Edgar Rice Burroughs's Tarzan, Jack London's adventurous heroes, and the irresistible Huckleberry Finn. There was an overarching theme in the stories his mother read to him and the books he read for himself, an outlook Reagan later summed up as "an abiding belief in the triumph of good over evil." Nelle was encouraging him to think in big, universal terms, implanting the belief that there were indeed these two great forces at work and that a hero would not stand by passively, but must fight for all that was good and oppose that which was not. Reagan's battle with "the evil empire" of communism was stamped by the same die, and his imagination, always lively, was stirred by the then-new realm of science fiction, stories of rockets and space travel, of other worlds and alien beings, tales that triggered his lifelong interest in space.

For Nelle, teaching the boys to read was only the beginning of her program to broaden their horizons. If there was a play or a band concert in a park, she and Jack, who shared her love of music, took the boys. Nelle and Dutch enjoyed special interests of their own, in particular the traveling Chautauqua program, a uniquely American movement that combined inspiration, practical education, and family entertainment for small-town folk bent on self-improvement. (A creation of the nineteenth century, Chautauqua continues, considerably changed, at its original grounds in upstate New York.) In Chautauqua week, local children had their first stage experience in educational playlets or playing simple instruments in the tiny tots' orchestra, and parents shared in this menu of learning and moral guidance. It was a way for small-town folks to stretch their minds and better themselves and their children. Nelle exposed her sons to the finer things and showed them the avenue to a life that would always be beyond her reach. She willed her boys to success.

But how could they ever manage college? In 1926 Neil, about to graduate from high school, declared that he would not attend the events, feigning indifference to conceal the real reason: the boys were supposed to wear tuxedos to both the prom and the commencement, and that cost money his father didn't have. Jack was generally indifferent to his sons' activities (a pattern Ronald Reagan would repeat), but he couldn't let his Moon, the first of the Reagans to finish high school, miss that milestone in his young life. Somehow he scrounged up enough cash or credit to rent the tux, and Neil, resplendent in black tie, danced at the prom and marched across the stage to receive his diploma with the others. Long after he had become an affluent executive, Neil spoke gratefully of his father's understanding and concern: "This was a sacrifice for my dad." He had always felt closer to his father than to his mother, shared his Irishness, his skepticism, his features, and his religion, and the commencement tuxedo episode lingered as a reminder that his father's good heart made up for his shortcomings.

Neil didn't want to go to college—he had a job at a cement plant that paid regular wages—but his kid brother had enough determination for the two of them. Ron virtually dragooned his older brother into joining him at Eureka College, a small liberal arts school of 220 students operated by Nelle's Church of Christ denomination. Dutch lined up a campus job, a scholarship, and even fraternity membership for his brother, just as he had done for himself the year before. With all of that, Neil acquiesced. It's safe to say that his successful career in advertising was possible only because of his brother's persistence; otherwise he might have risen to superintendent at the cement plant.

Nelle was thrilled that both sons were in college, proud that both of them had been awarded scholarships—partial scholarships that came not as a result of precocious reading and exposure to culture, it must be noted, but for their ability on the football field and swimming team. Dutch, with game-day cheers ringing in his ears and his first love, Margaret Cleaver, the minister's daughter, looking into his eyes, didn't bother much with studying, but he was into everything else on campus—sports, dramatics, college politics, every big event. Fortunately, he was also a first-rate crammer, a talent that squeaked him through exams and kept him eligible for his scholarship.

Some years in Reagan's boyhood actually were idyllic, yet anxiety about Jack's alcoholic binges made life uncertain and distressed Nelle, who tried to keep unpleasant matters from her boys. Dutch became aware of arguments in the night, of absences and a smog of worry that sometimes settled over the house. In his autobiography, *Where's the Rest of Me?*, Reagan, baring all as he entered the fray of California politics, related in wrenching detail an episode that took place when he was a boy of eleven: "I came home to find my father flat on his back on the front porch and no one there to lend a hand but me. He was drunk, dead to the world. . . . I wanted to let myself in the house and go to bed and pretend he wasn't there. . . . I felt myself fill with grief for my father at the same time I was feeling sorry for myself. Seeing his arms spread out as if he were crucified—as indeed he was—his hair soaked with melting snow, I could feel no resentment against him. . . . I managed to drag him inside and get him to bed."

Though those moments in his impressionable growing-up years were painful, the effect on him could have been far worse. Had it not been for Nelle's understanding and remarkably advanced thinking, he might have become bitter, rejecting his father. "With all the tragedy that was hers because of his occasional bouts with the dark demon in the bottle," Reagan wrote, "she told Neil and myself over and over that alcoholism was a sickness—that we should love and help our father and never condemn him for something that was beyond his control. . . . If he ever embarrassed us, she said we should remember how kind and loving he was when he wasn't affected by drink." Her explanation would help the boys accept their father's lost jobs and erratic behavior, even the time he drank up all the money Nelle had set aside for Christmas presents.

For a small-town wife with aspirations, a resolute teetotaler subjected to humiliation and fear of what might come next, to ask her sons "to help him and love him" was indeed an enlightened attitude in the 1920s, one that benefited both husband and sons. But Reagan also remembered "occasional absences [and] the loud voices in the night," suggesting that Nelle, despite the understanding she asked of the boys, was sometimes distraught beyond the limits of her own compassion. Yet she would never reveal those feelings or show her sons her heartbreak; she put on her cheerful face and reassured herself once again that there was a purpose in everything in life and things always work out for the best. But even Pollyanna herself couldn't have anticipated just how rosy life would someday become.

★   ★   ★

NEARSIGHTED as he was, Dutch, home from college for the summer, could see that the man sound asleep in Nelle's sewing room was a total stranger. Then he realized it was one of Nelle's wounded sparrows, a prisoner out on parole who was being eased back into life on the outside by the very personal assistance of Nelle Reagan. Welcoming a prisoner into her home was one of her many acts of mercy, her personal way of carrying out the Bible's mandate to comfort the afflicted. Nelle's faith was deep and abiding, her refuge and her strength in a life marred by her husband's alcoholism and the ever-present threat of hard times. While she never doubted the power of prayer, she also believed in the need for action. As her sons grew up, she impressed upon them the importance of helping people in trouble, and she did more than mouth platitudes—she was a hands-on good samaritan who set an example by doing. She literally followed Jesus' exhortation, as related by Matthew, to visit the imprisoned.

Nelle had been brought up in a staunchly religious home dominated by her mother, who, as an orphan of sixteen, had come from England to the American Midwest to work as a housemaid. She set a pattern of much churchgoing and serious Bible study. Shockingly, her husband, Thomas Wilson, a hitherto conventional second-generation Scotsman, walked out on the family when his youngest child, Nelle, was seven and was not heard from for a number of years.

At length Thomas Wilson returned home, in time to object to his daughter's intended marriage to John Edward Reagan, Irish and Catholic. Nelle could hardly have felt close to her father, given his long "sabbatical" from his family, and though she was usually a respectful girl, she defied his parental edict. She and Jack were married in the rectory of Immaculate Conception Church in Fulton, Illinois. If Nelle pledged, as was required, to bring up the children in the Catholic faith, she soon forgot it, and Jack apparently did not make it an issue.

Jack Reagan embodied both the charm and the thread of melancholy typical of the Irish—and the tendency to take refuge in drink as a way of coping with an inner sadness and a life marred by disappointments. He was proud of his Irish blood, and had he known then what British genealogists searched out after Ronald Reagan became president, he would have been even prouder: the roots were traced back to the brother of Brian Boru, an eleventh-century king who was the first Irish national hero and is claimed as an ancestor by a vast number of Irish Americans.

The blue in the blood faded as the centuries slipped by, and the 1800s found the O'Regans living in abject poverty in the dismal Tipperary village of Doolis, a place of dirt-floored cottages—huts—where, to guard against thieves, the animals were brought inside at night. Small wonder that Michael O'Regan, a laborer in some better-off man's potato fields, ran away with a local lass to London, where they were wed, in the process dropping the O and adding an a. The family managed to scrape together enough money to come to America, the land of hope, in 1850, homesteading in Illinois. Michael begot John and John begot John Edward, who was always known as Jack. In four generations the O'Regans went from stone cottage and dirt floors shared with the livestock to the White House. Only in America.

Jack Reagan's mother, Jenny Cusick, had respect for education and had taught him to read before she died when he was six, but that special attention ended when he went to live with his aunt and her much-older husband. In the absence of encouragement, he left school after the sixth grade and went to work in the uncle's general store. There he met and courted a pretty young lady making fancy hats in the milliner's shop.

Opposites attract, they say, and these young newlyweds—Jack, not yet twenty-two and Nelle, barely twenty—were indeed opposites. Their younger son once cataloged their differences: "If my father was Catholic, my mother was Protestant. If he rebelled against the universe, she was a natural practical do-gooder. If he was Irish, she was Scots-English. If he was occasionally vulgar, she tried to raise the tone of the family. Perhaps she never understood the reason for his week-long benders once or twice a year, any more than he understood her cultural activities, but they put up with each other." His choice of "put up with" was significant, for Nelle's close friends knew the marriage was rocky and advised her to leave him once the boys were away at college. But neither partner believed in divorce and so they stuck it out, settling into an accommodation.

★ ★ ★

IT seems odd that Ronald Reagan, the Republican president whose ship of state steered hard right under conservative colors, first became interested in politics as a strong supporter of FDR, the Republicans' nemesis. The whole Reagan family was staunchly Democratic, including Nelle, who saw Roosevelt as the savior of the little man beset by forces beyond his control. Jack Reagan was a Democrat right down to his bought-at-cost shoes, a no-apologies Democrat in a bastion of Midwest Republicanism. He had enthusiastically supported fellow Irish Catholic Al Smith, governor of New York, in the 1928 election and lamented the landslide victory of Herbert Hoover, a cautious midwesterner of high ideals and little charisma. With notable prescience, Jack dismissed Hoover's acceptance speech at the Republican convention as sheer malarkey: "We in America are nearer the final triumph over poverty than ever before in the history of any land," the nominee intoned. "The poorhouse is vanishing from among us. We have not yet reached the goal, but, given a chance to go forward with the policies of the last eight years, we shall soon, with the help of God, be in sight of the day when poverty will be banished from this nation."

Scarcely a year later, the bottom dropped out of the American economy; yet eight months after that President Hoover, in an exercise of whistling past the graveyard, pronounced the Depression over, as if by fiat. His ill-founded optimism did nothing to relieve the country's anxieties. It was a time of pervasive, gnawing fear. The vibrant, made-in-America democracy that had been a beacon to people everywhere seemed unable to rekindle its flame, and the future that had always been the repository of promise now seemed foreboding, empty of hope.

The fear first struck Jack Reagan when the Fashion Boot Shop in Dixon was forced to close its doors. With his experience and charm he landed a job with another shop, but the Depression was stalking all trade in those days. He was forced to try to eke out a living as a traveling salesman; then he was hired as manager of a down-market chain store, selling "bargain" shoes that pinched the feet and split in the rain, a sad contrast to the stylish footwear of the Fashion Boot Shop. The worst of it was that the store was two hundred miles away in Springfield, which split the family, but a job was a job as the Depression gathered strength like a hurricane heading toward landfall. Nelle found a job at a dress shop, selling and altering ladies' clothes for fourteen dollars a week.

As 1931 came to a grim close, Dutch, home for the holidays, put on his good suit and slicked his hair, headed for a Christmas Eve date with his one-and-only. There was an unexpected knock on the door and a special-delivery envelope for Jack, containing not a Christmas greeting or a little bonus check but a blue slip with a black message: the store in Springfield had closed. Jack Reagan was out of a job, at the end of the road. At that moment the Depression literally hit home for the Reagans, Nelle most of all. Now Jack was one of the thirteen million unemployed Americans—one third of the workforce—and the lucky ones were scrambling for whatever jobs they could turn up. One family in seven was on either public or private relief, a recourse Jack considered a humiliation that he would not inflict on his family.

Desperate times call for desperate measures, and the Reagans did the only thing they could do—they rented out all but one room of their apartment, and hung on by sheer endurance and a hot plate. Dutch sent home part of his own meager pay each week, and some of Nelle's good works came back like bread upon the waters as old friends brought in food when their credit was cut off at the corner grocery. In despair, Nelle wrote Dutch—without telling Jack—explaining their crisis, and somehow he managed to come up with fifty dollars so they could buy food, and Neil added his share.

In 1932 the country voted for a radical new direction with the election of Franklin Delano Roosevelt, a watershed in American political history, but by Inauguration Day, March 3, 1933, more than fifty-five hundred banks had closed, wiping out almost $3.5 billion, the savings of millions of Americans who had been taught to be thrifty. But the Reagans, and countless Americans like them, found reassurance in Roosevelt's confident voice on the radio, speaking directly to them, proposing an array of programs to get the country started again, programs he called "a new deal."

The Democratic victory brought Jack a personal new deal. For his hard work in FDR's campaign he was made head of the welfare office in Dixon, then head of the local Works Progress Administration. (The WPA, which put unemployed people to work on public works projects, was a favorite target of Roosevelt critics, the kind of government program President Reagan would assail.) Democratic patronage also extended to a job for Neil. In the simplest terms, it saved their lives, and Jack Reagan remained a Democrat for the rest of his days. Ron would always admire FDR and patterned his rhetoric after the master's, but somewhere along the way he took a distinctly different political turn.

The Reagan right turn began with the Eisenhower race in 1952 and hardened into total commitment by 1964. Reagan played a supporting role to Barry Goldwater, "Mr. Conservative," in his race against Lyndon Johnson, which was the most ideologically polarized campaign of the era. This may seem perplexing in a man whose first brush with politics was to organize a successful student strike at quiet little Eureka College, but it was consistent with the instinctive conservatism that underpinned his mother's Democratic compassion. She was always guided by the old values and time-honored standards, and she firmly implanted them in her sons.

★   ★   ★

IN the year before her second son was born, Nelle became deeply committed to religion through the Disciples of Christ and was baptized by immersion. In 1832 the Disciples, now known as the Church of Christ, an offshoot of the Presbyterian church, had emerged as a specific denomination, an indigenous American church that rapidly flourished, particularly in the Midwest. It was a denomination singularly suited to Nelle Reagan's inclinations that strongly supported education and literary pursuits and placed great emphasis on good works. Hers was an out-there, on-the-street faith; she ministered to those who needed a helping hand, sharing what skimpy resources she had with those who had even less. Reagan characterized his mother as "a natural do-gooder," open-minded and tolerant at a time when the country was blighted by racism, jingoism, and xenophobia. Long before her

son had considered entering politics, Nelle had written in her
Bible, "You can be too big for God to use, but you cannot be too
small." Nancy Reagan found that significant: "It's something
Ronnie always lived by."

Extending her hand and opening her home to people whose
plight was grimmer than the Reagan family's served another, unin-
tended purpose: it reassured her sons that they were not at the
bottom of the heap and instilled pride that the Reagans could help
others. (Almost sixty years later her example was still bearing
fruit. At a festive White House party for foster grandparents and
children, with Nancy's old friend Frank Sinatra singing "To Love
a Child," the First Lady chose to sit beside an eighty-year-old life-
termer out on parole, a touching gesture inspired by Nelle.)

Nelle's religious rebirth was undoubtedly heightened by the
near-death experience that stamped her life, an experience she
could never forget. In 1918 she fell victim to the epidemic of killer-
influenza that roared through the world like a tornado, striking
down young and old indiscriminately, decimating families and
threatening entire towns. In the United States alone, five hundred
thousand lives were snuffed out; worldwide, twenty million died.
In a desperate effort to contain the epidemic, many towns, like
Monmouth, where the Reagans were then living, shut down the
schools and townsfolk who had to go out in public wore protective masks. And still the virus found its way into their lives.

For days Nelle hovered on the edge of death. She later described
in vivid detail the phenomenon common to others who come back
from the brink: the memory of being bathed in a brilliant light, the
feeling of floating, the awareness that she was dying. A distraught
Jack lit candles in church and knelt in prayer at her bedside, and
Dutch, who was only seven, went to bed each night with what he
remembered as "a weight dragging at the pit of my stomach." In
those days, before there was even a thought of penicillin and its
connection with blue mold, the doctor told the anxious husband

to feed his wife moldy cheese, the moldier the better. Perhaps it was the prayers, perhaps it was the mold, but against heavy odds Nelle pulled through.

Some of Nelle's friends, observing her way with the sick and troubled, came to think of her as a faith healer. "She never laid on the hands or anything like that," one said to Anne Edwards. "It was the way she prayed, down on her knees, eyes raised up and speaking like she knew God personally, like she had had lots of dealings with Him before. Maybe she didn't always pray herself a miracle, but folks could bear things a lot better after she left."

Nancy Reagan could see the influence of Nelle's deep faith on her son. "Nelle believed strongly in the power of prayer, and instilled this in both her sons," she said. "Ronnie has always believed in the ability of prayer to make changes in his life. Nelle was a deeply religious person, and that rubbed off on Ronnie his whole life."

Nelle's church was central to her existence, for both spiritual and social sustenance. She was president of the Missionary Society, taught Sunday school, and every Sunday without fail she and the boys, in knee pants and long black socks, sat side by side well down front. She was inordinately pleased that Dutch was an active participant, occasionally teaching the young boys' Sunday school class and leading prayer meetings. Neil, however, decided in his midteens to begin attending Mass with his father. His decision might seem to be another show of preference for his father over his mother, but the fact that he remained a churchgoing Catholic until his death some seventy years later indicates something more profound. Nelle wept at his decision, oblivious to her early commitment to bring up the children in the Catholic faith. To be Catholic meant many things that conflicted with her strict beliefs, most particularly the church's tolerance of alcohol in moderation. Total abstinence was demanded by the Disciples, a denomination that had a strong affinity with the Woman's Christian Temperance Union and its most colorful exponent, Carry

Nation, who made a lasting name for herself by invading saloons and whacking the rows of whiskey bottles with the hatchet that was her trademark.

★   ★   ★

NELLE, in her service to the Lord, might qualify as a twentieth-century small-town sort of saint—but how many saints loved to perform? Along with roles in local plays, Nelle performed as an elocutionist. Drawing on nothing but her own innate talent, in church halls or living rooms she held audiences under her spell, even if they were only her friends in the Ladies' Missionary Society. As her actor-son described it, "Nelle was the dean of dramatic recitals for the countryside. It was her sole relaxation from her family and charitable duties; she executed it with the zest of a frustrated actress. She recited classic speeches in tragic tones, wept as she flung herself into such melodramas as *East Lynne,* and poured out poetry by the yard."

The magic of her voice had an impact on Dutch's responsive ear; in high-school and college productions, then in a lucrative film career, he used his voice as an instrument, a dramatic skill that originated with Nelle's teaching and example. As president, his smoky tones, his almost whispered intensity, his inspirational lilt will linger in the American mind, like JFK's staccato Boston cadence and FDR's aristocratic eloquence, long after the speaker has left the great stage.

In Nancy Reagan's view, Ronnie's love of acting unquestionably came straight from his mother. He retained an actor's pride that their joint mother-son performances were noticed by the local newspaper, from the time he recited a poem called "About Mother," when he was barely nine. Then Nelle came up with a new idea: add music to their act—she could pick a banjo quite acceptably—and play to wider audiences. Their little show for mental patients at the Dixon State Hospital was reviewed in the Dixon paper as "a short and enjoyable program," citing "two

entertaining readings" by eleven-year-old Dutch Reagan. A video-
tape of that performance of Reagan & Reagan would now be a
historic treasure, but as Mark Twain once commented, "Among
the three or four million cradles now rocking in the land are some
which this nation would preserve for ages as sacred things, if we
could know which ones they are." (Another historic performance
that was never reported, until it appeared in Nancy Reagan's
memoirs, took place at a state dinner given by Queen Elizabeth at
Windsor Castle: the American president and Britain's Queen
Mother spontaneously and jointly recited a favorite poem, "The
Shooting of Dan McGrew." "All eleven stanzas," Nancy reported,
"back and forth at the table!")

Dutch was emerging as Dixon's young star, though many
thought Neil was the better actor. It was that extra dimension
Dutch brought—his touching sincerity, his light touch and easy
manner, his rapport with audiences, whether he was reaching
them from a little church stage or bunting-festooned dais, in a
darkened movie house or on a television screen or from the East
Room podium. He learned early, and it would be a key to his election as president.

★   ★   ★

A year or so after Nelle and Jack had joined Ron in Hollywood,
Neil, too, succumbed to the lure of California. Though he was
doing very well in radio as a director and announcer in Des
Moines, thanks to his brother's connections, California beckoned
as the place of the future. Nelle was ecstatic. Now her family was
back together again, her boys successful, her ailing husband content in his new life. Yet even though the brothers lived within a
few minutes of each other, they led noticeably detached lives.
From boyhood, Moon and Dutch had a somewhat edgy relationship, the kind in which it's hard to tell where the kidding stops.

They were always very different brothers: Moon outgoing,
always ready for fun, slightly racy, a realist; Dutch more reserved,

mannerly, idealistic. Moon graduated from the plain-folks high school and shot pool with his pals; Dutch chose the more upscale school and met his friends at the ice cream parlor. From the time Dutch was fourteen, he saved his money for college; Moon had to be pushed to enroll. Their unlikeness, physically and temperamentally, mirrored their parents—Moon embodied Jack, Dutch reflected Nelle. (Journalist Lou Cannon, author of two outstanding Reagan biographies, was struck by the fact that in a day-long interview Ronald Reagan "talked nonstop about his mother for several minutes without even mentioning his father.")

Neil and Nelle never seemed to be in close harmony. Observed an old friend of Neil's: "He always seemed to be defying his mother. I had the feeling he really disliked her and preferred to think he was like his father." Lorraine Wagner remembers the scene she witnessed at a family picnic at Yearling Row, Ron's first ranch. In an argument over a trivial matter, sharp words were exchanged between Nelle and Neil—until Ron stepped in to put a stop to it, siding with his mother. At the big celebration in Dixon in 1950, Nelle had told Lorraine how hurt she was that Neil refused to attend, and she was sure it was because Ron, the movie star, would be the center of attention.

At Ron's first inauguration, in 1981, Neil and his wife, Bessie, refused to stay at the White House: "I'm no Billy Carter," he declared. There would be no "Moon's Beer." They chose to stay at Blair House, the president's guest quarters across Pennsylvania Avenue. At the second inauguration, with Blair House closed for remodeling, Neil was insistent on staying in a hotel, and only after much persuasion by his brother did he give in and stay at the White House. Why would anyone refuse to stay a night or two with family in that historic house? Perhaps it was Neil's reluctance to be compared with Billy Carter and Sam Houston Johnson or to play second string to his kid brother yet again, which loosened the long-buried feeling that, once again, his mother would have been proclaiming, "That's my Dutch!"

Nelle probably thought she was evenhanded with her two sons, but it seems clear that her Dutch was the apple of her eye, that she was closer to him, shared a deep empathy with him—and he with her. Neil could have sensed, perhaps unconsciously, the difference in his mother's attitude toward his brother, which raises a question: was his tilt toward his father the cause or the result of feeling that he took second place to Dutch in his mother's esteem? The distance between the brothers would have been widened by Neil's suspicion—justified or not—that his younger brother felt superior to him. In college Neil was the better student, which wasn't saying much, and the better football player, which was, but Ronnie was the big man on the little campus. Maybe Moon didn't mind; maybe deep down he did. How to explain charisma, magnetism? How did Dwight Eisenhower eclipse his older brother Edgar, who could never quite accept being second best? Some people call it star power, but there isn't a rational explanation. However, that kind of unformed resentment could gnaw at a boy who was older by two and a half years; it would have led him to turn to his father, who was also something of an outsider in the family's emotional life.

"There's a game we play back and forth," Neil half-joked to Reagan biographer Laurence Barrett. "I don't give him credit for taking a deep breath and he won't give me credit for taking a deep breath." He said this not when they were kids, but when both were long-established successes and might be expected to be big-hearted about the other brother. Still, when the final returns came in on the night of Reagan's first presidential election in 1980, Neil retreated to a quiet corner and gave way to tears as he absorbed the enormity of the event: his kid brother had been elected president of the United States. President Dutch! Neil must have had some powerfully mixed feelings than night.

In Los Angeles Neil rose to the top ranks of the influential McCann-Erickson Advertising Agency; he grew wealthier and had wider connections than Ron before his brother entered politics.

When Ron's career was in the doldrums, Neil returned the brotherly favor that had given him a start in radio in Des Moines. Pulling strings at General Electric, Neil landed Ron the job of host and company spokesman on *Death Valley Days,* a popular GE-sponsored television series that proved to be the turning point of the younger brother's life. He not only hosted the weekly show but gave speeches around the country on behalf of GE, an activity that converted him from movie actor to public figure and led him into politics. Both of Nelle's boys were sailing with California winds at their backs.

When Ron and Nancy were married in March 1952, in the Little Brown Church in the San Fernando Valley, Ron—as at his first wedding—did not choose his only brother to be his best man. The ties between the brothers would be further strained by this marriage, because Neil and Nancy just didn't hit it off. Despite Neil's outgoing personality, gift of gab, and undeniable success, they never seemed to find mutual grounds for easy friendship.

★   ★   ★

THE union of Ron and Nancy was made in heaven; their bond was so deep, so real, that forty years later they still had the glow of newlyweds. Nelle and Nancy got along well together (Nancy bears a strong resemblance to the young, fresh-faced Nelle of 1915), and the delighted grandmother was there, beaming and looking smart at seventy, for baby Patricia's christening. But she was growing ever more frail, more confused mentally.

Writing to Lorraine Wagner on New Year's Eve 1957, Nelle confessed, "My memory isn't so good now that I have reached the age of seventy-four." Three years earlier, she had told Lorraine, "I have hardening of the arteries in my head—and it hurts just to think. I know that this will be the last Christmas that I will be shopping for gifts. I'm getting too old to face the crowds of people—it's just too awful." Don't send gifts, she pleaded. "Let's have Christmas for children, but send cards so I know you are thinking of me."

It was sad that Patti and her brother, Ronald Prescott Reagan, younger by five years, could not have the chance to know the loving grandmother who was part of Maureen's and Michael's life. Parenting, as their children candidly state, was not a role Ronnie and Nancy starred in. When the Reagans were in the White House, Laurence Barrett observed that even accounting for the geographic scattering, "There had been for many years an emotional distance that was unmistakable." Others have used the more direct term: "dysfunctional family."

Ron called Nancy "Mommy," which a psychiatrist might make much of, in that Nancy moved to the center of his life as Nelle was fading. To spare her Ronnie every problem, Nancy shouldered the burdens of managing life in the spotlight both in Hollywood and Washington. In the view of many presidential scholars, Nancy quietly exercised more influence over the president's staff than any First Lady had since President Woodrow Wilson's wife, Edith, virtually ran the executive office after he was felled by a stroke.

Ronald Reagan, for all his geniality, was a loner who gave no indication that he wanted a closer relationship with Neil or anybody else, so long as he had the two women he loved, Nancy and Nelle. They were, each in her own time, the deep, influential forces in his life, the only persons with whom he let down the barriers and shared his inner self. Like Nelle, Nancy offered total support, adulation, protection against anybody or anything that might be troublesome. Sparkling and devoted, she shared his pleasure in the quiet life, just the two of them, upstairs at the White House or at their California ranch.

But long before Nancy made her entrance in the Reagan drama, it was Nelle who had shaped the man she would pass on to Nancy, like a runner handing off a baton in a relay race. Lou Cannon left no room for doubt when he asserted, "You and I would never have heard of Ronald Reagan if it had not been for his mother, Nelle."

*   *   *

ON August 15, 1962, nearly a month after Nelle's death, Reagan shared his personal thoughts with longtime pen pal Lorraine Wagner: "Mother's passing was peaceful and without pain. It was just a matter of going without waking. I'm sure it was what she wanted, too, because these past few years have found her unable to do any of the things that had always made her life meaningful." Lorraine agreed, for almost five years earlier Nelle had written her, "I still feel young—only my heart tells me I am wrong because I do have pretty bad heart spells. The doctor tells me I may pass on to the great beyond in any one of them. But don't mourn for me because I really want God to call me home."

In 1985 Ronald Wilson Reagan again took the oath of office on Nelle's Bible, crumbling and taped together, which to her was a living document. As he placed his left hand on one of her cherished verses he wished, again, that his mother could have been there for the event, a simpler ceremony this time, held in the East Room of the White House because of brutal winter weather.

On his way up in the world he used to say to Nelle, "Look at your son—he's making $300 a week. . . . Look at your son—he's featured with Pat O'Brien. . . ." How he would have loved to say, "Look at your son—he's president of the United States! For a second term!" Nelle was more than twenty years dead by then, but the Nelle who had brought up the boy and encouraged the man was gone years before that, lost in the gray mists of an illness that was then called "senility," for lack of a more precise name, but now would undoubtedly be identified as Alzheimer's disease.

Six years after his two terms as a highly popular president, those same mists would capture her son, stealing his memories of an Oscar-winning life, erasing the joys of a true-love marriage, robbing him of his final role as an elder statesman, leaving behind an amiable, unaware shell of the son whom Nelle had nurtured into history.

PLAY BY MOTHER'S RULES

Example is the school of mankind, and they will learn at
no other.

—EDMUND BURKE

DOROTHY Walker Bush was a competitor to beware on the
tennis court. Her second son, George, remembers that starting
from the time he was old enough to hold a racquet, she faced him
across the net and made him work for every point. When she won,
which was often, it left him with mixed feelings: whether to be
embarrassed that his mother could outplay him or to be proud
that he had a mother who, at an age when most women gracefully
retire to the sidelines, could still beat the socks off him and the rest
of her brood of five.

A little humble pie was good for them, in Dorothy's view; kids
who had everything should not boast about anything. She knew,
because she, too, had always had everything, and she hewed to
that standard for herself as well as for her children. Along with
instructing them in cross-court returns and slashing serves, she
taught them that it was not good manners to talk about yourself.
Applied to tennis court or playing field, good manners to Dorothy
were the equivalent of sportsmanship and thoughtfulness to
others, which were more important to her than winning.

"My mother had a tremendous influence on my life," said former President Bush, thinking back to his growing-up years. "It was about *values*." He ticked off a litany of his mother's exhortations as she implanted unchanging values and immutable standards in her growing sons. "Give the other guy credit. Be kind. Don't whine and complain. Count your blessings. Honesty. Your conscience will be your guide. Along the way she taught us values—and that conscience was important in later years."

Yet there was nothing about her that was preachy or overbearingly moralistic. This was a woman who, when she was a girl, lashed her brother to the gatepost when he damaged her tennis racquet. Her unflinching character meshed comfortably with her more aggressive side: powerful swimmer in frigid Maine waters, accomplished horsewoman, and fierce competitor on the tennis court. She could preach sportsmanship because she lived it. When she was seventeen, she was runner-up in the National Girls' Tennis Championship, back when they wore modest white skirts that flapped almost to the ankle. The bulky costume could get in the way of a running backhand, but it didn't slow down Dorothy one bit.

"I can vividly remember the bottom of my mother's feet," Bush said, marveling at her competitive drive. "Mother literally wore the skin off the bottom of her feet." In today's sports world she might have played on the women's tennis circuit, but her father refused to allow her to enter adult women's tournaments, even with amateur status—George Herbert Walker's daughter mixing with those hardened athletes? Out of the question. But Dorothy kept her competitive edge throughout her active life. In 1940, when she was thirty-nine and had recently had her fifth child, she played an exhibition match against a Polish champion, Jadwiga Jedrzejowska, and proudly won a set.

Growing up, when she wasn't on the tennis courts, she was in the water, in bathing costumes that were as revealing as a snowsuit. As a teenager, she took a young friend's challenge for a dis-

tance race, dived off the pier at the Kennebunk River Club, left her
friend in her wake twelve hundred yards out in the choppy open
sea, and kept swimming for more than a mile to the family's own
dock. Just meeting the challenge was not enough—she would go
the distance, which she had set for herself.

"She was a great athlete," the admiring son said of his mother.
"A fine women's tennis player, a good golfer, and an all-around
athlete." Heaven help the timid or uncoordinated when Dorothy
entered whatever game was being played. "She captained the
mothers' teams at our school. She was the fastest runner, the best
pitcher, the best football thrower." She was small and concen-
trated, not at all like the amazon women athletes of today, but her
determination more than made up for her size. Daughter-in-law
Barbara dipped into superlatives to declare her, without qualifiers,
"*the* most competitive human being." And she did it all without
trimming the generous sense of fair play that she demanded of her
children.

Her athletic prowess and competitive spirit came straight from
her father, Missouri's amateur heavyweight boxing champion in
his younger days (one of the few sports Dorothy didn't care to
compete in) and at one point co-owner of a racing stable with
Averell Harriman. But George Walker's passion was golf—he had
a six-to-seven handicap and an array of trophies—and he took
pride in his daughter's good game. In 1923, after founding his
own investment firm in New York, he followed the example of his
old friend, the industrialist Dwight Davis, who had set up the
Davis Cup as the most prestigious trophy in amateur tennis.
Walker, a former president of the U.S. Golf Association, permanently stamped his name on the game of golf, establishing the
Walker Cup, which remains the major biennial competition for
the best amateur golfers of Britain, Ireland, and America.

With those genes going for her, it was a very long time before
Dorothy began to slow down. In what would be called old age in
anyone else, she was still bicycling around Kennebunkport, the

colonial town on the Maine coast, the Walker family's summer gathering place for the better part of the twentieth century. Celebrated evangelist Billy Graham, a close friend of the Bush family, likes to tell about the time he and George were on their way to church, with a driver who seemed unaware of bicyclists: "You'd better watch that little old lady on a bike," cautioned the president. "That happens to be my mother!"

Tennis was such a part of her life that she tended to make judgments of people on the basis of their game. When someone predicted that Son X would probably marry the girl he'd brought for a visit at Kennebunkport, Dorothy's reaction was "Oh, no, she won't play net!" Barbara, in her middle years, realized she was slipping into Dorothy's way of thinking: the former First Lady confessed that when she looked in the mirror, what she saw was not the face of a woman losing the fight against crow's feet but "the face of a woman who needed to improve her serve."

Born to affluence and position, Dorothy rejected the superior mind-set that often accompanies elevated status, and when any of her children showed signs of a swelled head she could bring it down to size like a knife deflating a lemon soufflé. A good deal of thoughtful child psychology was probably involved, but her explanation was simpler: "I couldn't bear to hear them boast." The former president has never forgotten her lessons in sportsmanship: "The messages she would give us about sports resonated throughout my life," he said. He carries a particularly acute memory of the time he came home from school and triumphantly reported, "Mother, I scored three goals in soccer!" Her pointed reply was: "Fine, George, but how did the team do?" Nor did she brook excuses about poor performance. "Mother, I was off my game," George alibied after losing a match. Dorothy was not sympathetic: "You don't have a game. Get out and practice!"

Dorothy Bush would have had little tolerance for the current "self-esteem movement," which preaches feel good about yourself, don't worry about the rest. She was all about team play and effort, which left no room for egos. She taught her sons well. "I'd like to be known as a team player," George said as he campaigned for the presidency. It was an attitude that had brought him a succession of high-level appointments—U.N. ambassador, chairman of the Republican National Committee, U.S. representative in China, and director of the CIA—each a demanding post requiring the steady hand of a team player. It was a most appropriate quality when he was vice president—he never differed with President Reagan—though not altogether useful as president, when, as the captain, the chief executive must call the plays.

Dorothy's motherly criticism grew less pointed, a little more oblique, as George moved ever upward in the world of politics. Still, she felt it right to chastise him for reading from a piece of paper while President Reagan was delivering his State of the Union address. But, the vice president explained, he was following the president's printed text. She wasn't convinced: "I really can't see why that's necessary. Just listen and you'll find out what he has to say." She was dead right: to the viewer at home, it not only looks discourteous but also suggests that a vice president doesn't know in advance what his president plans to say.

"Sometimes Mother is more subtle in her suggestions about my deportment as Vice President," he noted in his autobiography, *Looking Forward*. " 'George, I've noticed how thoughtful President Reagan is to Nancy,' she once called to say. 'I've never seen him climb off a plane ahead of her or walk ahead of her.' " Her son added wryly, "I got the message." Such criticism, even if subtle, would rankle most offspring, and perhaps goad them into a retort, but George took it as intended—and remembered not to stride ahead of Barbara. Even when he was running for president, she monitored the I-factor: "You're talking about yourself too much, George." But the voters expect to hear something about his qualifications, he said in his own defense. Grudgingly she conceded, "I understand that—but try to restrain yourself."

The fact that Dorothy was a very special mother had not

occurred to George until he was about thirteen: "It was all summed up for me when an eighth grade classmate told me, 'I wish my mother was like yours.' I couldn't imagine a friend saying that, because I held my mother on such a pedestal I figured, way back then, that everyone else did that, too."

★　★　★

FROM July 1, 1901, until her death ninety-one years later, Dorothy Walker Bush was blessed with a life unmarred by tragedy, unsullied by scandal, unscathed by the Depression. Her family was wealthy and socially prominent in St. Louis and she was pretty and bright. She could have grown up as a self-centered, self-indulgent debutante at the center of the city's top-drawer Veiled Prophets ball. But her parents managed to keep the family's eye on the ball—in that family of athletes the clichés took on literal meaning—and keep their social position in perspective. Dorothy passed on these attitudes to her children, and they to theirs, on to the growing generation of great-grandchildren.

This midwestern girl was actually born in Maine, where the Walker family was summering. While their ancestral roots were in Maine—the first Walkers had settled there in the 1600s—the more prosaic reason for their annual sojourn probably had more to do with escaping the muggy river heat of St. Louis. The Walker forebears had migrated south from Maine to Maryland; then a later generation joined the country's westward push, with all its promise. In the midnineteenth century the Walkers settled in St. Louis, where Dorothy's grandfather went into the dry-goods business. Ely Walker & Company grew into the largest dry-goods wholesaler west of the Mississippi.

David Walker sent his son George to England for the best of Catholic education, and on returning home George fell in love with, and married, Lucretia Wear, who was known—in St. Louis, at least—as "the most beautiful girl in America." She was a staunch Presbyterian whose Bible was always close at hand, "the biggest Christian that ever lived," in the view of one of her sons. Lucretia brought up their children in unbending, but not oppressive, rectitude and tasteful, but not ostentatious, luxury.

Daughter Dorothy, a girl with a lively mind, wished she could go to Vassar along with her best friend, but her father had strong opinions about a young lady's education. "He didn't want any bluestocking daughter arguing with him," said Dorothy's daughter, Nancy Bush Ellis. "Grandfather wanted no disagreement from anybody!" And so she was sent East to Farmington, Connecticut, to join other young ladies of impeccable background at Miss Porter's, a prestigious academy that has been called a social training school.

Then a young man with charms that set hearts aflutter swept into St. Louis—Prescott Sheldon Bush, a handsome six-foot-four bachelor from a wealthy family (his father was a leading Ohio industrialist and a national figure in commerce), with the right pedigree (his forebears, Puritan dissenters, left Scotland for America in 1673, and he was about a twelfth cousin, if there is such a thing, of Queen Elizabeth II, according to *Burke's Peerage*, Britain's definitive guide to the aristocracy). He went to the right schools (St. George's in Newport, Rhode Island, and Yale, where he was a star athlete and member of Skull and Bones, the most elite club at the university) and summered at Watch Hill, Rhode Island, a sedate enclave of old money. And if that weren't enough, he was a gifted 'round-the-piano singer (Whiffenpoof level) and a terrific golfer (Ohio's schoolboy champion). With all the connections that accrue to such a breathtaking social résumé, he was the ideal match for Dorothy Walker.

They met in her own living room when, on a hot day in 1918, Pres (as everyone called him) was calling on her sister, Nancy. Dorothy, sweaty after an afternoon of tennis, her blond hair tousled, bounded into the room and was startled to see this newcomer. "Don't look at me," she implored, "I'm a wreck!" He looked. Their daughter, Nancy, finishes the story: "Dad fell head over heels in love with her." He was six years her senior, which Dorothy thought was too old for her until love bridged the gap. In the summer of 1919 they were engaged, and the following August their wedding in the Church of St. Ann was the social event of the season in Kennebunkport, the town that was to become famous through the Bush name.

On the family softball field behind the Kennebunkport house, Dorothy Walker Bush—nine months pregnant with her first child—stepped up to the plate, belted a solid home run, rounded the bases, tagged home, and continued to run right off the field to the hospital, where she gave birth to Prescott Sheldon Bush Jr. Nit-pickers say the story is a mite exaggerated. Oh, she hit the homer, all right, and ran the bases, but it was a day or two before the baby was born. But the story is so like "Dottie" that the family sticks with the tale as told, and any woman who has ever given birth would be agape at the fact that she could even swing a bat, much less run, with her obstetrician on alert. "Physically she is a small woman," her son said when he retold the story, "but she is made of mighty stuff."

Two years later their star-tapped second son arrived, in Milton, Massachusetts, where his father worked for a short period. He was given Dorothy's beloved father's full name, George Herbert Walker Bush, and even inherited the grandfather's nickname—"Pop," miniaturized to "Poppy" for the boy.

In short order Prescott Bush became a wealthy man, rising to partner at Brown Brothers, Harriman, a powerful Wall Street investment house, when he was only thirty-six. (The chairman was his father-in-law, the redoubtable George H. Walker, which was no small help in getting a foot in the door.) By then the Bushes had settled in Greenwich, Connecticut, the preeminent town in New York City's commuter belt, the symbol of success and a mag-net for achievers on Wall Street and in business. They bought a spacious old house with plenty of room for the children to play—necessary since the number had increased to five, as Nancy, Jonathan, and William came along.

With motherhood came the heavy responsibility of bringing up children who would do credit to their doubly distinguished lineage, starting with Puritan forebears and coming down through the years to their admirable grandfathers and, not least, their father, a natural civic leader. And then there was the matter of wealth, which Dorothy knew could easily erode moral standards and lessen ambition; she would be on guard against these dangers long after her children had left her immediate care.

After the children were grown, Prescott, broadening his sphere of public service, ran for the United States Senate; he lost his first election by a hair and then won twice. The imposing and personable Prescott Bush could have been the senator from Central Casting: John Kennedy once commented to the senator's daughter, Nancy Ellis, "Nobody looks as much like a senator as your father." He personified the eastern establishment Republicans, internationalist moderates who are now an all-but-extinct species, and with his wife's hearty support was one of the first senators to speak out against the intimidating Senator Joseph McCarthy and his communist witch hunt. Throughout her husband's two terms, Dorothy, a favorite in Washington's political and social circles, made a name for herself with her regular chatty column for the newspaper back home.

Like everything else in Dorothy's world, her marriage was a lifelong idyll. When Prescott died in 1972, she asked the family to "wear colors" to the funeral to show that it was a celebration of a remarkable man, and she composed a eulogy thanking him for giving her "the most joyous life that any woman could experience."

Asked how Dorothy Bush brought up something very close to a perfect family, the former president reflected, "Mother would always see the good in people. She was a devout Christian and her life was guided by the teachings of Christ." In the Bushes' sophisticated setting, Dorothy, a resolute Episcopalian, practiced her religion with the same direct simplicity as Hannah Nixon on her family's remote California lemon farm. Dorothy would read the daily lesson at mealtime: she not only read from the Bible over the morning bacon and eggs but also related the lessons to their daily lives. The parable of the Good Samaritan, for example, should remind them to step forward when anyone needed help.

"She set the example, she led by the way she lived her life rather than lecturing to us all the time," her admiring son recalled. "But when she did give advice, it was always sound—always about being thoughtful and considerate of others. And about being kind, and honest. She didn't lecture us about love—she lived that part of life and we saw it every day in our lives, love for my Dad, love for all five of us."

"Our upbringing," said daughter Nancy, "was a relatively strict one. Mother was a non-Freudian. She would drill into us, 'Think about other people. Put other people first.' We didn't want to put other people first—we were only interested in our wonderful selves. But she was a real Second Commandment person. Our family was very oriented toward religion." Along with concern for others, Dorothy emphasized self-discipline and good habits but laid down her rules in such a way as not to provoke resentment, much less rebellion, among the five children. In bringing up their five, Dorothy and Prescott Bush were never negative. "They were our biggest boosters," said a grateful President Bush, "always there when we needed them."

Dorothy's religion did not stop at the church door; like Hannah and Rose, Ida and Nelle, she brought it into the heart of the home. Billy Graham, who was devoted to the "delightful and devout" Dorothy and who has called George Bush his "best friend in the whole world" outside his own close associates, has frequently been their guest at Kennebunkport. Amid summer pleasures and family fun, Dorothy did not hesitate to ask the world-famous evangelist to talk about the Bible in informal conversations with the children and grandchildren. "Some nights," said George Bush, with a touch of amazement, "I've stepped into a room and found my children seated around Billy asking questions about the Bible . . . it was wall-to-wall grandchildren and their friends."

One particular summer evening in 1985 might one day be marked as a turning point in a presidential biography. Among the family members clustered around Dr. Graham was George W. Bush, the then–vice president's eldest son, who was drifting toward alcoholism in an unfocused life. There in the family living room, George W. found his road to Damascus. Billy Graham, he said, "planted a seed in my heart and I began to change." That evening he reevaluated his life and found his way, which would lead to two terms as governor of Texas and a run for the White House.

Governor Bush considers himself fortunate to have been able to spend time with his grandmother during summer vacations as a boy in Texas and more often when his father moved into Washington politics. Racing from one event to the next in his presidential campaign, he paused to think of her: "My grandmother was one of the most kind and thoughtful people I have ever met in my life. Everyone who ever knew her felt her compassion and was impressed by her grace." Dorothy's sterling spirit now influences a third generation of political leaders.

★   ★   ★

WHEN George Bush was on his way to becoming the forty-first president, Barbara, contemplating the influences that had shaped her husband, offered a decisive opinion: "His father had enormous influence on him, and his mother had ten times more. They held their father in enormous awe as young children." Jonathan Bush, the third of the four Bush brothers, acknowledged that "we were all a little afraid of Dad," and Prescott Jr. told an interviewer that when his father was away on business trips "it was like the Fourth of July as far as George and I were concerned."

Their father was a towering two-hundred-pounder with craggy good looks, a dynamic presence who could be a bit gruff when his home was overrun by a bunch of boisterous kids. He put in the long days of a Wall Street commuter, and on weekends he teed off early at the exclusive Round Hill Club. (His shelves were crowded with silver trophies, testament to his near professional level of play.) Barbara hasn't forgotten that "even when I married into the family, we tip-toed around his dad. As he got older, he got gentler." She could measure the difference in the generations: his grandchildren thought of him as "sort of scaryish," while his great-grandchildren thought he was "a sweet, wonderful, funny, huggable man. Time does that.

"But their mother had the most influence, and not in a bossy way," Barbara told biographer Fitzhugh Green. "As a young woman, just a great athlete, even-tempered, fair, loving. But I'm sure that her sisters-in-law heard their husbands asking, 'Why can't you be like Dottie?' " The president explained a fundamental difference in the lessons his parents passed on to their children: "Dad taught us about duty and service. Mother taught us about dealing with life on a personal basis, relating to other people." He dedicated his autobiography to his parents, "whose values lit the way."

Their mother was the one who was there, present in their lives every day when they were growing up in Greenwich, encouraging, correcting, settling disputes, teaching, disciplining. George and Pres Jr., two years older, did what brothers do—they gave their little sister, Nancy, a hard time. To this day Nancy is still a little sore about it. "George was a wicked tease. They left me out all the time—they'd shut the door and say, 'You can't come in.' " The former president admits that he and his brother bedeviled their little sister. "We teased her," he told me. "Mother bawled us out for that. If we got really nasty Mother would indeed discipline us. And if we got totally out of hand, then 'I'm going to tell your father on you.' That meant there was real trouble ahead for us—pain maybe, too."

Mary Walker, Dorothy's sister-in-law, agreed with Barbara: "I would say that Dottie did more to bring up the children than Pres because he was so busy in his own activities." Like many high-powered fathers, "he was more remote from us," recalled daughter Nancy. "Mother was there with us every day. If we left a tennis racquet or a bicycle out in the rain, she would say, 'Your father is working six days a week to give you these things.'" Dorothy was not above employing a little guilt to make her children shape up.

Prescott was proud of his children and, of course, loved them, but like many fathers of his time he may have been more devoted to them in the abstract or during the calm of the dinner hour (jackets and ties required at the table) than when they were rough-housing underfoot. Dorothy must have sensed that he needed to be a little closer to the boys, for she sometimes urged him to take his sons along with him to play golf, to give them a little coaching. "Oh, Dottie," he would grumble, "if they want to learn, they'll pick it up."

On evenings when he was not out at meetings of one kind or another, he loved to gather the family around the Steinway grand for a songfest. From his Whiffenpoof days at Yale, singing had been his joy, especially close-harmony quartets. Throwing back his handsome head, he belted out show tunes and old favorites in his rich, deep baritone, and he knew the words of songs everybody else had forgotten or had never heard. For George, the family musicales had one drawback: he couldn't carry a tune.

There were always kids running in and out and Dorothy, warm and outgoing, welcomed them all, whereas Prescott wondered why they had to pick on his house. (She sometimes quietly sent her children's friends up the back stairs so they wouldn't bother him.) A Ping-Pong table was set up in the front hall—dinner guests had to weave their way around it—and in bad weather the boys played a limited form of football in the long upstairs hall. Dorothy was quite relaxed about the fact that the house sometimes seemed to be a gym. "Mother was not much on houses and decorating," Nancy Ellis once said. "She didn't care about all that." Just as well. What mattered more was that the house was ideal for raising children.

It was a dark-shingled turn-of-the century style, with nine bedrooms on three floors, set on two acres with a stream in the back and a separate carriage house. There was a governess when the children were small, the requisite cook, two maids, a gardener, and leading the list was general factotum Alec Chodaczek, a Ukrainian immigrant. He kept the black sedan gleaming and insisted that the children leave their shoes outside their bedroom door for him to polish, as in five-star European hotels. (In gratitude, Prescott Bush helped send his two children to college.) On snowy days Alec took pride in getting his charges to school faster than any of the other chauffeurs, a vignette that brought gales of laughter from President Reagan, who depended on shank's mare to get him to school through Illinois snowdrifts. Years later Dorothy looked back on those times and reminisced, "Life was easy in those days."

Prescott bought the choice property in 1931, the year Jack Reagan lost his job selling shoes and Jerry Ford's father was struggling to keep his paint business out of bankruptcy. Life for the Bushes during the Depression was cloudless. Christmases were spent at Grandfather Walker's estate in South Carolina's horse country. (He had a keen eye for horseflesh and for a time was chairman of the New York Racing Commission.) They rode horses through pine woods, played tennis on his private court, and shot quail and dove. Along with Dorothy's other athletic skills, she was a great shot who could pop clay pigeons in doubles, two at a time out of the trap. Calamity Jane couldn't have done it better.

With her background and all that had come to her as daughter and wife, Dorothy Walker Bush might have cast herself as the grand lady, but she chose not to. She hammered her attitude of quiet modesty into her children to the point that her son, both as candidate and as president, found it uncomfortable to blow his own horn, something most politicians don't have any problem doing. Modesty in a man of power is a most becoming trait.

★　★　★

EACH summer, as soon as schools were out, the five Bush children and two dogs piled into the station wagon, with Dorothy at the wheel, heading for Kennebunkport. The place of their hearts was Walker's Point, so named by Dorothy's father and grandfather in 1919, when they bought seventeen acres on a neck of land jutting between the ocean and a small bay. For eighty years Walker's Point has been the family's private world.

At the most dramatic point of the promontory is a large, sprawling three-story "cottage" of shingle and stone, built right on the water, so close that the sound of waves splashing against the rocks lulls the sleepers in its ten bedrooms. "You feel as if you're on a boat," said Mary Walker, who owned the house with her husband, Dorothy's brother Herbert, until George Bush bought it when he became vice president. Until then the Bushes occupied one of a number of family houses in the compound.

At first Mary Walker hadn't wanted to take on the burden of the big house. "But then one day I saw a little cousin from the West playing with a little cousin from the East, and I said, 'This is a much bigger thing than I am.' " Barbara Bush also understood the importance of those family ties. When she and George were living in Texas, every other summer she piled their five children and dogs into the station wagon, just as Dorothy had done, and drove twenty-five hundred miles to Walker's Point. For many years, presiding over it all was the patriarch and his ever-beautiful wife, Lucretia, the beloved grandmother. Walker's Point was family, continuity, occasional squabbling without backbiting, acceptance without reservations, individuality without egoism. Walker's Point was for the Bush family what Campobello had been for Franklin Roosevelt and Hyannis Port for the Kennedys.

The minute the four boys (of either generation) leapt out of the station wagon, they were in action and in competition. Who could swim fastest, catch the biggest fish, hit the home run, serve the sizzling ace, sink the longest putt. And always there were boats, power boats in sailing waters, for the thrill of it. Nor did the Walker-Bush competitive spirit stop when the sun went down—tiddledywinks, peggity, and backgammon could be intense contests indoors. When the inevitable nor'easters bore down on the Maine coast, "Mother taught us bridge and thousands of games and organized incredible treasure hunts," Nancy fondly recalled. (But cards and dice were strictly forbidden on Sundays.) Cousins and aunts and uncles and nieces and nephews and friends all joined in.

"Maine in the summer was the best of all possible adventures," George Bush reminisced in later, more pressured times. "We'd spend long hours looking for starfish and sea urchins, while brown crabs scurried around our feet. . . . Then there was the adventure of climbing aboard my grandfather's lobster boat, *Tomboy,* to try our luck fishing. . . . When Pres was eleven and I was nine, we got permission to take the *Tomboy* into the Atlantic by ourselves."

Once again, even for a boy living in such privileged circumstances, there was the opportunity for independence and adventure, something that runs through the boyhoods of all of these modern presidents, whether wealthy like George Bush, FDR, and JFK or of modest means like most of the others. It made for self-reliance and was a building block of confidence. Dorothy, like the other presidential mothers, encouraged this spirit, especially since she harbored an adventurous streak herself.

Summertime at Walker's Point was remarkably like summer-

time at the Hyannis Port compound, and the Bushes were every
bit as competitive as the Kennedys. Mary Walker will never forget
the afternoon when George, about nine, was in the finals of the
River Club tennis tournament. Watching from the porch, "I
started to laugh at something. He looked up and said, 'Aunt Mary,
would you please get off the porch immediately!' And so I walked
off." John McEnroe couldn't have been tougher in his hotheaded
days at Wimbledon. But George lost the match, and his mother
made him apologize to Aunt Mary.

One can only be glad, for young George's sake, that Grandfather Walker was not there that day, as he often was, in his customary white flannels, navy blazer, and straw boater. Years before,
when Dorothy and her brother Louis were scheduled to play doubles in a club tournament, Louis, who had been enjoying more
than a few drinks with friends, had forgotten about the match. He
was chased down—and stepped onto the court obviously tipsy. As
he lurched about and even missed the toss on his serve, the specta-
tors tittered. Back at the cottage, his father, furious at such behavior by a Walker, roared at his son, "You are not going to college.
You're too stupid to go to college!" That very night Louis was
ordered to pack up and was promptly banished to work for a year
in a Pennsylvania coal mine. No son could embarrass Grandfather
George Herbert Walker and get away with it. Expectations in the
Walker-Bush family were high and transgressions were not tolerated—the line you had to toe was not the chalk markers of the
tennis court but the invisible line set down by Grandfather.

As the years passed, a new generation of children was learning
to handle boats and getting tennis tips from their grandmother
and snuggling in heated towels after splashing in the swimming
pool. It was—and is—a snug island where family roots grow deep
and kinship blossoms in summer, and it engendered strong feelings
in George Bush: "Family is not a neutral word for me. It's a powerful word, full of emotional resonance. I was part of a strong

family growing up, and I have been fortunate to have a strong family grow up around me."

★ ★ ★

THE Bush brothers were generally very well behaved, but now and then they were not. More than sixty years later the former president confesses to egregious conduct that even his close associates hadn't heard about, a naughtiness so unlike George Bush that one can only roar with laughter at the story he revealed.

"Once my brother and I, sixth and seventh graders at the time, did a bad thing involving the sister of one of Pres' classmates. She 'told on us.' And all hell broke out at home. The matter was grave enough to go from Mother directly to the father of all discipline— my Dad. We were summoned into his presence. He had a squash racket in hand. He was a big man—very big, and at that moment intimidating. He confronted us with our sin, shoved us out the front door and yelled at us to walk to the girl's house. It was a two-mile trek. We went fast at first to get away from Dad, but then slowed noticeably as we drew near the house. There we had to ring the bell. My brother Pres took the lead: 'We are very sorry for what we did.' I mumbled, 'Never should have done it.'

"And indeed we were sorry—we never should have given that girl and her friend ten cents each to run across the room naked while we Bush boys and the offended girl's brother watched. It was a tawdry thing. Parental discipline worked. Mother lectured us. Dad gave us hell."

Back on the straight and narrow, George was dependably good as gold and everybody's favorite. Though Dorothy never admitted to favoring any son over another, George was "funny and fun," said his sister, "and Mother was always amused by him." He was even known to do or mumble something funny in church that would make her break up—with the Patriarch glowering sternly at the wife and son from down the pew.

At home, at Andover, and at Yale as a married student after the war, George was a natural star. With his mother and father both outstanding athletes, making the team was predictable, and being captain followed. Dorothy's ban on boasting, when he had much to brag about, enhanced his popularity with his peers at every age, and their approval made him a student leader from the beginning. At Greenwich Country Day School and at Andover, he was an indifferent student (like most of the other presidents), but by the time he enrolled in Yale he had experienced four hard years of war; he returned as a seasoned man with a wife and, after his freshman year, a son, christened George Walker Bush. His changed status as a battle-toughened veteran and new father produced a changed student: majoring in economics, he made Phi Beta Kappa and captained the baseball team into the College World Series.

Years later, as George Bush's campaign in his race against Massachusetts governor Michael Dukakis was taking shape in the summer of 1988, Dorothy talked about George and his siblings and ruminated on the fact that her family had always been free of internal rivalry: "They accepted from the start that George was going to be the best in whatever activity. Someone asked me, 'Wasn't it hard for Pres that his younger brother was able to do everything so well?' But George never boasted, so it was all part of the family performance."

For his part, the former president, thinking back, says, "I never felt competitive with my brothers. We were too close for that. Had I ever done so, I expect Mother would have corrected that inclination very fast." Dorothy had nurtured a team spirit in her five children—as Rose Kennedy had in her nine—which made them share the pride of George's achievements rather than resent the success. That, plus the parental formula spread evenly among the five—as the shining son put it, "generous measures of both love and discipline." With much more of the former than the latter.

Aunt Mary, whose husband provided financial backing for

George's oil enterprise in Texas, remembered her nephew as "a real star from the day he was born." In school "he was always the most popular boy . . . and very, very considerate to everybody." His early schoolmates remember that even as a kid George was different from the others. There was the time at parents' day when a fat child got stuck in a barrel in the obstacle race. While the other boys laughed, George, then eleven, dashed from the sidelines and helped extricate the embarrassed boy, then ran the rest of the race at his side. Dorothy had the satisfaction of seeing her training in action.

At Andover, his prep school, George saw a Jewish student being unmercifully hazed and commanded, with some inborn authority, "Leave the kid alone!" The bully slunk away. Bruce Gelb would later say, "At that point, he became my hero and has been ever since." Indeed, Gelb never forgot: almost forty years later, then a very wealthy man (his father founded the Clairol company), Bruce Gelb returned the favor, raising almost three million dollars for the 1980 Bush for President campaign. Dorothy Bush would have seen in that story the biblical message of the rewards of kindness.

Throughout his life George Bush, with that instinct instilled by his mother, coupled with his easy charm, has collected friends the way other men collect stamps. "Close friendships are my under-pinning," he has said. "People who really care are with you when you're up and with you when you're getting kicked. Friends are really what make life worthwhile."

"The most consistent characteristic in his life, the characteristic I could see in him as a young child and that has stayed with him over the years," his mother said, "is his kindness to others. He wanted everyone to have a fair chance, and he was always looking out for the underdog." Every biography relates how little George was called "have half" for his generous way of offering—or accepting—half of a toy or a goodie. Nobody told him to do those things, but the former president attributes his concern for people to his mother: "She showed the way by her innate kindness."

Dorothy explained her teaching very simply: "I taught them love. I taught them to love everybody, no matter what their background, and I taught them to be unselfish." Her text was a favorite Bible verse, "Though I have the tongues of men and of angels and I have not love, I am nothing" (1 Corinthians 13:1). Nor did her influence stop with her own five. "She's had a really strong influence on all of us," said her grandson Marvin. "First on her children, and then kind of trickling down to our generation. She is one of the most spiritual people I have ever met . . . she's one of those very good and special people that you meet once in a lifetime." She had the deep beliefs of a clergyman and a wicked tennis serve.

Along with millions of other American mothers during World War II, Dorothy worried night and day about her son, who had entered the U.S. Navy as its youngest pilot. (He was her only son to serve: Pres Jr. had bad eyes; the other two were too young.) Her faith was put to the test when he was reported missing in action, shot down on a bombing mission. With a streak of luck—she probably saw it as an answer to prayer—he was rescued by an American submarine and came through unscathed. "I had faced death, and God had spared me," an introspective Vice President Bush said later. "I had this very deep and profound gratitude and a sense of wonder. Sometimes where there is disaster people will pray, 'Why me?' In an opposite way I had the same question: Why had I been spared and what did God have for me?" It was a question that would deepen the commitment his mother had drilled into him: from those to whom much is given, much is expected. The gospel of Saint Luke was instilled in him, just as Rose Kennedy had instructed her children.

Young Americans went to war as boys and came home mature men. The rush to settle down, to go to college on the unprecedented G.I. Bill of Rights, and to catch up on the living they had missed led to overflowing university classes, produced the fat and happy fifties, along with cookie-cutter suburban developments, and gave birth to the baby boom that continues to have an impact on every aspect of American society.

George Bush was among those who came home a different man: "It wasn't so much that I knew what I wanted as that I knew what I didn't want. I didn't want to do anything pat and predictable. I'd come of age in a time of war, seen different people and cultures, known danger, and suffered the loss of close friends. I was young in years, but mature in outlook. The world I'd known before the war didn't interest me. I was looking for a different kind of life, something challenging, outside the established mold. I couldn't see myself being happy commuting into work, then back home, five days a week."

His description of the life he didn't want was precisely his father's Wall Street life. The entrepreneurial bent of his mother's family had led him to chart a new course and make it on his own. With financial backing from Uncle Herbie's investment firm, he and Barbara headed for Texas (Barbara none too ecstatic at the prospect of life in Midland) and the expanding oil industry, reminding themselves, when things got dicey with the offshore rigs, that "to achieve something, you've got to be willing to take risks." After about fifteen years, George stepped into another risky business: politics—Republican politics in then-Democratic Texas. It proved to be a risk worth taking.

★   ★   ★

"IT must be so amazing to have your daughter the First Lady of the land," Dorothy Bush, Senate wife, wrote in a column for the hometown newspaper in 1962. She was ruminating about young Jackie Kennedy, but she would have been amazed had she known that in the years to come she would not only have a daughter-in-law as First Lady but also would set a record as the only woman to be wife of a senator, mother of a president (and two-term vice president), and grandmother of not one but two governors—with the possibility of adding grandmother to a president to the list.

Only Rose Kennedy could claim as many titles for her sons and grandchildren, but she had more to draw from.

At the time Dorothy made that observation, the Bush family would not have predicted a three-generation political record but would not have been surprised that it produced a president. "I just knew he would be doing something important some day," said brother Jonathan during Bush's 1988 campaign. "Even when he was a small boy, we all thought he would be famous." Candidate Bush, the sitting vice president, revealed that he had been pointing toward the White House for years. "It was almost evolutionary," he explained. "You have conviction in what you feel for the country—I guess realistically you look at the field and think, Why not me?"

Later, when he was President Bush, Aunt Mary Walker recalled a moment that left a lasting imprint on her mind. Not long after Barbara and George were married, several of the women of the family, chatting on a summer afternoon at Walker's Point, idly pondered the hypothetical question: How would you like to be First Lady? Young wife Barbara spoke up firmly, "I'd like it, because, you know, I'm going to be the First Lady sometime." The ladies were astonished, because she seemed to be serious. George had not yet said anything about going into national politics, but Barbara, said Aunt Mary, felt "there wasn't anybody in the world like George. She worshipped him." And when her prophecy came to pass, Barbara did indeed like being First Lady and became a national favorite in the role.

The two strong women in George Bush's life were not ruffled by the tenuous relationship between the daughter-in-law who feels the husband is hers and the mother who sees the son as the product of her efforts. They became immediate and lasting friends.

It could have been otherwise. The romance between George and Barbara started in 1941 at the Greenwich Country Club's Christmas dance, where charming girls of affluent WASP families met attractive boys from the same. George, seventeen, spotted tall, dark-haired, sixteen-year-old Barbara Pierce, asked a friend to introduce them, danced with her, talked with her, fell in love with her. Just like that. The music never stopped. "I married the first man I ever kissed," declared Barbara Bush when she was a white-haired mother of five and a grandmother. "When I tell this to my children, they just about throw up."

Any romance that was launched in December 1941—only days after the attack on Pearl Harbor—was heading for rough sailing as the lives of all Americans were changed. George rushed to enlist immediately after graduating from Andover; on his eighteenth birthday, he signed on with the navy, postponing college, postponing life—but not postponing Barbara, who was by then the central fixture in his plans. He was off to preflight training at the University of North Carolina in Chapel Hill; she was finishing up at fashionable Ashley Hall in Charleston, South Carolina. But before she was to enter Smith College in the fall, there was a weekend of fun—and young romance—with him in Chapel Hill.

If the Bush children were amazed that their father was the first man their mother ever kissed, they would have been astounded to learn that it had been the same for George, until that summer of 1942. In an unusual letter to mother from son, published in his collection *All the Best, George Bush*, he confesses, somewhat defiantly, "I kissed Barbara and I am glad of it. I don't believe she will ever regret it or resent it, and I certainly am not ashamed of it. . . . I have never kissed another girl." He then went on at length to deliver rather preachy views on sex, inveighing against "intercourse before marriage," deploring the carryings-on of his fellow cadets. "To think," he wrote, "all this was brought on by your asking me what I thought about kissing."

The following summer Barbara spent three weeks with George and an intimidating array of his family at Walker's Point, aware that she was under the clan's microscope, especially since she was only eighteen and George, barely nineteen, had already decided to marry her. But first there was a war to fight.

If his mother thought they were ridiculously young to be planning marriage, she—unlike Sara Roosevelt—never let on. In 1943 mothers of boys preparing to go overseas had more urgent worries than a son's romance, and Dorothy must have been pleased that George was so happy in that anxious time. She welcomed Barbara with an open heart and told her mother that her "most wonderful daughter" had made great peanut butter sandwiches for everybody, hardly a culinary feat. Barbara later speculated, with her unfailing sense of humor, that because Dorothy would say only good things about people, there wasn't much she could say about George's girl—other than the peanut butter sandwiches—at that stage of her life.

Barbara adored "Dottie" Bush from the start, and later admitted that she was more comfortable with her than she'd ever been with her own mother. ("That's a sort of chemical thing," she explained.) She sang Dorothy's praises as "the most supportive, wonderful, loving . . . she just was the perfect mother and the greatest mother-in-law." Dorothy's relationship with her other daughters-in-law was also warm, to the point that Nancy Ellis recalled, "I remember being jealous—Mother would tell me how wonderful they were, but I felt I was the one she should be saying that about." As the family moved into the White House, Barbara talked about her relationship with her husband's mother. "I had to decide early on as a daughter-in-law that you can't beat her—you have to sit back and enjoy her. When I was a new bride, she beat me in paddle tennis with her right hand, then with her left. She is an extraordinary woman. She brings out the best in us."

There was nothing but approval by both families when Barbara and George were married on January 6, 1945, at the First Presbyterian Church in Rye, New York. It was like a Jimmy Stewart movie: George, the young flyer, had cheated death when his bomber was shot down over enemy-infested waters; he arrives home on Christmas Eve amid what he would remember as "tears, laughs, hugs, joy, the love and warmth of family in a holiday setting"; he and the girl whose name had been emblazoned on his TBM Avenger are together at last. And the reality was as bewitching as Hollywood could have made it—eight smiling attendants declared the season in holly green dresses and red carnation bouquets; George could have been a recruiting poster in his lieutenant j.g.'s dress blues; dark-haired Barbara was radiant in white satin. Her new mother-in-law smiled with particular sentiment at the veil her son's bride wore: it was Dorothy's own, the cascade of rose point lace she had worn when she was the bride of Prescott Bush, so many wonderful memories ago. From that day forward Barbara looked to Dorothy as her model as wife and mother, even repeating the family pattern of four sons and a daughter. (Tragically, their first daughter died of leukemia when she was not yet four.)

With son Jeb the governor of Florida and George W. the governor of Texas, a Bush political dynasty is in the making. It is possible, after the election of 2000, that a historical footnote may be entered in a presidential facts website: Barbara Pierce Bush could become the first woman since 1825 to be both First Lady and First Mother, a distinction held only by Abigail Adams from the time her son, John Quincy Adams, followed his father's footsteps into the White House.

Should that occur, historians and journalists should take note of the commencement address Governor George Bush—"W" to the press, to distinguish son from father—gave at Southern Methodist University as he began his national campaign in 1999. "Remember that no matter how old you are or what your job is," he admonished the graduates, "you can never escape your mother. I know." His tone was light and loving, reflecting the special bond between him and his mother, forged in his early childhood years in Midland, when his father was away a great deal of the time developing his oil business.

The son recounted a civic event in a small Texas town where he was to introduce his parents: "I said, 'Mr. President, welcome to

Fredericksburg,' and there was a nice round of applause. And, then I said 'Mother . . .' and the place went wild. 'Mother,' I said, 'it's clear the people of Texas still love you and so do I, but you're still telling me what to do after fifty years.' And a guy in a big cowboy hat cups his hands and screams, 'And you better be listening to her, too, boy!'

"I do listen to her," he assured his college audience, "and one of the things she has taught me is the most important fixed star of all: God exists today and forever. . . . We need a renewal of spirit in this country, a return to selfless concern for others, for duty and for country. We must let faith be the fire within us." He was speaking of Barbara, but his message of selfless concern stemmed from the grandmother who had been her daughter-in-law's model and the guiding light for three Bush generations.

Like Dorothy, Barbara keeps a sharp eye on details. Before an important debate with Senator John McCain, the governor had an urgent e-mail from his mother—"She told me to stand up straight." And he divulged to Jay Leno that when his mother learned he would be on the *Tonight* show she admonished him, "You make sure your socks are pulled up!" He then ostentatiously tugged at them, as if to say, "See, Mom, I did what you said." The governor's wife, Laura, says her husband "is more like his mother—both are feisty."

★   ★   ★

TOURISTS being conducted through the White House amid the inaugural festivities of January 1989 were incredulous. There on the portico was the frail, eighty-seven-year-old mother of the new president, parked in her wheelchair, chatting and shaking hands with the line of public visitors, and when she moved inside, she took up a spot by a window where she could wave to them. The legs that had raced over so many tennis courts were no longer steady, and illness had left her fragile, but nothing could keep her from participating in this grandest moment of her life. It took a bit of doing—she was flown from her winter home in Florida in an ambulance plane, accompanied by two doctors and a nurse—but a mother could not miss such a day.

In an "Inauguration Exclusive" for the *Houston Chronicle,* her son Prescott Jr. mused that their mother may have privately viewed the inauguration "with mixed emotions—bursting pride that her second son was about to become the forty-first president of the United States, mixed perhaps with some concern for his welfare in a dangerous world."

Now, swearing on the Bible on which George Washington's hand had rested at the country's first inauguration two hundred years earlier, George Herbert Walker Bush began his chapter in the history of American presidents. In his inaugural address he hoped to set the tone for a less strident America, appealing to all Americans to help one another, to draw upon their "goodness and courage" to do what government cannot. Obliquely decrying the greed that marked the decade of the eighties, he declared, "We are not the sum of our possessions. They are not the measure of our lives."

Dorothy could hear her personal articles of faith going out to the thousands on the Mall and millions around the world. The new president was proclaiming the values she had passed on to him as those which would define his conduct of the presidency. After the ceremony he bent to kiss her and, speaking for the family, said, "We all worship the ground she walks on."

In the crowded days of the celebration, Dorothy, the matriarch, basked in her personal heaven, surrounded by family, four generations of her own Bush tribe and the remaining members of her generation of Walkers; her tiny great-granddaughter was asleep on the Lincoln bed. Two hundred and fifty Bushes and Walkers attended the family luncheon—cousins, nieces and nephews, this-and-that relatives and in-laws, reaching to the outer fringes of kinship.

Dorothy had sampled every stage of George and Barbara's variegated official life. When he was the American representative to China, she bicycled through the streets of Beijing at seventy-three. ("You should have seen the people stare at old momma on a bicycle," George wrote to a friend. "At each traffic light a little group would stand around, nudge each other, the kids were openly incredulous, but she cycled majestically off at each stop, doing beautifully in her teenage looking ski outfit.") She helped in his campaigns, enjoyed the vice president's residence with its old-fashioned Christmas-card charm—but nothing could match being in the White House, the *Bush* White House.

A few months earlier, as George Bush was embarking on his first presidential campaign swing, he relaxed on the sun-washed lawn of the Kennebunkport house and talked about his family with Donnie Radcliffe, who was writing a biography of Barbara. In his view, a big family gave the advantage of being "love squared." "We're very, very close," he said of the scattered and ever-expanding Bush clan. Then he motioned toward his mother's weathered cottage. "The lady in that house—my eighty-seven-year-old mother—keeps it all together. She is still the anchor to windward for all family members of whatever generation—she is their idol."

Three summers later, the family gathered en masse at Walker's Point to celebrate Dorothy Bush's ninetieth birthday, a joyous event made more precious by the awareness that her supply of birthdays was running very low. Some months earlier, George had confided to his diary that his mind that day was not on world problems or the budget: "It's on Mother and our love for her. . . . It's Mum's words: do your best, try your hardest; be kind; share; go to church—and I think that's what really matters on this evening."

Growing ever more frail, Dorothy lasted through the four years of her son's presidency, then, less than three weeks after his traumatic loss to Bill Clinton, another in a series of strokes left her semi-conscious. With his daughter, Dorothy, her namesake, the president flew to her bedside in Connecticut. He held her small hand, embraced her, and told her that he loved her, hoping his love would find its way through the shuttered portals of her mind.

At her bedside he saw her Bible, worn from long and regular use—and letters he had written to her as a boy at Andover and a birthday card he had sent when he was eighteen and going off to war. She was ninety-one and had lived a golden life, but there are moments when rational thought gives way to deeper emotion. He was losing the mother who had been his polar star throughout his life, and he wept.

Presidents must always put out statements, even on such personal occasions, regardless of private pain. His was heartfelt: "She was the beacon in our family—the center, the candle around which all the moths fluttered—she was there, the strength, the center, the power but never arrogance, just love was her strength, kindness her main virtue. How many times she taught us to be kind to the other guy, never hurt feelings, love." He had been dealt two terrible blows, one upon the other, and now he must hand over the White House to the man who had defeated him. He performed that wrenching task with the grace and goodwill he had been taught by his mother. Her hallmark on her favored son was indelible.

MY LIFE IS LIKE A
COUNTRY SONG

Sons are the anchors of a mother's life.

—SOPHOCLES

*I*T was early morning on May 18, 1946, before breakfast, when the telephone rang in the Cassidy residence in Hope, Arkansas. Virginia Cassidy—Mrs. William Jefferson Blythe Jr.—answered and was puzzled when she heard her husband's good friend saying to the operator, "No, I don't want to speak to Mrs. Blythe, I want to talk to one of her parents." Her mother took the phone, listened for a moment, turned pale, and burst into a flood of tears.

In those few seconds Virginia, not yet twenty-three years old and six months pregnant, felt her life pulled out from under her. Her handsome husband Bill Blythe, who had come home unscathed after two years of war in Italy, was dead, the victim of a freak one-car automobile accident. He had been heading west across Missouri, traveling fast on Route 60 from Chicago, eager to join his wife in Hope. A tire blew. Hours later Bill Blythe's body was found facedown in a drainage ditch, his hand still grasping weeds, as if to save himself. The young veteran of twenty-eight had drowned in a muddy ditch on a Missouri prairie.

Three months later, on a night riven by an electrical storm of terrifying force, Virginia gave birth to a son and named him William Jefferson Blythe III in honor of the father he would never know. That August night in Hope marked the beginning of a mother-son relationship of singular devotion, one in which the son would become her friend and protector and virtually a surrogate father to his young half brother.

"Even when he was growing up," Virginia said proudly, "Bill was father, brother and son in this family. He took care of Roger [Jr.] and me." In giving him responsibilities beyond his years, she gave him the confidence that is the basic building block of leadership. But psychologists suggest that casting him as the supportive husband who was absent, by death or character, in her life imposed too great a burden on the adolescent boy; the fallout was negative as well as positive. Clinton later commented, not happily, "I was forty years old by the time I was sixteen."

Forty-six years after that stormy August night in Hope, on the day her fatherless infant, now the forty-second president of the United States, moved into the White House, Virginia, ensconced in the queen's bedroom, let her mind roam back over the incredible highway they had traveled together: "What a journey it has been. Lord, the hills we've climbed!" And as she looked down the road, she recalled in her memoirs, "I wondered what Bill's election would mean for all of us. I knew one thing for sure: None of us would ever be the same again." Her son was part of history—that comes with the election. But how will he be remembered?

She had witnessed his finest hour, the culmination of dreams beyond dreaming and years of unceasing effort. She did not live to see him reelected; he would become only the fourth Democratic president to win a second consecutive term, joining the illustrious company of Andrew Jackson, Woodrow Wilson, and Franklin Roosevelt. But it can only be seen as a kindness of fate that the doting mother was not forced to endure the ignominy that would indelibly stain his presidency and make him the butt of raunchy jokes on late-night talk shows.

That Bill Clinton's personal character is flawed is beyond argument. If other presidential mothers can be credited with building their sons' characters, how much of the responsibility for his flaws can assigned to his mother? The impact of Virginia's influence is magnified by the dual role she was forced to play in her son's life, functioning as both mother and father to an extent not required of the others. Thus, for better and for worse, the imprint she made on her son is overwhelming.

The flamboyant Virginia set some examples unlikely to figure in a parenting book. If there was a slot machine within reach, she was quick to pull the lever; if a roulette wheel was spinning, she wanted chips on the board. If there was a friendly guy around, she would flirt. If drinks were being poured, her glass was ready. If a party was going on, she would be the life of it. Oaklawn racetrack and Hot Springs nightclubs were her natural habitat, and her heavy hand with makeup was a signal that she was a fun girl who liked to show off.

When any lingering vestige of inhibition was loosened by cocktails and jollity, she didn't hesitate to leap onto the nightclub stage and sing along with the celebrity performer. At Saturday-night gatherings with friends she would climb up on the kitchen counter and belt out a song of her own composition, "I'm the Hempstead County Idiot." "Which," she chuckled in hindsight, "I obviously was." The lovable thing about Virginia was that the most outrageous stories were those she told on herself; she unfurled her less-than-motherly behavior with much laughter and no remorse. Her autobiography, *Leading with My Heart* (with James Morgan), is an exuberant, no-holds-barred memoir that reflects her zest for living life at full gallop.

She summed up her life story with the boast "I was not one for rules." That attitude was not lost on her son, who came to feel that rules were for other people, not for a governor, much less a president. His disdain for other people's rules was reinforced by his luck in rarely being caught, in not having to face the consequences, in getting away with it.

Virginia's anything-for-fun exterior tended to obscure her kind and engaging real self. "I'm a character, a cut-up, a kook," she said of herself. "That's me, all right, but it's not all of me." She was a warm heart her friends could always count on, a cockeyed optimist who triumphed over more adversities than most women would encounter in two lifetimes: tragic death, an abusive husband, a son in prison, and the one thing she could not conquer—breast cancer.

How did she survive all of those battles and stay the same jaunty, irrepressible Virginia? In the first place she taught herself never to worry about the what-ifs, but when bad things happened, she had a system to deal with them: "Inside my head I construct an airtight box. I keep inside it what I want to think about, and everything else stays beyond the walls. Inside is white, outside is black. Inside is love and friends and optimism. Outside is negativity, can't do-ism, and any criticism of me and mine."

In passing on her airtight box system to her son, Virginia enabled him to compartmentalize his life to a degree that has astounded the people around him. Throughout the long year leading up to his impeachment by the House of Representatives and trial in the Senate, Clinton seemed able to distance himself, mentally and emotionally, from the accusations that were directed at him. Embracing his mother's dismissal of "any criticism of me and mine," Clinton just kept going, as she always did, never outwardly revealing his humiliation. In keeping with her practice of denying unpleasant facts, he locked the episode that jeopardized his presidency outside his airtight box.

★　★　★

VIRGINIA was already a registered nurse when her son was born. She was a go-getter, a single mother who could see a guaranteed future for herself and her son if she could qualify as a nurse-anes-thetist. With a year's specialized training at New Orleans' large Charity Hospital, she figured she could be an independent woman for life. Independent. That was her goal, to get away from the stul-tifying confines of her unhappy home and the constant irritations of her cantankerous mother. Though it hurt to leave her son a few months past his first birthday, her parents were delighted to take over the jolly, chubby toddler.

It was a move that gave Billy Clinton a second mother in the strong and ambitious Edith Cassidy and a kindly substitute father in Eldridge Cassidy. They adored him—all the more because they showed little love for each other—and Billy dearly loved "Mam-maw" and "Pappaw." Far from being an abandoned child, he was swaddled in attention; it was Virginia who suffered the pangs of separation.

By the time Virginia returned to Hope in 1948, nurse-anes-thetist diploma in hand, her mother had come to feel that two-year-old Billy was her own, setting up a competition in which the precocious boy had two women fighting over him—and the real mother was the weaker contestant.

From childhood Virginia had lived at sword's point with her mother, an intense, angry—yet oddly flirtatious—woman. "She met every day with anger," Virginia said, "and took it out on me and my father." Edith's "nightly screaming fits," her shrieks and jealous accusations against her husband, horrified their young daughter, lying awake in the next room, her heart aching for her father. "I loved him as much as it's possible for a daughter to love her dad," she declared, an echo of Lillian Carter's statement about her father. "But the truth is, he was too good for his own good. He was kind and gentle and he loved laughing and fishing and story-telling and people—especially me." His "Ginger" was every bit his daughter, except for the flash and tough-mindedness that were every bit Edith.

Eldridge had been an iceman in Hope, a cotton market town of about five thousand situated on the main rail line, a town where tent shows would set up now and then. Barnum & Bailey Circus even played there once. In those days, the iceman let himself into the kitchen and with huge tongs placed the frozen block in the icebox, an oak cabinet with a pan underneath to catch the melting ice. With his easy access, the iceman was the counterpart of the farmer's daughter in risqué stories, which fueled Edith's dark suspicion of affairs with other women.

Later, he operated a little grocery that was one of the few places in Hope frequented by both blacks and whites. Both races were welcome to buy a bottle of illegal liquor from the stash he kept well hidden to evade the county's "dry" laws. (It was easy enough to bring in a regular supply from Texarkana, thirty-five miles away, where booze was sold lawfully.) Eldridge gave his black customers the same fair treatment he gave the whites, extending credit as well as cordiality, and Edith, a practical nurse (certified through a correspondence course), shared that tolerance. On her own time she cared for black families in need, much as Lillian Carter did in Plains. The Cassidys' attitude, which was neither conventional nor comfortable in a rural southern town in the days of segregation, inspired in their observant young grandson a deep commitment to equal rights, which, many years later, won solid support for him among black Americans.

Though Virginia saw her mother as "negative and cynical," with her grandson Edith was imaginative and loving. She poured her abundant energies into him, making flash cards to drill him on the alphabet and beginner's words, teaching him numbers with playing cards, enrolling him in the First Baptist Church when he was only three. "They taught me to count and read," their grandson remembers. "I was reading little books when I was three." Family friends were astonished to see Billy reading the newspaper headlines in the first grade, and a junior-high chum retells the running joke "Let's go over to Bill's and watch him read."

His grandparents, the president said, "didn't have much formal education, but they helped embed in me a real sense of educational achievement." Both as governor and as president, he pressed for better education for young children, never forgetting how that early teaching by his mother and grandmother had instilled in him a love of learning.

Much as Virginia adored her son, she was a vivacious young woman not cut out for widowhood. She used to brag a bit that "I've always loved men"—and they were likewise attracted to her. Her track record suggests that she judged them less by the content of their character than the cut of their clothes, the flash of their car, their line of chatter, the fun they provided.

Less than a year after Bill Blythe's death, she began going out with Roger Clinton, a car dealer twelve years her senior, and after she returned home from New Orleans they began to "get serious." He was a two-toned-shoes kind of guy, a sharp dresser whose friends called him "Dude," a gambler, a drinker, and a heavy partyer with a repertoire of off-color stories and a keen eye for thoroughbred horses and less-pedigreed women. "You've never seen such strutting in all your life," Virginia said many years and much pain later. "But the thing was, he made you feel like strutting, too."

Edith Cassidy despised Roger Clinton, and her brother, Buddy Grisham, gave his niece a measure of Dutch uncle advice when she told him she planned to marry Roger: "You're fixin' to marry a bunch of Buick cars—and you'll have hell from then on." Virginia, headstrong as always, ignored the warnings.

"I accepted Roger Clinton at face value," she explained in her memoirs, trying to justify why she would marry this alcoholic gambler, a high-school dropout not yet divorced from his second wife when they met. Whatever went through her mind as she made that decision, it involved ignoring such things as the high-stakes crap games in his apartment, his first wife's divorce on grounds of physical abuse, the garnisheeing of his salary to pay court-ordered child support, his vicious attack on a young Puerto

Rican in a gambling dispute. But Roger always seemed to have money (until he lost his Buick dealership through mismanagement), always drove shiny new cars, and was always ready to party. Though he was not much of a model for her bright young son, he was crazy about him and little Billy was excited to have a father. For four years Virginia had been a widow, living uncomfortably with her parents, sharing her son with them, longing to be a wife, to build a family of her own.

They were married on June 19, 1950, in the parsonage of a church Virginia could not later recall, except that it was convenient to Oaklawn racetrack. No family members were present; Edith was threatening to seek legal custody of Billy, arguing that the child, soon to be four, should not be in the same house with Roger Clinton. "Over my dead body!" roared Virginia, and the law was on her side. A new chapter of mother and son's entwined life was beginning—and Edith Cassidy's instincts proved to be sound, Uncle Buddy's prediction on the mark.

★   ★   ★

THREE years after their marriage the Clintons moved to Hot Springs, where Roger, who had lost the car dealership, managed the parts department in his highly successful brother's automobile agency. Virginia was overjoyed at the move. Hot Springs was her kind of town, a Las Vegas on training wheels. Gambling, nightclubs, prostitution, a mid-America stop for big-name performers—Liberace, Tony Bennett, Peggy Lee, a roster of pre-rock stars. "I loved Liberace," she admitted. "I loved his diamonds, I loved his furs, I loved his *makeup*!"

All manner of entertainment, much of it illegal, flourished in Hot Springs, a town of only thirty thousand, but the town fathers and godfathers worked together in perfect harmony. In his heyday, mobster Al Capone and his bodyguards used to nip down from Chicago—for the restorative mineral baths, of course. It took a governor who was immune to the influence of money—Winthrop

Rockefeller—to take the hot out of Hot Springs, ostensibly putting a stop to the blatant illegal gambling in 1967.

Virginia's favorite club was The Vapors, all red velvet and sparkling chandeliers, "as plush and glittery and showy as anything Las Vegas ever dreamed of," she boasted with local chauvinism. Of herself she declared, with endearing candor, "I obviously was born with a flashy streak inside me just waiting to burst free, and Hot Springs let me be me with a vengeance." But Hot Springs had another side: it was also a conventional southern town of traditional values and strong Baptist dominance that cast a cold eye at the rhinestones and sequins element. In those circles, "being me" generated considerable gossip about the well-known nurse-anesthetist. That she was "a lightning rod" for gossip did not bother her in the least, and she saw no need to change her style: "When you go out of your way to look and act the way I do, it comes with the territory. I think it's one of the trade-offs for being who you want to be." With her ability to live in denial and to go on the offense against critics, she turned the gossip into a compliment: "I've always known that people speculate about strong people. People speculate about leaders."

Now and then she took Bill to a club to hear such celebrated jazzmen as Jack Teagarden play, to encourage him to think beyond the marching band, to take up the saxophone and form a group of his own. Though her intentions were well meant, it was hardly appropriate to take a young teenager to a nightclub where the ambience was booze-fueled and his stepfather would most often be gambling in the back room. Bill, uncomfortable in such a setting, always wanted to leave earlier than his mother did; she had placed her son in the husband role as her escort, and he didn't relish the grown-up part.

Of all of Hot Springs' charms, the most alluring, to Virginia, was Oaklawn Park Racetrack, which came to be her second home (she calculated she spent one-third of her life there, except during

Bill's campaigns). Sometimes she won, sometimes she lost, but it was always exciting.

She loved the wild side of Hot Springs and apparently never gave a thought to the gamble she was taking in raising her son in an environment rich in temptation.

★   ★   ★

"I, Virginia Cassidy, will to Mary Jo Monroe my magnetic attraction for boys," read the jokey "Last Will and Testament" of Hope High School's class of '41. To which the young editors added, "Help us, please, if she turns it on full force."

"High school was fun," she reminisced fifty years later. A boy could take her out on a date for fifteen cents—two Cokes, and five cigarettes at a penny apiece. She flirted with boys, jitterbugged to Tommy Dorsey on '78s, and shared a few kisses on the porch swing, but there was another side to Virginia Cassidy. She was in plays, worked on the school newspaper, was elected class secretary, and made grades good enough to get her into the National Honor Society, all of it done with her rollicking good humor. Her quote under her class picture pronounced her philosophy of life: "I'd like to be serious but everything is so funny."

Yet it was an uncertain time. The pall of the Depression had not lifted from Arkansas, which has always ranked among the poorest states, and everyone feared that the war raining destruction on England would soon involve America. Virginia never considered going to college, even if her family could have afforded it, but set her sights on nursing, her mother's profession.

One look at Virginia as an adult and you knew she was not destined for a quiet, sedate life. In her book she detailed the intricacies of getting her face ready for the day. She would start with wildly false eyelashes—"some people have said I look like a spider with them . . . but I think they're pretty." Then came layers of powder, rouge, lipstick, liners, red nails—the whole bag of tricks.

Later, as her hair turned gray, her hairdresser devised her trade-mark "racing stripe," a three-inch streak, front and center, left white while the rest of her tresses were colored dark brown. (She later married the hairdresser.) "I like people to notice me," she said, dismissing timid souls who made do with nature's allotment. "I hate them not to notice me."

Whether it was the cyelashes or her pheromones, there were always fellows around, and she married four of them, one of them, Roger Clinton, twice. But she always thought of Bill Blythe as her heart's true love, from the moment she first saw him at the hospital in Shreveport, Louisiana, where she was a student nurse. He had brought a woman friend to the emergency room and "when I saw him standing there, I was stunned," she said, still a bit breathless half a century later. "He smiled, and the only way I can describe it is that he had a glow about him. I was weak-kneed. There was some strange and powerful attraction." As he was leav-ing the hospital, halfway down the steps he turned and came back to the desk where she was working. So many lives would be affected by Bill Blythe's split-second decision to go back up those steps.

First he asked about the ring she was wearing on her fourth fin-ger. In a flash she lied that the almost-engagement ring meant nothing really, nothing at all. He suggested that they go out for something to eat. There was almost half an hour before the nurses' curfew, time enough for a Coke and kisses—though kiss-ing on the first date, she said later, was "something you just didn't do in those days." (Since the very shy and innocent Rosalynn Smith confessed doing the same with Jimmy Carter, perhaps a first-date kiss was something you just didn't admit to rather than just didn't do.) In midwar 1943, everything was on fast forward, and Virginia acknowledged that soon there was more than kisses—and within two months Virginia, twenty, and Bill, twenty-five, a new recruit in the army, were married in Texarkana by a justice of the peace.

Five weeks later he was headed overseas, to North Africa. It was all so fast, yet three months allowed ample time for him to mention a few relevant details about his life. He could rightly claim to be a hard worker: at thirteen W. J., as he was then known, had been obliged to drop out of school to help support his dirt-poor family in rural Texas (nine brothers and sisters crowded into a four-room house). Then he got a chance at a future, a job as a traveling salesman for heavy machinery, working the Midwest region. Impressed by his hustle and charm, a salesman's best tools, Virginia envisaged a solid life ahead, blissfully unaware that his moral road map had had many detours and that she was, in fact, W. J. Blythe's fifth wife, which he had neglected to mention.

Court records and family memories of his life before Virginia reveal a complicated story of a marrying kind of man. He had been married four times before; he had a son by his first wife, born two years after their divorce, and a daughter by his fourth wife, born eight days after the marriage (the father's name, Wm Jefferson Blythe, is clearly inscribed on the birth certificate in the flowing handwriting of a Kansas City courthouse official). His second marriage had ended in a rancorous divorce after nine months, in which he was judged guilty of "extreme cruelty and gross neglect of duty." His third wife was his first wife's younger sister, this five-month marriage is generally explained as W. J.'s way to avoid marrying another girl who claimed she was pregnant by him. The fourth marriage had a special relevance because the subsequent divorce was granted in Kansas City in April 1944, while he was overseas—seven months after he had married Virginia. The sequence of events points to bigamy, though the charge was never brought against him. Blythe had accrued this multiple-marriage record, which few men would accumulate in a lifetime, by the time he was twenty-three. Not until his son and namesake was elected president did the details of his reckless marital history come to light. Bill Blythe was obviously a ladies' man, and his DNA is woven into Bill Clinton's.

It is inevitable that a young father cut down by war or accident will be mythologized by his children and his widow; in fantasy he becomes the wonderful husband, the perfect dad. Bill Clinton, in his first presidential campaign, told TV journalist Bill Moyers that he had once driven that fateful stretch of Missouri highway to see for himself where his father's life had ended in a rain-filled ditch. "I was wondering what it might have been like," he said wistfully, "and wishing he'd landed the other way."

Nothing in Bill Blythe's short and turbulent life story suggests that he would have been much of a father. Virginia, stunned when all this came out after her son's election, could still find it in her generous heart to be forgiving, certain that she had been "the love of Bill Blythe's life." She doubtless was, for the few months they were together, and possibly their marriage would have fared better than his previous several. A dream in retrospect, a dream unattainable, is better than no dream at all, and she would consign his less than respectable past to the black space outside her box of happy thoughts.

The shadow of the father he never knew was always there in his subconscious, a grim reminder that life can be short. "I always felt that I should be in a hurry in life, because it gave me a real sense of mortality," the president reflected. "Most kids never think about when they're going to run out of time, when they might die. I thought about it all the time." With that sense of urgency, he leapt into a race for Congress when he was only twenty-eight—his mother was first to encourage him to go for it—and at thirty-two he became the youngest of the nation's fifty governors. "I thought I had to live for myself and for him, too," he said in a moment of self-analysis. "I sort of had to meet a very high standard of conduct and accomplishment, in part because of his absence." The fact that his dead father had demonstrated neither high standards nor accomplishment adds a touch of irony; nevertheless, Bill Blythe's son felt a need to prove himself to his father.

It was his mother Bill Clinton thanked first of all for his success, with her example of unwavering competitiveness, unbeatable optimism, and her philosophy of maximizing your luck. What better tools are there for a politician? Virginia and her son were alike in one other way—their sheer delight in entering a room of strangers and captivating them. She and her son were alike, she said: "When we walk into a room, we want to win that room over. Some would even say we *need* to win that room over. If there are one hundred people in a room and ninety-nine of them love us and one doesn't, we'll spend all night trying to figure out why that one hasn't been enlightened."

★   ★   ★

DESPITE the distractions of Hot Springs' lively nightlife, the Oaklawn races, and her thriving career as a nurse-anesthetist, Bill was always at the center of Virginia's life. Her living room was a cross between a shrine and campaign headquarters, its walls covered with his awards, mementos, band contest medals, photographs. This son, she was certain from his birth, was something special. Bill had "a sense of destiny" about him, and she would see that he met it.

He was the golden boy at Ramble Elementary and at Hot Springs High—class officer, the lead in school plays, in every kind of club and the band, a National Merit semifinalist. He excelled in calculus and Latin and was "one of those students," recalled classmate Paul Root, "that history teachers pray to have in their class."

And he was popular with his schoolmates. Running for office years later, he described himself as having been the stereotypical fat kid, teased and self-conscious; that could only have been an appeal to the formerly-fat-kid vote, because Billy Clinton was a star, though not on the football field like Jerry Ford or as a lifeguard like Dutch Reagan. He was a leader, more mature, more focused, harder-working than his peers. One black mark: in the fourth grade he got a D in conduct for talking too much—a politician in the making.

His growing-up years were, by his account, as happy as those of presidents from simpler times and less troubled homes; anxiety about his stepfather's uncontrolled drinking was kept strictly separate from the lighthearted side of his life. There was touch football, shooting hoops in the driveway, hiking on the mountain close by. For a different kind of fun he and his pals piled into his convertible (an uncle with a car dealership was a great asset) and crowded into a booth at Cook's Ice Cream, grappling with the hard decision of which of the thirty-one flavors—all homemade—to choose.

They were hooked on music, spending hours transfixed by every Elvis record—Elvis fanatics, they knew every word, every beat of The King's hits. Virginia scored high with them as a mother who actually loved the gyrating icon and knew the words, too. Bill and two pals who were also music-makers formed a trio, "The Three Kings," who were good enough to play at school functions. (They adopted the gimmick of dark glasses for their gigs and predictably were dubbed "The Three Blind Mice.") Bill and Carolyn Yeldell (now Staley), the Baptist preacher's daughter who was his next-door pal and sometime girlfriend, spent hours together, Carolyn at the piano and Bill on tenor sax. Virginia had enough enthusiasm to spread it around. "She was cheerleader for our dreams," Carolyn recalled fondly.

Virginia, happy that Bill's home was the after-school hangout for his friends, would join them when she came home from her day at the hospital. After she produced cocoa and peanut butter and banana sandwiches for everybody, they pulled up chairs around the dining room table and talked. "Bill and his mother did most of the talking," Carolyn chuckles. Virginia arrived overflowing with tales from work, indignant about some controversy at the hospital, about patients' problems, about some unfairness she had seen or experienced. Her opinions were never mild. Carolyn calls these sessions Virginia's "kitchen table teach-ins," discussions about "whatever was the news of the day," including racial issues, which were extremely volatile in Arkansas at that time. "She felt strongly that it wasn't right for people to be looked down on, or made second-class citizens."

The kids around the kitchen table were made to think beyond the Saturday-night dance to issues of social justice. Carolyn (who now works for the National Institute of Literacy in Washington) has no doubts that "Bill got his sense of justice from his mother. She was early in understanding that people without a voice must have someone to speak up for them. She would talk about people who were denied access to the best medical treatment because they didn't have money, the unfairness of it. She was a champion for those who have not, always looking out for people who by some quirk of fate didn't have money, position or rank." Her views resonate in his presidency. Up until Bill and the others were about twelve, they mostly listened, but when they got to high school, Bill often challenged his mother and questioned her views; these arguments, Carolyn stresses, "were always good-natured," a kind of sport.

Carolyn was impressed by the way Virginia treated Bill as an adult, an equal. "She respected him, discussed issues with him, and trusted him with a great deal more freedom than most of us had. She recognized, I believe, that Bill was a uniquely gifted young man and wanted to help him have the fullest range of opportunities possible in life."

Those kitchen-table debates and their impact on him were the genesis of the priority Clinton, as president, would place on health-care issues and his early, though unsuccessful, effort to address the problems of the system. Her stand on racial issues was reflected both in his policies and his warm personal relationships with black friends; her strong opinions, reinforced by his wife's abilities, led him to appoint more women than ever before to policy-level jobs. Virginia's professional problems made him understand the inequities women face in the workplace. At that kitchen table the food for thought was more nourishing than peanut butter and banana sandwiches.

Virginia was consciously training Bill to think about issues larger than the Arkansas Razorbacks' next game, to marshal his facts and argue his points, just the way a candidate would have to do in a presidential debate.

*　　*　　*

LITTLE Roger Clinton was running as fast as his six-year-old legs could carry him to the house next door, shouting, "Bubba! Bubba! Daddy's killing Dado!" Bubba—Bill—was off like a flash, because in his little brother's vocabulary "Dado" meant mother. Bill found them in the laundry room. Roger Clinton, raging drunk, had pinioned Virginia over the washing machine, brandishing scissors at her throat. Bill pulled his mother and brother to safety behind him and confronted his stepfather: "You're not going to hurt them anymore. We're not going to take it anymore. If you hit them, you're going to have to go through me." After more shouting, Roger slunk away, the old lion displaced by the young alpha male.

Roger had always been a mean drunk, quick to pull off his belt and threaten to lash his wife or sons. More often he hurled jealous taunts at Virginia, accusing her of carrying on with other men. Once, not long after they were married, inexplicably furious that she was taking little Billy for a last visit to her dying grandmother, Roger whipped out a pistol and shot a hole in their bedroom wall. The shot went over her shoulder—was it a display of anger, or of more sinister intent? Virginia chose to take it as blind anger. More than once police were called when Roger was beyond managing.

Bill was only fourteen the first time he stood up to his stepfather; matching roar for roar, he dominated him on that night when the violence reached such a pitch that he burst into the bedroom to protect his mother.

"I'll never forget the scene as long as I live," Virginia recalled, still aching that their family life had sunk to such a low. "I'll never forget how straight Bill looked him in the eye. 'Hear me,' he said to Roger. 'Never . . . *ever* . . . touch my mother again!' " The son who had always been her friend and confidant had now become her protector. For Virginia it was a traumatic yet affecting moment. For Bill, it was the moment he was no longer a kid. The roles had reversed. "I was the father," he said, looking back on the brutal episode from the distance of years. "My mother just put the best face on it she could. In later years she dealt with a lot of stuff by silence."

As a result of responsibility thrust on him prematurely, teachers and school friends saw Bill as more mature than other teenagers, "a young man older than his chronological age," observed band director Virgil Spurlin, who was deeply impressed by this serious and talented boy. The family traumas that had forced him to grow up too soon did, however, give him an advantage over his peers— the confidence of maturity, an ease with leadership, and a certain arrogance that he knew more than the others.

Despite his promises to reform, Roger took no positive steps and remained an alcoholic given to unbridled rage. On such nights, Virginia said, "Our house was just bedlam." When she could stand it no longer, she asked for a divorce. She agonized over taking that step: how much easier it had been to say she would marry him than to say she would leave him. "Breaking up a family," she observed, "is a frightening decision." When Roger refused to move out of the house, Virginia, undeterred, moved into a motel with her sons while she looked for a place to live.

She found a nice one-story rambler on a quiet street, which she bought with money she had squirreled away out of her own earnings, which had always outstripped Roger's. "It was *my* house, earned by my own tears, the home I made for *my* family." A few years earlier she had left Roger, only to discover that her sizable income, which she had turned over to him without question, had been swallowed up, used to pay off his gambling debts. What she thought were payments on their house were, in fact, rent paid to Roger's well-to-do brother. That time she had had little choice but to go back to him, a humiliation endured by many women without funds of their own. Like Scarlett O'Hara, her role model, she declared, "Never again will I be without money to protect us!"

But when Roger moped outside her house in his car day after day, she relented and, despite Bill's plea not to make the same mistake twice, remarried the spouse who had bruised her physically and emotionally. Twenty-five years and two husbands later, Virginia tried to explain why she had changed her mind: "I took Roger Clinton back—I remarried this man who had abused me and my children for twelve years because I felt sorry for him." She no longer loved him but acted out of duty. With no feeling for her husband, she channeled her enthusiasm and attention into her firstborn, the pride of the family (little Roger was still very young), its "hero." Oddly, it was during the brief period of divorce that Bill chose to have his name legally changed from Blythe to Clinton (which he had used from the time Virginia married Roger). He wanted to make his records consistent with those of his little brother, who was soon starting school, "for family solidarity," as Virginia put it.

Life with the violent, alcoholic Roger Clinton had permanent and widely divergent effects on the characters of both of her sons. His own son, Roger Jr., would assert flatly, "I always thought of my dad as a cruel, mean, frightening man," whom he hated, which was understandable in light of Virginia's contention that her husband "enjoyed inflicting the fear [of violence], especially on little Roger." Still angry twenty-five years later, the biological son would say, "I have no pleasant memories, just memories of pain, violence, confusion." Bill, the stepson who was seemingly the better loved of the two boys, had the opposite reaction. When Roger Clinton was undergoing weeks of treatment for advanced cancer at Duke University Medical Center, Bill, then a student at Georgetown, drove the two hundred and fifty miles to North Carolina each weekend to be with him, and the two reached a reconciliation.

With commendable generosity of spirit, Bill remembered him as "a marvelous person who was very good to me. . . . It really was a painful experience to see someone you love, that you care about, in the grip of a demon." Like Ronald Reagan before him, Clinton had thought his way through the suffering his father's alcoholism caused the family and could be forgiving. (Unlike Roger Clinton, however, Jack Reagan was never abusive.)

Both mother and son armored themselves with denial, and Clinton, as president, said of those difficult years, "I had a normal childhood. I had a good normal life. . . . I had to learn to live with the darker side of life at a fairly early period. But I wouldn't say it was a tormented childhood. I had a good life." He may not have felt his childhood was tormented, but it requires his gift for compartmentalizing reality to describe it as normal.

Analyzing himself, President Clinton has said that growing up with an alcoholic father had lasting effects: "When you grow up in a dysfunctional home, inadvertently you send mixed signals. . . . I think I have to be acutely aware that I grew up as a peacemaker, always trying to minimize the disruption. People underestimate your resolve because you go out of your way to accommodate them before you drop the hammer."

His tendency to recoil from confrontation and to send mixed signals has been interpreted by his critics as the mark of a man who tells everybody what he thinks they want to hear. But to psychologists who work with the families of alcoholics, these are modes of coping with a disturbing situation. Clinical psychologist Paul M. Fick, a specialist in the problems of adult children of alcoholics and author of *The Dysfunctional President*, sees Clinton's "clearly dysfunctional upbringing" as the root of his "tendency to lie, to waffle on significant issues, [to be] energized by self-created chaos." Fick identifies blocking out hurtful memories, denying, and lying or "reshaping the truth" in stressful situations as symptomatic of the child of an alcoholic parent, and Virginia's way of dealing with a difficult marriage, he asserts, "was the epitome of denial."

★   ★   ★

HOW would Virginia have dealt with the Monica Lewinsky scandal? It is likely that she would have defended her son as vigorously as she did when the Gennifer Flowers affair became public just before the 1992 election. This time, though, even she would have found it hard to turn aside the charges with the glib dismissal that many women found her son sexy. Would she have counseled Bill to bull his way through by denial and evasion—and would she have recognized, in his behavior, some of her own traits and the examples she had set?

She might have recalled the advice she had given him years before: "Be who you want to be and don't worry about what people say about you." That chin-out attitude was fine for a woman thinking in terms of flouting Hot Springs conventionality, but taken too literally, it could be construed as "do what you want to and thumb your nose at your critics." That translation could be dangerous for an overly confident president who had relentless political enemies, a man who relished risk and exercised little restraint in his sexual behavior. The choice of being "who you want to be" was coupled with a warning: "You have to have a thick hide." In the year of ugly revelation and threats to his presidency, he demonstrated just how thick his hide really was.

Virginia was inordinately proud of her little man, the good son, her young achiever, and fulfilled his every wish. As a child he was rarely punished and any little transgression was always quickly forgiven. And, a child treated this way by a doting mother might very well expect the same leniency from everyone else. "I wasn't much of a disciplinarian," Virginia conceded. "He would do anything to avoid punishment. It offended his dignity." That also suggests that he was adept in talking his way out of trouble. Virginia's reaction to the Monica crisis would almost certainly have been to defend him, no matter what, and blame it all on other people.

During the months leading up to the impeachment, psychiatrists, depending on their specialties, analyzed him (from afar) as a sex addict, a reckless risk-taker stimulated by danger, the untreated adult child of an alcoholic, a habitual dissembler, an egocentric who disdains rules and expects to charm his way out of any tight corner, an insecure man needing sex to bolster low self-esteem. Geneticists would add to that catalog the influence of his DNA.

Though sexual addiction is not accepted as an illness by the psychiatric profession—the variables are too great and the specific boundaries too vague to establish diagnostic guidelines—many practitioners share the view of Jerome D. Levin, a New York psychotherapist who maintains that sexual addiction is a real and "surprisingly prevalent" problem. In his book *The Clinton Syndrome,* which addresses the "self-destructive nature of sexual addiction," he argues that for Clinton to endanger his political agenda, his presidency, and his place in history "for the sake of having oral sex [with Monica Lewinsky] bespeaks of either recklessness bordering on madness, which simply is not Clinton, or the irrationally compulsive behavior of an addict seeking affirmation and reassurance." If even half of the charges against Clinton are true, Levin concludes, "he has undoubtedly lost control—he has become powerless over his sexual impulses." In short, he's a sex addict.

David Maraniss, the preeminent Clinton biographer, summed up the president's sexual behavior in a journalist's concise terms: "narcissism, arrogance, stupidity, and cynicism."

Clinton has long been concerned with the subject of addiction, an interest heightened by his younger brother's cocaine habit, which led to his arrest in 1984. During that bleak time, in a serious discussion with his old friend Carolyn Yeldell Staley, Governor Clinton ruminated, "I think we're all addicted to something. Some people are addicted to drugs. Some to power. Some to food." Not insignificantly, at that time he was involved with Gennifer Flowers, according to her account, and talk of his womanizing had become a topic on the grapevine of national politics.

Fifteen years later the White House intern who will be a scarlet footnote in history volunteered that the president, among his many personal comments to her, said "he didn't want to get addicted to me." And he flattered her with his highest compliment: "You are like my mother—full of piss and vinegar." In those circumstances it is doubtful that Virginia would have liked that comparison.

Addiction spawns addiction, as many studies show, and Bill Clinton's life was scarred with addictions. His stepfather was an alcoholic and "a rageaholic," his brother a cocaine addict, the grandmother who partly raised him fell prey to morphine, his natural father's record points to what is called sex addiction, and his mother was an inveterate gambler, small-scale and never out of control, but happily hooked. Any psychiatrist or geneticist or sociologist might wonder how he reached the heights he did and was able to stay there.

★　★　★

WHO *is* this girl Bill has invited for a visit? Virginia was not prepared for Hillary Rodham, the Chicago girl, Bill's friend from Yale Law School. Apparently she was a very special friend. Since his high-school days he had brought lots of girls home to meet his mother, but none of them had been like Hillary; she wore thick glasses, had mousy brown hair that had not benefited from the attention of a hairdresser, and used no makeup. No makeup!

It was a shock. Virginia maintained that she and Roger Jr. were "polite" when they were introduced, but Bill could read their eyes. While Hillary unpacked, he took them into the kitchen—figuratively, the woodshed—and delivered a stern lecture: "Look, I want you to know that I've had it up to *here* with beauty queens. I have to have somebody I can talk with. Do you understand that?" (Virginia had always approved of his penchant for girls she called "strikingly attractive, a bent I wholehearted agreed with"—except for the Arkansas beauty queen who talked too publicly about Bill to suit her.)

Virginia never forgot the way his eyes bored through them, the chastened mother, the sulky brother. Recalling that cold first meeting, a contrite Roger told *Paris Match* years later, "In retrospect, I think I didn't want to get along with her. I wasn't doing very well at the time, and I refused to listen to her advice. I was stubborn. I didn't want her to play a role in my life, even though what she was telling me was for my own good." After he grew up—the hard way—he developed "the best relationship in the world" with Hillary and, he declared gratefully, "It's thanks to her that I have my head on my shoulders." Virginia, customarily the friendliest of women, must be faulted for setting a mean-spirited tone in that first meeting, which the teenage Roger emulated. But she had the grace to be ashamed of herself, writing a letter of heartfelt apology to the young woman who would become her devoted daughter-in-law.

Nor was Hillary all that taken with them during that first visit. And when Bill spent a week at her home in Park Ridge, a Chicago suburb, the Rodhams' reaction was much the same as the Clintons'. Her mother was distant and cool to her daughter's new beau, despite his Oxford and Yale Law School credentials. Clinton recognized "a kind of cultural tension" on both sides of this still-tenuous relationship, but since Arkansas, and his mother, were to be the launch pad for his career, Hillary's reaction was crucial. "She was cool, quiet, unresponsive," Virginia recalled. "Just think about what she was seeing: a mahogany brown woman with hot pink lipstick and a skunk stripe in her hair, and that woman's son, a budding rock musician."

One of Bill's friends later said bluntly, "Virginia loathed Hillary then. Anything she could find to pick on about Hillary, she would pick on. Hillary did not fit her mold for Bill." (In a very similar way Rosalynn Smith had not fit the mold Miss Lillian had in mind and Eleanor Roosevelt did not fit Sara's.) The only one who liked Hillary from the start was Virginia's then-husband, Jeff Dwire, who saw two strong, self-assured women "so much alike it was funny." And both of them would expend the last iota of that strength to stand up for Bill Clinton.

During a law school break at home later that year, Bill and his mother talked about what lay ahead for him after graduation. Thinking back, Virginia remembered the conversation clearly. "I'll never forget his words as long as I live. He said, 'Mother, I want you to pray for me that it's Hillary, because if it isn't Hillary, it's nobody.' " Carolyn Yeldell Staley remembers Bill defining for her the kind of woman he would marry: "The woman I marry is going to be very independent. She's going to work outside the house. She needs to have her own interests and her own life and not be wrapped up entirely in my life." He had not yet met Hillary Rodham, but the description was custom-tailored for her.

In time, Virginia came to understand and love Hillary; she realized that they were products of different worlds—the mother spawned by the Depression-stricken rural South, the wife, a postwar, antiwar suburban Yankee with her sights set on the fast track. And, Virginia admitted in the privacy of her memoirs, "I might've been intimidated a little bit [because] this was—and is— the smartest woman I've ever encountered." Ruminating on her life, Virginia was very frank about the rocky beginning in the relationship: "Oh, Lord, yes, I confess that for me, Hillary has been a growth experience. I love her dearly now."

As the Clinton presidency began, Virginia proudly portrayed Hillary as "a remarkable woman who will bring a new perspective to Washington. I'm convinced she will do great things both on her own and as the president's wife." On her own? Virginia showed great prescience—but would she have predicted that this daughter-in-law with the new perspective would run for the Senate in a state she had never lived in? Historic firsts are not easy to foretell, but Virginia would have championed Hillary every step of the way—so long as Bill agreed that it was a good idea.

Virginia Clinton was a learning experience for Hillary as well, but she wisely kept her negative impressions to herself for the very good reason that she had fallen in love with the son who was the mother's pride and joy. She and Bill had met in the Yale law library, after significant eye contact. Once she had gotten to know him, Hillary was impressed that he had broken away from backwaters Arkansas to go east to Washington, to the prestigious School of Foreign Relations at Georgetown University. Its campus crowns a high hill overlooking the serene Potomac River, with the great presidential monuments in the distance; its spires soar above the ten square blocks of Georgetown, an eighteenth-century enclave that was a bustling river port well before the American Revolution. This fashionable neighborhood had become the social center of Democratic politics.

Much of the credit for that crucial fork in Bill's life must go to Virginia. She might have insisted on sending her son to the low-cost University of Arkansas, the traditional breeding ground for the state's leaders, instead of expensive Georgetown, the jewel of Jesuit higher education, an eastern establishment university. Then there was the hard fact that Georgetown would take him twelve hundred miles from home, meaning a painful separation from the son she depended upon, her life-brightener. But if Bill felt that Georgetown was the right course, then Georgetown it would be.

From the minute he unpacked in his dormitory room Bill began campaigning for freshman president and was duly elected, then sophomore president, and then and then . . . Along with his campus politicking, he more than held his own academically. When a fellow student bet Bill would not rank as high as third in the class, the future president was confident that he would; Virginia—the inveterate betting woman—put a hard-earned twenty dollars on

Bill Clinton to show. Virginia sent the twenty; Bill came through
like a thoroughbred. She didn't always pick such sure winners at
Oaklawn.

In his senior year, Bill took a giant step with even longer odds,
applying for a Rhodes scholarship at Oxford University. Only
thirty-two of the hundreds of applicants would be chosen for the
two-year grants to that legendary British seat of learning that for
centuries has been synonymous with scholarship in the classic tradition. After making it through the first cut, Bill was summoned
for the final round of interviews. Back in Hot Springs, Virginia
would not leave the telephone—it was the only time in more than
thirty years as an anesthetist that she refused to respond to an
emergency call from the hospital. At last the phone rang. It was
Bill: "How do you think I'll look in English tweeds?" She
whooped with joy. Never could she have dreamed of such an
honor: Virginia Cassidy of Bodcaw and Hope, with a son at
Oxford.

From the age of ten, when he was riveted by the 1956 Democratic convention, Billy Clinton had been inclined toward politics,
running for office in school classes and clubs, and winning. Most
important was his election as a "senator" to represent Arkansas at
the 1963 Boys Nation, a highly competitive leadership conclave
sponsored by the American Legion and held in Washington each
summer. Virginia was thrilled at his success. A "senator"! She
took him to the airport to give him a good send-off for this first
independent journey of his life. While waiting for the plane, the
other young Arkansas winner was so struck by Virginia Clinton's
intense interest in her son, almost hovering over him during their
animated conversation, that he could clearly recall the vignette
thirty years later. The special bond between this son and mother
was plain to see, and their easy closeness made other kids feel a
touch of envy.

In that first trip to the nation's capital Bill drank in the sights
and had lunch in the Senate dining room with Arkansas Senator

J. William Fulbright, the powerful chairman of the Foreign Relations Committee. The self-possessed teenager made a strong impression on Fulbright, who would give Bill a job on Capitol Hill when he entered Georgetown and a strong recommendation when he applied for the Rhodes scholarship. Then came a defining event in Bill Clinton's life. The boys were taken to the White House, to meet President Kennedy in the White House Rose Garden. And who was first to break ranks and shake the hand of his political idol? The White House photographer snapped the moment.

When Bill returned to Hot Springs clutching that photograph, Virginia "could see in his eyes" that he had decided to go into politics; that brief meeting had left him "bubbling over with idealism." It is a historic photograph now: the charismatic young president smiling at the palpably eager boy from Arkansas, a freeze-frame moment made magic by the realization that we are seeing the idea of the presidency take shape in Bill Clinton's head. Thirty years later, the forty-second president took his mother into the Rose Garden, to show her the exact spot where he shook John Kennedy's hand.

★   ★   ★

FOR Hillary, settling in Arkansas was a different road from any she had in mind at Yale, but Bill Clinton, who was teaching law at the University of Arkansas in Fayetteville, was persuasive, and she was madly in love with him. In 1974 she had been a young lawyer on the staff of the House of Representatives committee that voted to impeach President Richard Nixon; when Nixon resigned in order to avoid trial by the Senate, her work abruptly ended. With Bill Clinton as the lure, she went to Fayetteville, joined the faculty of the law school, and began working in the first of many Clinton campaigns. Enthusiastic about politics as well as Bill, she was a key member of the inner circle in his bid to unseat the Republican incumbent in Congress. (Whenever she arrived at campaign headquarters, a staff member was alerted to whisk away the student volunteer with whom Bill was having an affair.) Though he lost, Clinton made a good showing that attracted attention across the state; Arkansas voters nodded knowingly that this young man would be heard from again.

Without too much trouble, Bill overcame Hillary's reluctance to live in Arkansas. He bought a house she had casually admired— without consulting her—with a nice fireplace that served as a background for their wedding in October 1975. Virginia was tearful when she learned that Hillary was keeping her own name; on the other hand, her new daughter-in-law was at last wearing a little makeup.

A few years later, while relaxing in the garden of the governor's mansion, Hillary said to their friend Carolyn Staley, "I wonder how history is going to note our marriage"—a conundrum that continues to be puzzling.

The wedding was the beginning of a great adventure that took them to the governor's mansion in Little Rock in 1979, where Bill garnered national attention as the nation's youngest governor, and fourteen years later, in his fifth term, they made the leap to the White House. It sounds storybook easy, but it wasn't all smooth sailing. After his first two-year term, young Governor Clinton was defeated, a crushing blow that could have ended his career, but for his mother's example.

"She taught me to never give up, never give in, never stop smiling," he told his supporters in that dark hour. "I remember when some pretty awful things would happen to her—she would get up the next day and go to work, always smiling. She would never show it. She never put her problems on other people." Many years later he followed her example through the long slog in which his sullied private life was laid bare for the world to judge—like his mother, he was always smiling, never showing the anxiety that must have been gnawing at him.

In the wake of his shattering defeat, the chastened youngest ex-governor, for whom admitting error does not come easily (one of his mother's less laudable traits), reluctantly apologized for getting too far away from the people of Arkansas, for talking too much and not listening enough. Arkansas reciprocated by reelecting him three more times, and never again did he lose an election.

There are worse things than a political defeat, and it happened to Bill, then governor, and Virginia in 1984. His half brother Roger, ten years younger than his supportive big brother, was arrested for using and dealing cocaine; he worked as a direct contact for a Colombian drug ring based in New York. At twenty-eight Roger had dropped out of three colleges; he aspired to be a rock musician but was a layabout whose only income came from shooting pool, with stakes as high as sixty dollars a game—and drug trafficking.

Nothing could be worse for a mother, and it was devastating for the governor who had to authorize a state police sting to net his brother, and then wait, lips sealed, as undercover agents set up a series of deals, all secretly recorded. One conversation revealed Roger assuring a drug cohort that with the name of Clinton he was arrest-proof in Arkansas. It was a cruel way to repay the brother he called "my friend, my guardian, my father, my role model."

The governor himself announced the arrest, which was painful both personally and politically. He and Virginia loyally stood by the errant son, who had never shown the slightest trace of his brother's strengths. Roger, sentenced to two years in prison, served thirteen months. "We were sort of the two prototypical kids of an alcoholic family," Bill mused later, after learning about the damaging psychological effects a parent's alcoholism imposed on the children who live with it. One son had become the family hero, the other, a wastrel. In the governor's mansion that night a sorrowing Bill Clinton reflected, "There, but for the grace of God, go I." And a distraught mother continued to replay the past, agonizing over what she had done, or failed to do, that had made her second son so sadly different from his brother.

★   ★   ★

"HERE we were," as Virginia told it, "alone in a truckers' motel in Texas on the night before Christmas, eating Snickers bars and drinking whiskey and getting ready to go see my son in prison. That night I realized why I had never liked country music. My life was too much like a country song." That Christmas Eve in 1985 was another low point in a life that had been a harrowing soap opera.

She had grown up in country-song territory, geographically and emotionally, in the benighted time of the Depression. They had a car but no money for gas; they bought groceries one meal at a time. Her father had wept when he didn't have two dollars to buy his daughter a new dress for Easter—a big thing for a ten-year-old girl growing up in the small-town South.

By the time Virginia was fifty-one she was three times a widow: the first, much married husband was killed in an accident; the second was an abusive alcoholic gambler; and the third, Jeff Dwire, her gentlemanly hairdresser, had been in prison for stock swindling. "He was duped by others," she insisted, in her way of blaming others for everything that went wrong. The fourth, Dick Kelley, a food broker, was the only solid figure among them.

Her flourishing practice as a nurse-anesthetist, which had kept the family afloat, had been destroyed by two malpractice suits. Her insurance company settled one out of court; an autopsy cleared her in the other, amid gossip that the coroner was influenced by his friend the governor. But the damage ended her career. "Losing my work had broken my heart," she said in retrospect. "You need something to cling to. Work was it for me."

Her philosophy was expressed on a needlepoint sampler at her bedside: "Lord, help me to remember that nothing is going to happen to me today that You and I can't handle." It was a belief she shared with Rose and Ida and Nelle. (Although none of them would have understood another of Virginia's rules of life, spelled out in cross-stitch: "A Racetrack Is a Place Where Windows Clean People.") Elvis, she felt, had had her in mind when he sang, "This Time You Gave Me a Mountain."

Fighting deep depression, she took positive action by organizing her friends in a Birthday Club. "It's more than a bunch of rowdy old ladies who like to drink at noon," she explained. "It's a life support system, and I don't know what I would have done without it." She didn't take adversity lying down; she whipped it with a good drink and a lot of laughs over meat loaf.

★   ★   ★

IN Madison Square Garden, packed to the rafters for the 1992 Democratic convention, Virginia Clinton Kelley stepped up to the Arkansas delegation's microphone. "Madame Secretary," she trumpeted, "Arkansas proudly casts our forty-eight votes for our favorite son—and my son—Bill Clinton!" With those votes he was over the top, the party's official nominee for president. The hall erupted. Never before had a candidate's mother been given the thrill of declaring the vote that nominated him.

In his acceptance speech the young and vigorous standard-bearer, hoping to rekindle the party's zeal after twelve years of Republicans in the White House, addressed the convention and the nation. "You want to know where I get my fighting spirit?" he asked. "It all started with my mother." Looking up to the family box, her son said simply, "Thank you, Mother. I love you." Spotlights and television cameras focused on Virginia and an ovation rocked the hall. This was the ultimate gift of the son who, from boyhood, had wanted to protect and please her. "I think that Bill thought I hadn't been dealt the best hand in life," Virginia observed, "and he was determined to make up for it."

Throughout the primary campaigns, he had stressed the need to help families, and there was no better model than Virginia: "She held steady through tragedy after tragedy. And she held our family, my brother and I, together through tough times. As a child I watched her go off to work each day at a time when it wasn't very easy to be a working mother. As an adult, I've watched her fight off breast cancer, and again she taught me a lesson in courage. And always, always, she taught me to fight." Said Virginia: "I don't give up easily. Like never." Her son learned her lesson well, albeit over certain issues of less honor.

Her spreading cancer did not stop her from throwing herself into her son's campaign, as she had in every election since his first run for Congress. From that first race, in 1974, she came to feel that she was "no longer just a mother and a protector—I was now a partner in my sons' lives." She would load her car with campaign paraphernalia—buttons, pins, flags, bumper stickers—and when people waved at her, "I'd screech to a halt and get out my wares." Then they got the full picture—the candidate's mother plastered with Clinton campaign trappings. She shook every available hand and talked to everybody who would give her a minute. If they weren't friendly, she tried to persuade them, but when one farmer set his dogs on her, she figured he was a Republican and raced for her car, hounds snapping at her heels. They were probably Republican, too. Canny tactician that she was, she would affix the stickers on the bumpers herself, "just in case they had a notion of sticking it in a scrapbook somewhere." She prided herself in being "one crackerjack of a cold-call campaigner—I loved knocking on those doors and telling those good people about what a wonderful governor my son was going to make."

In the presidential campaign, Virginia found herself in constant demand. TV networks and publications from everywhere fell all over themselves to get to this bright, funny woman with her striped hair and five rings and her tendency to say the most amazing things. For Virginia it was a little like her old days of jumping

onstage to share the spotlight with the star—and this time the star
was her son and she was singing in his key.

Bill Clinton has repeatedly said he could never have been president without his mother's rock-solid support, her example of
strength and backbone. James Carville, Clinton's astute campaign
strategist, noted that "anytime he was asked who was the most
influential person in his life, he would say, without a doubt, 'My
mother.' " Thomas Caplan, Bill's roommate at Georgetown University, was equally certain: "She's the root of all this."

Her influence was both general and specific. "She gave me a real
democratic—small 'd' democratic—outlook on life," Clinton said.
"She taught me that everyone's life has equal worth." That lesson
might be the key to Clinton's fabled ability to connect with "plain
people," conveying genuine understanding of their problems and
never condescending. And it was Virginia's brave battle against
breast cancer that moved him as president to press Medicare to
cover the cost of yearly mammograms.

She had encouraged him to go for it, whatever "it" was at the
moment, long before that initial political effort. From girlhood,
Virginia Cassidy was competitive to the bone. "I'm proud to say
that Bill inherited my competitiveness," she boasted. "I play any
game to win—Bill is the same way." (He is competitive to the
point that, even when playing golf with good friends, he has been
known to fudge his score.) "He always had to be the class leader,"
a high-school classmate recalled in a collection of memories of
their president-friend. "He had to be the best in the band. He had
to be the best in his class."

In 1992, with his future on the line in his most important race,
Bill had to tap every ounce of that competitive drive. Virginia was
pleased that he said "he consciously thought of my life and tried
to draw strength from it." Bill left no doubt: "During those really
tough days in the campaign, I would think about how so many of
Mother's difficulties had gone on for years and years. In my case,
the voters would simply make up their minds and everything

would be resolved with the election, so I might as well get up each day and do my best, do what I believed in and try to have a good time."

Easier said than done, but along with her fighting spirit, her son had absorbed another of his mother's maxims: Pretend you have confidence, even when you don't—and then you may actually get it. (That advice echoed President Eisenhower's admission that even when the famous Ike smile camouflaged a flagging spirit, it could convince other people.) As president, Clinton made fullest use of her counsel during the year of investigation and revelations that led to the dubious distinction of being the only elected president to be tried by the Senate. Richard Nixon had resigned rather than risk the Senate's verdict.

A longtime Nixon associate, former White House counsel Leonard Garment, pointed out the contrast in the reaction of those two presidents under fire: "[Clinton] seems infinitely elastic, positive and resilient. He turned in a four-star performance as Mr. Cheerful, aggressively counter-punching, ever resourceful. On a comparable occasion, Nixon looked like Mr. Gloom, sweaty and defensive." Both presidents had supportive mothers, but Virginia was the defiant optimist who met life on the doorstep while Hannah Nixon, the noncombative Quaker, stayed well inside.

Optimism is not to be minimized, and Virginia had an endless supply. Gamblers are by nature optimists—the wheel will stop on their number on the next spin, a royal flush will come with the next deal—and Virginia was an inveterate gambler. Experts who track and analyze such traits say that optimists tend to report having good luck while pessimists more often report bad luck, suggesting a built-in link between attitude and external circumstances, or perhaps the way in which luck is viewed. And increasingly the medical profession finds that an optimistic outlook has a beneficial effect on a patient's recovery. Bill Clinton inherited optimism from his mother—and he is lucky. Throughout his first presidential campaign, when things looked dire, he remained optimistic, proclaiming himself "the comeback kid." Back in Arkansas, in every dark moment Virginia braced him with her motto: Never give up.

Then came the victory; the final votes were tallied and the American people had elected her son as the forty-second president of the United States. Virginia Kelley was only the sixth mother to witness the simple yet awesome ceremony of the inauguration, the thirteenth mother to have lived to rejoice in her son's triumph. It was a day when Virginia's makeup alchemy took even longer than usual, because she had to look her best on world television. Forswearing her usual bright colors, she wore sophisticated black, remembering that her third husband, the style-conscious Jeff Dwire, had insisted that "black is my color," but among the home-folks at the Arkansas inaugural ball she could let go with the spangles that were the real Virginia.

For the president's mother, the second most memorable event of the inaugural festivities was meeting Barbra Streisand, an enthusiastic Clinton supporter. Barbra had long occupied a top pedestal in her pantheon of idols (along with Elvis) and hearing the incomparable Streisand sing "The Way We Were" always turned her "weepy." And here they were, not only meeting but becoming instant friends.

Moving into the White House, Virginia and her husband, Dick Kelley, took possession of the queen's suite: the same elegant chambers that Sara Roosevelt and Rose Kennedy had found quite suitable, the bedroom from which Dorothy Bush had cheerily waved to tourists outside, the room with the canopied bed that Martha Truman had bypassed as too fancy and too high; the setting that Lillian Carter had found too restrictive. (And in these same quarters Winston Churchill had startled the White House domestic staff by strolling about naked.)

Like Cinderella, Virginia was late getting home from the ball, but not too late to relive the day: "I lay back on the bed. Even for an old girl who's pretty much seen everything that goes on after

dark, sleeping in the White House for the first time is quite a feeling. I wished my daddy could've been there to watch his grandson
become President—or even to see his daughter stretched out on
the queen's bed. I lay there a few minutes saying a prayer of
thanks for all we had endured."

In November she and Dick came back to Washington for a family Thanksgiving at Camp David. By then her health was deterio-
rating visibly. Chemotherapy had taken its toll and she was
wearing a wig, specially designed to duplicate her "skunk stripe."
Again at Christmas the Kelleys joined the family in the White
House, which was a picture of holiday joy, aglow with lights, with
a big tree on the official floor and another in the family quarters
upstairs. Thinner, weaker, but relentlessly upbeat, Virginia was
eagerly planning her New Year's jaunt to Las Vegas: Streisand was
making her first personal appearance in twenty-two years and Virginia would be at ringside. In the few days between Christmas and
Barbra, Bill and Hillary and her darling Chelsea would be visiting
her in Hot Springs. What better way to start 1994?

But as the family celebrated, the editors of *The American Spec-
tator,* a right-wing political journal, were putting final touches on
a big story that would be on the stands in a matter of days, a
scoop involving then-governor Bill Clinton and a woman named
Paula. It would be the beginning of five years of questions and
charges and denials and affidavits and, finally, ten votes that saved
Bill Clinton from the ignominy of being the first American president to be thrown out of office. Virginia had bowed out at just the
right moment.

★   ★   ★

"AN American original," the minister called her as family and
friends said farewell to the valiant, never-give-in, one-of-a-kind
Virginia. Always a party girl, she would have loved this final bash.
The president had booked the Hot Springs Convention Center for
the event, the biggest hall in town. It was crowded with twenty-

five hundred devotees, some of them personal friends, others who had never met her but admired her breezy style and scrappy spirit in facing down adversity. Hundreds of others waited outside for more than an hour in the eighteen-degree January morning.

The governors of three states who had known her—Kentucky (where she loved the Derby), Nevada (where she loved the casinos), and Arkansas (where she loved almost everybody)—were among the mourners. But more important to Virginia would have been the dramatic figure in a sweeping black coat—Barbra Streisand. She had arrived by private jet the day before to join the family and close friends at the funeral home, and at the service her million-dollar voice joined with the tuneless and the grief-choked in singing "Amazing Grace."

The presence of the megastar was a testimony of her great affection for Virginia, who only one week before had traveled to Las Vegas for Streisand's long-awaited onstage performance, the showbiz event of the year. Starting at breakfast Virginia played the slots, hoping for a little luck with her morning coffee, then on two successive nights she donned her fanciest gowns for the show that brought Streisand face-to-face with her public once again. For this inveterate nightclubber, the MGM Grand was the ultimate, eclipsing the glitz of even her beloved Vapors club in Hot Springs.

At their choice seats down front, where she and her husband shared the evenings with Streisand's mother, Virginia sat transfixed, transported, holding tight to every poignant note, and between acts all heads swiveled when the star came out and visited with them. It was beyond fantasy: Virginia Cassidy of Bodcaw hobnobbing with the Legend and all those other Names. As champagne flowed at the VIP party, she knew that for her this was more than a New Year's celebration—it was her farewell-to-life party. Exit laughing.

The Reverend John P. Miles, one of three Methodist ministers who officiated at the funeral, knew just what Virginia would have wanted. A somber good-bye would not reflect her life, and so he kept the crowd chuckling amid tears. She would have led the laughter when he talked about her passion for makeup and horse-flesh and his notion to hold the service at Oaklawn Park, where the two-dollar window was now short a regular customer. At the track that day they expressed their sentiment with a large sign— OAKLAWN MOURNS THE LOSS OF RACING'S NO. 1 FAN—and one of her favorite jockeys was among the mourners.

All around Hot Springs flags were lowered to half-staff, and hand-lettered signs of sympathy were posted in tribute to the president's unquenchable mother. Speaking of her unconditional love for her sons, the minister noted lovingly, "If you ever crossed those she loved, you knew she was not without fault." Virginia would have nodded that he got it right. She had dedicated her memoirs—which she had not completed, as she had not completed her life—"For my family, the reason I've run the race."

She had fought the cancer in her breast for four years, but at the end, when she was seventy, it overcame her like a rampant kudzu vine strangling a flame tree. When the cancer was discovered in 1990, Virginia, ever the optimist, was sure that a mastectomy and chemotherapy would beat it, but in 1992 it invaded her bones from skull to legs. She demanded that her doctors "tell my boys I'm fine," adamant that she would not burden her son in the most important year of his life—and hers—as he challenged the sitting president, George Bush. Her son raced to victory unaware of the extent of her illness and only months later, when her condition could no longer be concealed, did she reluctantly tell him the inescapable verdict.

Virginia, everybody knew, was not a stained-glass mother. "She didn't have a pious bone in her body," said the Reverend Miles, "but she had faith." Never much of a churchgoer—Billy Clinton, when he was only eight, would get himself dressed in his Sunday best and walk to Park Place Baptist Church, all on his own—she engaged in a kind of direct communication with God, a kind of no-ritual belief in which she looked out of her window each morning and said, "Thank you, God, for letting me see another morning."

On the family's final visit with her in Hot Springs following their Christmas in the White House, Bill, a member in good standing of the bare-your-soul generation, had wanted to talk about death. "She just shoved it off," said the minister. "She wanted to talk about life, she would say, 'Let's play cards, let's talk, let's celebrate.' "

As the mile-long funeral procession took her the ninety miles to the Rose Hill Cemetery, knots of people gathered on the overpasses to bid her farewell, hands over their hearts. She would lie beside William Jefferson Blythe Jr., the husband she hardly knew but would always love best, the less-than-admirable Bill Blythe who had given her the son who had succeeded beyond her fondest dreams. The cemetery was in Hope, the name so right for the woman who never ran short of it.

"Too many people seem to think life is the tablecloth, instead of the messy feast that's spread out on it," she stated in her memoirs. "They want to keep the cloth clean and tucked safely in a drawer. That's not life. Done right, life leaves stains. That's why I don't judge Bill Blythe for the things I found out about him. That's why I feel sorrow, not hatred, for Roger Clinton. That's why I love my mother, even though many a day she made me feel like murdering her. . . . It's called resilience."

★ ★ ★

THE *Washington Post* headlined her death, "The Woman Who Shaped the President Dies in Her Sleep." Shaped him. Supported him. Loved him unconditionally. A week after Virginia's funeral, the president, on a state visit to Moscow, quietly slipped into a restored church on Red Square. He lit a candle and with tears in his eyes confided to an aide, "For my mother." The passing of years has diminished the pain but not the loss. "I miss her terribly," the president said years after she died. "What I miss most is just being able to pick up the phone and call her. Every Sunday, wherever I was, I would call her. Still, sometimes I almost impulsively reach for the phone." (Tough old Harry Truman had said the same about his Sunday calls to his mother back in Missouri.) Photographs of Virginia are scattered throughout the White House living quarters as reminders of happy moments in their intertwined lives. She still hovers over him. When he broke the gender barrier with the bold appointment of the first woman secretary of state, Madeleine Albright, he beamed with pride: "My mama's smiling down at me right now."

Without being asked twice, Bill Clinton will recite—or sing, with a few chords of accompaniment—the old song he associates with his mother:

> One bright and shining light
> That helped know wrong from right
> I found in my mother's eyes
> Those fairy tales she told
> Those pathways filled with gold
> I found in my mother's eyes.

The eyes, Blake said, are the windows of the soul. That hers were fringed with unabashedly false lashes and overarched by brows drawn in the mood of the day only strengthened the contrast with the generous soul they mirrored. "She is a celebration," said John Miles, "on the way to heaven."

12

A NEW DYNASTY

A mother's pride, a father's joy!

—Sir Walter Scott, *Rokeby*

NOVEMBER 7, 2000. Election Day in America. But as the clock labored past midnight without a clear result in Florida, there was still no winner. Hours stretched into days, days dragged into weeks. The political vocabulary added a strange new word, chads, dangling or dimpled; Americans became sidewalk experts on arcane twists in the electoral process; partisans rooted for or against a recount of the disputed votes in Florida, which put the nationwide results in question. Not until the Supreme Court ended the marathon with its five–four decision, after five contentious weeks, did the country have a new president: George Walker Bush, governor of Texas, son of a president, brother of the governor of Florida, grandson of a senator, and, on his mother's side, lateral descendent of the fourteenth president, Franklin Pierce. The Bushes now rival the Kennedys as a political dynasty.

As the nation agonized through this history-making process, no one was more perturbed by the challenges brought against Bush's apparent victory in Florida than the new president's mother, Barbara Pierce Bush. More was at stake than victory for her son; it was vindication of her husband's defeat by Bill Clinton eight years earlier, robbing him of a second term in the White House. It was the restoration of Bush dominance, the Bushes back where they belong. Before this stunning new chapter in political history occurred, the puckish Barbara had joked, "Did you know that one out of eight Americans is governed by a Bush? That ought to scare you!" Now it was total—and double Bushes for the residents of Florida.

The family had confidently anticipated that a Bush son would follow the father to the White House, but the most likely successor seemed to be Florida governor Jeb, the more serious, more polished second son, who had told the family he intended to run. But the tobacco-chewing, fun-loving, reformed boozer, gifted schmoozer older son, who took forty years to settle down, then jumped in and won the prize. "He's a late bloomer," said Barbara, "like his mother." She did not underestimate his gamble—six years before, when he told her he planned to run for governor of Texas, her response was an unvarnished, "You'll lose." She was to learn that his breezy Texas style and outrageous charm, far removed from the correct reserve of his family's Eastern establishment roots, connects with the voters.

That was one of the rare times when Barbara's political antenna failed her. As early as 1989, George W., a young man in a hurry, had talked about running for governor. His mother openly opposed the idea. She sensed that, with a President Bush in the White House, it would be bad for both father and son; his move would be attacked as a blatant exercise in coattails and nepotism. And it was also too soon—George had not yet established himself as somebody other than his father's son. She effectively scotched his presumptuous notion by telling a Texas journalist that George should stick to his baseball enterprise for a time. George's reaction was less than affectionate joshing. "Mother is worried about Daddy's campaign affecting my race," he snapped. "Thank you very much. You've been giving me advice for forty-two years, most of which I haven't taken." But deferring to her better judgment, he built a Texas base for himself as part-owner and front man for the Texas Rangers baseball team. Then, in 1994, two years after his father lost his re-election bid, the son made his bold run for governor—successfully, as his own man. His victory was all the sweeter for the unseating of Governor Ann Richards, who had won the undying enmity of the Bush clan with her unforgettable line at the 1992 Democratic convention: "George Bush was born with a silver foot in his mouth."

From the outset of George W.'s run for the White House, Barbara was her son's most ardent campaigner, speaking all over the country—in as many as three states in one day—making the rounds on TV shows, sending thousands of letters to women voters in Florida, taping messages for his phone banks. And bringing in big money all along the way. She was out front everywhere, while the former president stayed in the background. Barbara, it is said, is the best politician in the family, and perhaps the most aggressive. In 1999, when her son was equivocating—or perhaps being coy—about running, she told an audience, "If he doesn't run, I'll kill him!" Earlier that year, at George W.'s second inaugural as governor, the minister at a private prayer service for family and close friends spoke about duty, citing the story of Moses, who flinched from the sacrifices God demanded of him. Leadership, declared the pastor, calls for sacrifice. At that, Barbara leaned over to her son and whispered, audibly, "He's talking to you, George." She had raised her children not to stand on the sidelines but to get into the game.

On a Florida vacation after the election was finally settled, Barbara discovered that her role as a principal player had changed. Entering an exclusive club for dinner with the president-elect and the family, she was chagrined when a Secret Service agent—obviously new—stopped her for a metal-detector search. "Is every club member being searched?" she asked with a tinge of indignation. "No," he said. "Only those who look like they might be reporters or look dangerous." One suspects that the hand of her impish son was at work.

In her four years in the White House, Barbara Bush was one of the most popular first ladies ever, and she continues to rank high on the list of most admired women. (She consistently outpolls Nancy Reagan, which must give her a little rush of satisfaction since she and Nancy were never a comfortable team.) In and out of the White House, Barbara's popularity consistently topped her husband's. With her fluff of unapologetic white hair, size fourteen (or so) dresses, her easy, believable smile, exquisite needlepoint, dogs and grandchildren romping about, she became "America's grandmother." Her lighthearted books about life in the White House, "written" by Millie, her spaniel, were immediate best-sellers. Coupled with all that Hallmark lovability is a lively sense of fun that breaks through official formality, as when she appeared at a Gridiron dinner (the press event attended by Washington's headline makers and publishing powers) in a strawberry-blond wig, just to tweak all those journalists who were fixated on her white hair. Her husband was flabbergasted, and uncertain, but she went right ahead and wore it.

Her tireless work to promote family literacy, to encourage children to read, added substance to the lovable image. Literacy is more than a project, it is a passion with Barbara Bush. "Focusing on the family is the best place to start to make this country more literate," she explains. "And I believe that being more literate will help us solve so many of the other problems facing our society." Building on her White House prestige, she established the Barbara Bush Foundation for Family Literacy and continues to make scores of speeches each year to advance its goals. She sprinkles her informal talks generously with tidbits from her First Lady days and self-deprecating quips about her weight and wrinkles. "You could plant a whole potato field in the rows of wrinkles," she told one audience, and she draws an appreciative laugh when she recounts her mother saying to her older sister, a sylph who was once featured on the cover of *Vogue,* "Eat up, Martha—not you, Barbara."

Like the blurry snapshots taken at a family reunion, where half the head is out of the frame, the picture of Barbara is only partly in focus. This grandmother is also irrepressible. She speaks her mind, vents her displeasure, and on occasion deploys a withering stare or caustic wit toward anyone who dares criticize or vote against any member of her family, who, in her protective mother-view, can do no wrong. George W. tells of her powerful impact on her husband's campaign staff during his 1992 presidential race. Her special targets were self-aggrandizing aides who undercut his authority by claiming credit for his ideas or leaking stories to the media. "I would then go and talk to that person," said her son, who had been brought in to ride herd on the campaign, "and explain to them that if they weren't careful, the wrath of the Silver Fox would fall upon them. A lot of those people lived in fear of my mother." The threat produced an immediate effect, as did the son's own abrasive manner. "Like his mother," observed an old friend, "instantly likable and instantly able to instill fear."

With her keen mind and forthright nature, it follows that Barbara would be a woman of strong opinions. She differs with some key positions of the conservative right, which her husband—and now her son—espouse. In the past she has taken a moderate view on the powder-keg issues of gun control and abortion. (She was an early supporter of Planned Parenthood and was quietly pro-choice.) In 1999, when a television interviewer asked about her views on abortion, Barbara stated, "I don't think it should be in the [Republican] national platform. There's nothing a president can do about it anyway, in all honesty." She called it a matter of states' rights, a phrase that is a conservative commandment. Republicans for Choice seized on her remark for a television commercial with the catch line, "Listen to your mama. That's a family value we can all agree on." (In his campaign her son avoided the issue.) Among family and close friends, Barbara remains her totally unfettered self, but as a political wife and mother, she has learned to muzzle controversial opinions, careful not to add to a president's problems. Nobody can take issue with literacy.

When the Bush children were growing up, the house was a place of the happy chaos of kids, pals, and pets. Through it all, she was the stabilizing center. "She did a good job of keeping the family intact," her son Jeb reflected, likening her to an executive of a large corporation. "She is a supermom—Cub Scout den mother, carpool driver, and Sunday school teacher. Dad was the chief executive officer, but Mother was the chief operating officer. We all reported to her."

Barbara is an awesomely organized manager—the kind who has her Christmas presents bought and wrapped by midsummer—but Betty Crocker she is not. She cheerfully acknowledges, "I never loved cooking, but I can light a fire and throw something on it." But in their twenty-three Washington years the easy warmth of the Bush hospitality made their Sunday cookouts a magnet for men of power as well as old-shoe friends—and her chili did indeed win kudos.

Her firstborn, the president, is the child most like her. "It's like you cloned Barbara to get George," an old family friend told George W.'s biographer, Bill Minutaglio. First Lady Laura Bush readily concurs. Mother and son are so much alike that when he was growing up, an obstreperous West Texas kid, they used to have regular shouting matches, while his brothers would never cross her. The younger George likes to say, "I have my daddy's eyes and my mama's mouth." In his campaign-oriented book, *A Charge to Keep*, he wrote, "My mother and I are the quippers of the family, sharp-tongued and irreverent. I love her dearly and she and I delight in provoking each other, a clash of quick wits and ready comebacks." Barbara relishes the verbal sparring: "We fight all the time—we're so alike in that way. He does things to needle me, always."

"I sometimes get in trouble for jesting with reporters," George admits. "And who can forget Mother's infamous quip, 'rhymes with witch'?" (Though her acerbic sally, in 1984, had been aimed at Geraldine Ferraro, her husband's vice presidential opponent, her target is often mistakenly remembered as Hillary Clinton, whom she abhors.) When the *New York Times* interviewed the Bushes a few days before their son was nominated, the former first lady frequently interjected what the *Times* termed "acidic interruptions" about the Clintons. At one point her husband admonished her, "You're not going to be in this interview if you're going to start talking like that. George will be furious." But he didn't really seem to mind.

Though she is quick to unleash her missiles against multiple political targets and will wrangle with her senior son, there is one untouchable exception—her husband. "I've never heard George and Barbara Bush utter a harsh or ugly word to each other," says the new president. Steeped in their love, they are mutually protective of each other and their family, reaching to its outermost connections.

From the night she met George Bush at a country club dance—she was sixteen, he was seventeen—Barbara has led a clearly focused life as a career mother. That was the role expected then of daughters of affluence. Barbara's father, Marvin Pierce, son of a wealthy Pennsylvania family that lost its iron foundry fortune in the 1890s, was president of the McCall Corporation; her mother's father served on Ohio's Supreme Court. Barbara could claim President Franklin Pierce—a Democrat—as her great-great-great uncle. Instead, learning that he was "one of our weakest presidents, I was humiliated."

Barbara, the third of four children, grew up in Rye, a fashionable New York suburb, in a five-bedroom house with live-in servants, country clubs, and private schools. After her first six grades in the local public school, she entered Rye County Day School, and soon she was old enough for Miss Covington's dancing classes—with boys! For her last three years of high school she was sent south to Ashley Hall in Charleston, South Carolina.

It was a carefree life, especially in summer, when Barbara, an outdoorsy girl, and her friends happily frittered their days at the Manursing Club, on its tennis courts and private beach on Long Island Sound. And at last the boys were tall enough that at dances for the young set she didn't have to telescope her five-feet-eight to accommodate them.

Like the other mothers of the presidents in this book, Barbara was a daddy's girl. Marvin Pierce was "a smiling man" with a "great sense of humor" who had a special bond with his younger daughter. He encouraged her to compete in sports and sided with her when she was being picked on, "I think because Mother never took my side," Barbara said. In her memoirs she wrote, "He was a hero of mine."

Her mother, Pauline, was a beautiful woman and an uncontrollable spendthrift, without a gram of humor. She was the disciplinarian, scolding and spanking, usually with a hairbrush or wooden clothes hanger. (Barbara acknowledged that she spanked her own children, "but not as hard as my mother did." But to a Bush cousin, Elsie Walker, "she was scary with her kids when they were little.") The distance between mother and daughter doubtless contributed to Barbara's becoming 100 percent Bush after marrying George, with special affection for her mother-in-law.

In 1943 Barbara entered Smith College. By then the country was deep into World War II, fun times were limited, and the young man she was in love with for life was headed to the South Pacific, a Navy pilot with his heart on his sleeve and "Barbara" on the side of his bomber. A year and a half later she happily left Smith to marry George Herbert Walker Bush the minute he returned home, a decorated hero.

They settled down, if that term could be used in relation to the makeshift apartment they moved into as George picked up his life as a student at Yale. There was never a question that he would go to Yale; he was following his father and brother, and furthering the family tradition for the son who would be born in New Haven in 1946. (Barbara's mother dosed her with castor oil to hasten the process.) But their little George would grow up to be a full-blooded extroverted Texan, a far cry from a reserved, blue-blooded Connecticut Yankee.

After graduating, George shunned his father's Wall Street life, which was his for the asking, to follow the old American challenge: "Go West, young man!" He would seek his fortune in the oil lands of west Texas. In that single decision he set the course for his life and his son's. In his new Studebaker, he headed for Odessa to join a family friend's company; Barbara and two-year-old Georgie followed by plane, a twelve-hour flight. She was less than keen about this new adventure, and her eastern heart must have sunk at the sight of Odessa, a bleak, unpaved, trailer park Saturday night kind of town inhabited mostly by the rough workers from the oil fields and maintenance shops. The right place for Bush to get started in the oil business, yes, but the executives and entrepreneurs and money men lived with their families twenty miles down the road in Midland.

Barbara was a good sport about their cramped apartment and the bathroom shared with two questionable ladies and the large number of gentlemen they entertained in the evenings. Barbara's family thoughtfully sent her care packages of niceties and necessities they were sure she couldn't find in untamed west Texas. In 1949 George was sent to California for a year, and on their return to Texas they wisely settled in Midland.

Midland, a boomtown that had exploded to a population of about 30,000, was as benighted topographically as Odessa, but its residents were upwardly mobile, many of them transplanted Easterners also seeking their fortunes in oil. Midland was a town all about money.

The Bushes moved into a street of little bungalows painted in bright hues, dubbed "Easter Egg Row." And it had a feature no eastern kid could boast—a buffalo wallow, a remnant of the days when millions of the great beasts roamed the West. Later, after the oil started flowing, the Bushes moved (one of thirty times, by her count) to a more spacious house with the first swimming pool in town and by a park that was the gathering place for all the kids. Meanwhile, the family had grown to include four boys, George, Jeb (John Ellis), Neil, and Marvin.

George, like the other president-sons who went before him, enjoyed the independence offered by a small town—rambling at will, bicycling anywhere, taking shortcuts over neighbors' fences, and making occasional visits to the emergency room to stitch up the results. The senior Bushes quickly became leaders in every civic organization and social club, central figures in the life of Midland.

A fellow den mother in Midland remembers George as "a typical little boy—sort of self-confident, blue jeans with the knees out, rough and tough, all boy. Not a bully—I don't think Barbara would have stood for it. He was cocky, but not rude or anything. He was having to assert himself in a family that was very strong, with a lot of other children in it, and he was the oldest."

Most of all, there was baseball, the passion that was to become young George's serious business, the gateway to his governorship and the presidency. His dad, a star at Yale, had made the college All-American team, a model to aspire to. Young George wasn't much of a hitter, but as catcher he was in on every play. His mother was the parent who was always there to root for him— and the team. Surprisingly, his father rarely attended a game.

For all its friendly attributes, Midland was a harsh kind of place, a world removed from the verdant East that Barbara had always known. Trips back to Kennebunkport, to the old Walker-Bush compound on the Maine coast, brought a return to the extended family fold of entwined siblings and cousins and in-laws; the ocean mists and cool summer evenings brought respite from the searing Texas heat. It would always be Barbara and George's true home, the one filled with the spirit of generations past and memories of her first visit, at seventeen, with the family that would become her own.

Governor Bush reminisced years later that "Midland had a

frontier feeling; it was hot and dry and dusty. I remember giant sandstorms blowing in. You could look out the back window but not see the fence because the sand was blowing so thick and hard." Storm windows were to keep the sand out; every morning schoolchildren had to brush fine coatings of sand off their desks. "Tumbleweeds blew into our yard," he recalled, and "once it rained and frogs came out everywhere, like the biblical plague, covering the fields and front porches." The boys probably thought that was great.

By 1959 the senior Bush already had made several million and with a political future taking shape in his mind, he moved the family to Houston—the four boys newly joined by baby sister Dorothy—Doro—named for much-loved grandmother Dorothy Walker Bush. A world away from Midland, Houston was the realm of big oil, a place of splashy designer gowns, custom-made alligator boots, private schools, and showplace mansions, a burgeoning metropolis ripe for turning Republican. Transplanted Yankee George H. W. Bush would be a catalyst in nudging it in that direction, and he soon was elected to Congress—a Texas Republican, an anomaly that was the first wave in a sea change for that state's politics, which in time elected and re-elected his Republican son as governor, and now president.

★   ★   ★

IN the long history of the American republic, only one other woman shares Barbara's status as both First Lady and First Mother: Abigail Adams of Massachusetts. Her husband, John Adams, was the nation's second president; her son, John Quincy Adams, was the sixth. The two women have much in common. Both can be described as strong and outspoken (President Bush says his mother "will just let it rip if she's got something on her mind.") with sharp intellects and lively personalities that charmed their compatriots—and both were the dominant figures in their families.

In a singular parallel, their husbands were former vice presidents who became presidents but failed to win a second term; their first sons took office after bitterly contested elections in which they lost the popular vote to a Tennessean (Adams to military hero General Andrew Jackson). The election of 1824 was a cliff-hanger equal to that of 2000. It was thrown into the House of Representatives, where the the dynamic speaker Henry Clay convinced a wavering member of New York's delegation, deadlocked in a tie, to cast the crucial vote that gave John Quincy Adams the victory. In what was lambasted as an unsavory deal, Adams then named Clay secretary of state, the stepping stone to the presidency. The tainted election cast a permanent pall over Adams's presidency; four years later he lost to Jackson, and Clay forfeited any chance to be president.

Barbara and Abigail could have compared notes on their disdain for the press ("the media" in the television age). Barbara blamed the media for her husband's re-election defeat and for bias against her son. Abigail's target was Benjamin Franklin Bache, grandson of founding father Benjamin Franklin, who constantly attacked John Adams, "His Serene Highness," and, even more unbearable, her son. "He is never at a loss for a lye," she charged, deploring his "lies, falsehoods, calumny, and bitterness—his malevolence is unbounded." Abigail once said of her beloved husband, "When he is wounded, I bleed." Barbara, too, would bleed—and fight back.

In one particular, Abigail and Barbara couldn't be more unalike: Abigail once declared that she would rather see her son "a log thrown on the fire" than to be president. Her husband had been brutally castigated as president and her four short months in the raw, unfinished White House, following his defeat in 1800, had been miserable. Barbara, to the contrary, reveled in her four years as first lady—in her view, "anybody who couldn't be happy in the White House couldn't be happy anywhere"—and she threw herself into helping her son follow in his father's footsteps. His victory would ease her lingering bitterness.

Like Barbara, Abigail was the daughter of a prominent family; Abigail's father and grandfather were ministers, and her mother bore the distinguished Quincy name—the family had arrived in America in 1633, accompanied by six servants, and her father was for fourteen years speaker of the Massachusetts House of Representatives. (Abigail married John Adams over her mother's objection that he was not good enough for her.)

As young wives, both Abigail and Barbara were struck by the greatest tragedy a mother can know—the death of a child. Each lost a daughter, Abigail's little Susana, "Suky," at fourteen months and Barbara and George's second child, Robin, was diagnosed with leukemia when she was only three. Seeking the most advanced treatment, the distraught parents took her to Sloan-Kettering hospital in New York but despite every effort and Barbara's constant vigil, after seven months Robin's frail little body slipped away.

George W. remembers it as if it were yesterday, the day his parents came to his school to pick him up. He had not been told that his little sister was facing death—it was the wrong decision, his parents later realized, but they had hoped to spare him anxiety. "I was sad, stunned," he wrote. "Forty-six years later, those minutes remain the starkest memory of my childhood, a sharp pain in the midst of an otherwise happy blur."

Barbara, who had been so strong through the long ordeal, not allowing any tears, sank into a deep depression when it was over. Young George did all he could to cheer her up. Barbara overheard him tell a friend that he couldn't go out to play because he couldn't leave his mother—she was sad, she needed him. "That started my cure," Barbara said. "I realized I was too much of a burden for a little seven-year-old boy to carry." Her reaction, her son reflected, "was to envelop herself totally around me. She kind of smothered me, and then recognized it was the wrong thing to do."

Abigail and Barbara both found themselves de facto head of the family as their husbands were away for long periods. George's oil business kept him on the road as much as three weeks at a stretch;

John Adams was away, with only rare visits home, for ten consecutive years during the Constitutional Convention and eight years as a diplomat in Paris. (Later she was able to join him when he was minister to Great Britain, and she was thrilled by London's cultural riches.) Meanwhile, Abigail, alone, with the flames of the revolution burning within sight of Braintree, capably managed their farm and house, homeschooled their five children, and wrote hundreds of thoughtful, literate—and amorous—letters in a legendary husband-wife correspondence.

Barbara was clearly the parent in charge of the Bush family, settling fraternal disputes and patiently tutoring both George and Neil with flash cards and phonics to help them overcome reading difficulties. On Saturday afternoons George could not join his pals for a baseball game until he had mastered ten new words from his mother's flash cards. With Neil, her teaching was to help him overcome dyslexia. "Barbara was always the guiding hand of the family," observed an old family friend. "If they did make a mistake, she would see they didn't make it again." George W.'s most egregious violation (not counting the time he stole a Christmas wreath from a New Haven lamppost) came when he was twenty-six. Quite drunk and with his fifteen-year-old brother in the car, he drove into a metal trash can, which clattered noisily down the quiet street. He then challenged his father: "I hear you're looking for me. You want to go *mano a mano* right here?" Cooler heads prevailed. (He forswore all alcohol in June 1986, whereas Abigail's second son, Charles, died of alcoholism when he was thirty.)

Jeb Bush recalled that "even when we were growing up in Houston, Dad wasn't at home at night to play catch. Mom was always the one to hand out the goodies and the discipline. In a sense it was a matriarchal family." Barbara, in her self-mocking way, said that when the boys (daughter Doro was more amenable) got out of hand, "I would scream and carry on. The way George scolded was by silence or by saying, 'I'm disappointed in you,' and they would almost faint." In his book Governor Bush noted,

"Mother has always been the front line of discipline in our family, something my own children and the other grandchildren are learning quickly. If she sees something she doesn't like, she makes sure you know about it. She quickly blows off steam and clearly lays down the law." Not that she is overly restrictive—as Jeb put it, "They gave us enough rope to hang ourselves without killing ourselves."

Son George has always tended to behave circumspectly around his father, but is the leader in the family style of banter and teasing with their mother. In that vein Marvin says, "We tried to be disrespectful on a regular basis." Jeb adds, "Out of disrespect we called her 'the gray fox.'" And Barbara would give as good as she got.

Abigail would have been horrified. In her day parents commanded obedience and respect. When John Quincy was only eleven years old, he went to live in Paris with his father. Abigail, concerned about that city's racy reputation, sadly (it was like having "a limb lopt off") sent him off with ample motherly advice—curb your temper, obey your father, acquire useful knowledge—culminating in an all-inclusive plea: "I would rather see you find a grave in the ocean you have crossed than see you an immoral, profligate or graceless child." Barbara also spoons out advice—her president-son has commented, "Dad gives me advice when I ask for it; Mother gives it even when I don't." Barbara's counsel tends toward the practical: "Stand up straight," she directed George before his first presidential debate.

On occasion Abigail's exhortations hardened into commands. When John Quincy was a young lawyer, just out of Harvard, he fell in love with the charming young Mary Frazier, but Mary was only sixteen, and marriage did not fit into Abigail's ambitions for her son. She simply prohibited it. (She also blocked her elder daughter's engagement, which led to a later miserable marriage.) Years later John Quincy, then a seasoned diplomat of twenty-eight, wrote his mother that he was engaged to a young lady he had met in London, Louisa Catherine Johnson. Abigail's reaction was negative: Not now, think of your future, stay single, some day you will find true love. When she learned that his marriage to a *British* girl was set, Abigail, ever the fervent revolutionist, dispatched a sharp note: "I would hope for the love I bear my country that the Siren is at least half-blood." She was reassured that though "the siren's" mother was English, her father was the American consul in London.

Barbara's reaction in meeting Laura Welch was very different. Only five weeks after George met Laura, they were engaged, and two months later they were married. Barbara immediately liked this calm, intelligent young librarian—perhaps sensing that she would be a settling influence on the not totally tamed George. Since he was then running for Congress, Barbara gave Laura one piece of advice: Don't ever criticize his speeches. But after an event that hadn't gone well, George said, "I didn't do very well, did I?" Abrogating her mother-in-law's counsel, Laura conceded, "No, it wasn't very good." As George tells it, "I was so shocked I drove the car into the wall of our house." Since then Laura has hewed closely to Barbara's warning.

If Abigail and Barbara shared much in common, John Q. and George W. do not. John Q., a lawyer who spent virtually his entire life in public service, was the only president ever to serve in Congress after his defeat. He was near genius with languages, not one of George W.'s strong points. But with all his background, Abigail's son was devoid of social graces; the stiff American diplomat was a dud at state occasions in Europe. He described himself as "an austere and forbidding" man; his enemies, he said, described him as "a gloomy misanthropist and unsocial savage." In contrast, George W. enlivens any gathering and makes friends at first meeting.

★   ★   ★

ABIGAIL saw her son ascending the ladder of power, as a successful diplomat in four world capitals and as secretary of state (after turning down an appointment to the Supreme Court), but did not live to see him as president. At her death in 1818 the grieving John Quincy wrote in his diary: "Had she lived to the age of the Patriarchs, every day of her life would have been filled with clouds of goodness and of love. There is not a virtue that can abide in the female heart but it was an ornament of hers."

As governor of Texas, one of George W.'s first acts was to introduce a resolution in the state Senate: Resolved, "The most famous mother in Texas, beloved and admired throughout the nation, Barbara Bush is a special treasure to the citizens of Texas, and Mother's Day is an appropriate opportunity to express our gratitude to her for her many contributions to our state."

The resolution was adopted unanimously. Barbara Pierce Bush is enshrined in Texas law and in the heart of her feisty, fun-loving, firstborn president-son, who is just like his mother.

PASSIONATE ATTACHMENTS

> I remember how I clung to her till I was a great big fellow,
> but love of the best in womanhood came to me and
> entered my heart through those apron strings.
> —PRESIDENT WOODROW WILSON

THE easy assumption is that men of such achievement as the presidents of the United States would, of course, be devoted to the mothers who nurtured them. But do some notable exceptions prove the rule? Consider the mother-son relationships of two of the nation's most-honored presidents, George Washington and Thomas Jefferson. Acclaimed for his exemplary character and sense of duty, Washington was a loving and devoted husband to his wife, the much-admired Martha. Yet historians agree that his mother, Mary Ball Washington, was a lifelong source of irritation to her firstborn son. "The strangest mystery of Washington's life was his lack of affection for his mother," Douglas Southall Freeman asserted in his definitive biography. "Added years and understanding brought no improvement in his relations with her . . . apparently he did not write her even once during the war. He who had so much magnanimity and patience in dealing with human frailty was so much like his mother, in most money matters, that he felt she had been grasping and unreasonable." Historian Margaret Bassett tartly characterized Mary Washington as "a woman of small intelligence and great complaints [who] gave him a hard time all her long life."

In happy contrast, Martha, wrote Washington scholar James Flexner, "created for George what he had yearned for but not possessed since he was a little child; a happy home." Things might have been different had his father not died when George was eleven. Augustine Washington was a prosperous widower with three children when, in 1731, he married Mary Ball, a comely young lady of twenty-three, possessed of a fine figure, pleasant voice, and notable skill as a horsewoman. In quick succession she produced five offspring, then at her husband's death in 1743 was left to raise them and manage the family plantation, Ferry Farm, alone. In those sad circumstances George, whose father had been a distant, often-absent figure, might have grown closer to his mother, but it proved to be quite the opposite. Saddled with responsibility for the four younger siblings, the boy was hostage to her whims and steely will. (At fifteen, George wanted a career in the navy, which would have removed him from his mother, but Mary forbade it, in her own interest. Unwittingly, she affected his place in history and possibly the course of the American revolution: a seafaring Washington would not have become the victorious commander in chief in a war decisively conducted on land, and thus would not have been considered in the choosing of the first president.)

Again she thwarted him when he turned twenty-one, refusing to turn over Ferry Farm, which had been willed to him by his father, and she compounded the affront by badly mismanaging *his* property. A family dustup was averted, temporarily, when George inherited from his beloved half brother a splendid site on the Potomac River south of Alexandria, Virginia. There George built Mount Vernon, now a national shrine. After much acrimony, Washington and his brothers eventually settled their mother in a comfortable house in Fredericksburg, Virginia, but she constantly complained that she was being forced by her famous son to live "in great want."

In the midst of the war, she publicly humiliated George with a whining letter to the Virginia House of Burgesses, seeking financial assistance because he—a man known as rectitude itself—had failed to provide for her properly. In a sharp reply to her, a much-vexed Washington protested, "I am as a delinquent and am considered perhaps by the world as an unjust and undutiful son." She continued to grumble. Nor did it help the tattered relationship that Washington's wife, the charming and well-heeled widow Martha Custis, was never comfortable with her rebarbative mother-in-law. Add to that Mary's unkempt appearance, which was a constant embarrassment to her fastidious son.

A more profound offense would have been his mother's lack of support for independence, the great cause he was willing to die for. If Mary, a Tory at heart, felt pride in the son who to this day is revered around the world, there is no record of it; to the contrary, in letters she denigrated his achievements. Small wonder then, as Douglas Southall Freeman observed, "From her renowned son in his manhood Mary Ball Washington never elicited the warm love a man usually has for his mother."

The second icon on Mount Rushmore, Thomas Jefferson, was, like Washington, the compleat Virginia gentleman—and distant from his mother, London-born Jane Randolph Jefferson. Though he shared her home until he was twenty-seven years old, she is scarcely mentioned in his voluminous personal records. The literate, intellectual Jefferson, a mirror of the Age of Enlightenment, wielded the most prolific pen in public life: letters by the hundreds express his every mood and idea, observations of nature, ledgers listing trivial expenses and grand plans—yet not so much as a sentence of introspection about his mother. "The only remark he is known to have made about her influence was negative," observed Jefferson scholar Dumas Malone. "He probably did not value her counsel very highly."

Like Washington, Jefferson lost his father in boyhood; like Mary Ball Washington, Jane Randolph Jefferson, a young widow of thirty-seven, was left to bring up eight children and, like Mary, she placed the family burdens on her elder son, Tom, her third child. Late in life Jefferson wrote to his grandson: "When I recollect that at fourteen years of age the whole care and direction of myself was thrown on myself entirely, without a relative or friend qualified to advise or guide me, and recollect the various sorts of bad company with which I associated from time to time, I am astonished I did not . . . become as worthless to society as they were." So much for his mother's care. That she lived on for nineteen years after his father's death apparently did not figure in his recollections. Like Mary Ball Washington, Jane Randolph Jefferson never expressed support for the colonies' cause, and her own family was sharply divided over the issue.

On March 31, 1776, Jefferson wrote a flat, unelaborated entry in his journal: "My mother died about eight o'clock this morning, in the fifty-seventh year of her age." Two months later, in a letter to Jane's brother in England, Jefferson commented fully on the war's effect on trade before mentioning, "The death of my mother you have probably not heard of. This happened on the last day of March after an illness of not more than an hour. We supposed it to have been apoplectic." Not a word about her admirable traits or what she had meant to him. Even so, her death apparently triggered deep mixed feelings in Jefferson, who was stricken by one of the disabling migraine attacks he regularly suffered in times of personal upheaval. After six weeks, declaring himself a "renewed" man—perhaps freed of his ambivalence about the cold mother-son relationship—he immersed himself in a torrent of activity in the cause of American freedom. Within two months he produced both the constitution of Virginia and that great landmark of history, the Declaration of Independence, blessed with inspiration and graced with glorious language.

John Adams, the president who came between Washington and Jefferson, had no such difficulties with his mother, who shared his New England fervor for independence. And eighty-six years later

Abraham Lincoln reached the White House on the wings of "my angel mother," Nancy Hanks, an unschooled woman from the Kentucky backwoods who died when he was ten. "All that I am or hope ever to be I get from my mother," he declared in tribute and in oblique defense against malicious gossip about her. Though he was devoted to his stepmother, he coldly rejected his father, refusing to go to his bedside in his final illness. "Say to him," he instructed his stepbrother, "that if we could meet now, it is doubtful whether it could not be more painful than pleasant."

No rule or precedent governs the bond between president and mother; the mother herself forges the link. As historian Flexner, examining the fractious relationship between the first president and his mother, observed, "History does not always draw noble men from noble mothers, preferring sometimes to temper her future heroes in the furnace of domestic infelicity."

These mothers of modern presidents belonged more to the America that Jefferson had struggled to create, an agrarian patchwork of small towns and farms, than to a nation dominated by great urban centers and industry. And apart from Bill Clinton's mother, they were shaped by the nineteenth century rather than the twentieth. (Though Dorothy Bush was born on the cusp, in 1901, her conventional midwestern upbringing was rooted in the previous era.)

In 1831 Alexis de Tocqueville, a French jurist who later became foreign minister, spent a year traveling throughout the young American nation, recording his insights in a remarkable volume, *Democracy in America*. American women in particular fascinated him. "She has scarcely ceased to be a child, when she already thinks for herself, speaks with freedom," he marveled. "She is full of reliance on her own strength. . . . I have been frequently surprised, and almost frightened, at the happy boldness with which young women in America contrive to manage their thoughts and their language. Even in the independence of early youth, an American woman is always mistress of herself; she indulges in all permitted pleasures . . . and her reason never allows the reins of self-guidance to drop, though she often seems to hold them loosely." He suggested that such independence came at a cost: "It tends to invigorate the judgment at the expense of the imagination, and to make cold and virtuous women instead of affectionate wives and agreeable companions to man."

But her independence, he observed, "is irrecoverably lost in the bonds of matrimony," as the nineteenth-century American wife willingly subjugated herself to her husband and whatever life he chose. Tocqueville told of meeting "in a comfortless hovel in a forest" frontier wives who had grown up amid the comforts of New England, yet the near-unbearable hardships of their new life "had not broken the springs of their courage," an "inward strength" he credited to education. "If I were asked to what the singular prosperity and growing strength of [America] ought mainly to be attributed," he told his European readers, "I should reply, To the superiority of their women."

A few decades later, he could have cited these mothers of presidents as paragons of the strong, self-assured women who so deeply impressed him. Martha, Sara, and Ida, in the vanguard of the next generation, enlarged the profile of the indomitable American women he delineated.

Though these highly individualistic women grew up in different places and led different lives, common threads run through their stories. In a time when young ladies expected to be married before they turned twenty, most of the twelve married later, some by a number of years, an unwillingness to be rushed into matrimony that suggests an independent turn of mind. Several married against their fathers' wishes, and several "married down," joining their lives to men whose background was not quite up to their own social standing—in their families' views, at any rate. A particularly poignant example was Hannah Milhous, who defied her well-to-do Quaker family to marry Frank Nixon, a knockabout who in their eyes was a nobody. Never again would she enjoy the status she had known in her father's home. This sense of position surely heightened each mother's determination to elevate her children to the standards of her early years and, most particularly in Rebekah Johnson's case, to mold her son in the image of the father she so admired. None "married up" into families of higher social status than her own.

The lives of most of them were darkened by trials and tragedies: the deaths of children, poverty, illnesses, dysfunctional marriages, homes lost, alcoholism, abuse. But they got through it all unbowed, fueled by their early dreams, spurred by disappointments, challenged by limitations, bolstered by unwavering faith. From the time they said "I do," they sought personal satisfaction in the homemaker's accomplishments and, most of all, through the achievements of their children. Each of these presidential mothers poured her energies into the vessel that was her son, vicariously sharing in his triumphs, finding her fulfillment in his. She wanted people to see that she was a successful mother. She wanted respect, and got it.

They were diverse women who led conventional lives, exercised their keen intelligence, and didn't shrink from controversial opinions—which they dispensed without hesitation to their political sons. As a group (with the exception of Sara Roosevelt) they were concerned about social issues, troubled by racial inequities and other injustices. Lillian Carter, the most defiantly independent-minded member of this little club, showed a tolerance toward blacks in her attitude and actions that shocked her south Georgia community and had a crucial impact on her son's future. Ida Eisenhower, in her later years, expressed only one regret about her life: she wished she could have worked for the fair treatment of black Americans. Such a strong conviction against racial discrimination was notably advanced for a woman of the Kansas frontier, who died years before the civil rights movement emerged. She traced her views to her Virginia family's opposition to slavery, a concern that grew stronger as she witnessed the mistreatment of black Americans. Rose Kennedy, stung by discrimination against Irish Americans, taught her children compassion for the outsider. Decades later, Virginia Clinton, in kitchen-table discussions with her son and his friends, railed against social injustices she saw in her work and read about in the papers.

Nor were they parroting their husbands' views, which were often quite different and less admirable; their moral stand was formed by their own consciences—and they instilled in their sons the responsibility to think beyond themselves. The penchant for independent thinking shows up among mothers of exceptionally successful sons in fields other than politics. In her study *Mothers and Sons,* author Carole Klein found that such women are not of the neutral image traditionally assigned to mothers. "An intelligent mother with strong opinions stirs more than her son's intellect," Klein asserted. "She stimulates both his curiosity and his energy so that he will one day be an opinion-shaper himself."

If these presidents' mothers had done nothing more than give their sons the underpinning of confidence, it was an accomplishment, for all around them were mothers whose sons did not shine, boys who did not reach beyond mediocrity, young men whose lives were not inspiring tales of success, much less the stuff of history. What was it that stirred these women to pour their lives and hopes into these particular sons and set them on the path to achievement?

Knowing nothing of each other, they were of much the same mind about bringing up sons. Without child development experts like Benjamin Spock and T. Berry Brazelton ladling out child-rearing advice in books and on television, they relied on common sense and maternal instincts, on God and gumption, encouraging and disciplining as they saw fit. Undergirding those natural instincts was unconditional love, the most important element in developing confidence in a child. That confidence is thought to have its roots in infancy, starting with the close emotional bond between mother and child, the secure and comforting presence who inspires trust, which matures into self-confidence and in turn builds trust in others.

Self-confidence, psychiatrists suggest, is fundamental to achieving success in any career, whatever the choice. Personalities may be unlike, motivations may differ, but the one who makes it to the top, the achiever, first believes in himself. Nowhere is this self-assurance more crucial than in present-day politics, where the candidate must stand, exposed and vulnerable, before the public and be judged by a handful of villagers in a cold town hall in New Hampshire and by millions critiquing the presidential debates. But unbridled self-assurance can balloon into hubris. The boy whose mother made him feel that he was indeed very special usually had his head kept to its hat size by his unimpressed—sometimes resentful—siblings, by a critical father, and by the mother's own instinctive balance of praise and discipline.

Where did the self-confidence of these mothers come from? Their fathers. Grandfather to mother to son. In her book *Fathers' Daughters*, psychotherapist Maureen Murdock states that "a father's daughter is a woman who identifies with her father and imitates men in pursuit of success." For the mothers of these presidents, success meant achievement through nurturing a successful son. Murdock sees such women, when they were young, as "a 'daddy's girl' . . . the apple of her father's eye [who] receives special treatment and attention from him." With that father-given confidence and drive, if these women were their granddaughters they would very likely be pursuing ambitions of their own today as well as managing a family. As it was, each settled for the reflected glory of her son.

In any study of mothers and sons Sigmund Freud, the father of psychoanalysis, cannot be overlooked. Freud, who held that sexuality was at the core of every relationship, was the first to theorize that a daughter would heap her earliest affection on her father, while the son would direct his infantile desires toward his mother. Freud interpreted this as the rivalry the child feels with the parent of the same sex in seeking the affection of the other parent—all elements of Freud's theory of the Oedipus complex. "It is a natural tendency," he wrote, "for the father to spoil his little daughter and for the mother to take the part of the sons." Freudian theories, which scandalized the sexually constricted late nineteenth century, no longer shock, nor do they dominate psychiatry as they once did; however, they remain a tool in understanding child-parent relationships.

The daughter-father bond is evident in the fact that six of these twelve gave these future presidents her family name—her father's and her own—compared with only two of the thirty-one mothers who preceded them. (Only four bestowed her name on a son born later.) It was as though the mother had stamped her name on that son as her special child. She couldn't have known that this newborn would come to have her personality traits, be closer to her, and in most instances even look more like her, or, when there were only two sons, that the other brother would tend to "take after" his father and have a stronger relationship with him. What explains this phenomenon? Quirk or pattern? DNA? Cause or effect?

Current research indicates that individuals are born programmed with specific personality traits as well as physical characteristics. The brain's hardwiring is far more important to a child's individuality than had been previously understood. The mother, too, has innate traits that affect her bonding more closely with one child than another; for example, consider the connection between Nelle Reagan and her second son, Ronald.

Research on the brain, a field that is virtually exploding, has produced evidence that a human being is at least 50 percent the product of heredity, not just in mental ability and physical characteristics but even in behavioral traits such as temperament and "emotional intelligence." The new findings wouldn't have mattered greatly to these mothers—they would just have made sure that their 50 percent was the half that counted most. Working on instinct, they followed a course that today's child-development experts preach to modern mothers: stimulate your baby's mind. Reading to children should start in the crib. They prized books and education with a passion that began in their own childhood, and from the time their sons were toddlers, the mothers had passed on to them a love of words, reading, learning, knowledge.

Again and again, these presidents boasted that their mothers taught them to read before they entered school, before they were four or five, to recognize letters at two. Taking their era into consideration, the mothers' own educational backgrounds are impressive: seven attended college, two were given a high gloss in European finishing schools, and two received nursing degrees. Only Nelle Reagan had no formal education beyond elementary school—and she made up for it with her natural quest for self-improvement.

In 1870 a total of 11,126 female students were enrolled in colleges and academies throughout the country, most of them instructing young ladies primarily in "suitable" pursuits such as embroidery, art, poetry, and the social graces. In ten years the number of female students more than tripled, reaching almost half that of male students, and the 1900 census indicated a continuation of this pattern. College at that time was for wealthy boys who were going into a profession; the boy of average means was expected to take over the family farm or get a job in a local business. Even high-school diplomas were not taken for granted, which makes these girls' hunger for advanced education all the more noteworthy.

In those early years young ladies pursuing higher education were eager to acquire a patina that would take them beyond their homespun backgrounds. "Culture," intoned an uplifting speaker to Rebekah and her classmates at Baylor Female College, "purges from the soul the coarse. . . . Culture betters not in worldly advantages but in true nobility of soul." So they read great works of literature, found solace in poetry, and dabbled in writing. They turned their hand to sketching and painting china plates, and with a tolerant instructor and a metronome they mastered the piano to varying degrees. A favorite course was elocution, in which the girls memorized and recited dramatic readings, with exaggerated expression and careful vowels. There were, however, female institutions aimed at the "blue-stockings"—serious students who would drink deeper of the springs of knowledge—with stiff courses in history, foreign languages, and mathematics. Women studied, but "women's studies" was not even a concept.

When they were mothers, their introduction to things cultural at college served as an incentive to bring up their sons to appreciate life's finer side. Rebekah Johnson would send Lyndon to dancing classes and (hopelessly) violin lessons; Martha Truman envisaged Harry as a concert pianist (realistically, he judged himself at the level of a music-hall performer); Hannah Nixon sent twelve-year-old Richard away for six months to study piano with her sister, a trained musician. They had such dreams for these sons—only to have them grow up to be president.

The other brothers' stories follow different lines. The problem brothers were younger, the second of two sons, born to be runner-up in the family and overshadowed by the hugely successful big brother. Sam Houston Johnson, Don Nixon, Billy Carter, and Roger Clinton, each in his turn, joined the long line of younger brothers who embarrassed their president-brothers. (From their early teens Dick and Don Nixon realistically were older and younger brothers since Harold, the eldest son, was mostly away from home, under treatment for tuberculosis until his death six years later.) This line stretches back to Washington's brother Charles, who was the first of a striking statistic: among the younger brothers of presidents, from Washington to Clinton, alcoholism and drug abuse have occurred at a much higher rate than the national average.

Billy Carter played ball with the pariah nation of Libya, his brother's antagonist, by accepting large fees as a "consultant," and in the Nixon White House Donald was considered such a potential problem that the president ordered his brother's telephone tapped. When the chain of restaurants owned by lighthearted and popular Don ("Nixonburgers" were the specialty) slipped into crushing debt, presto!, one of eccentric Howard Hughes's companies stepped in with an unsecured loan of $265,000: the result was a deluge of headlines. Edward Nixon is sympathetic toward his likable, much older brother: "Don was generous to a fault, taken in by trusting people too much. Mother and Dick would warn him, but the good advice was sometimes not heeded." Don Nixon wound up bankrupt, as did Sam Houston Johnson and Billy Carter. Roger Clinton wound up in prison.

The Truman brothers got along comfortably, but Vivian—his father's favorite—preferred his farmhouse to Harry's White House. He shunned Washington and was so outspoken in his disdain for the place that, if listening closely, one might hear the younger Truman disparaging his older brother's career and success. Ronald Reagan's brother, Neil, caused no problems. To the contrary, in a relationship that was always a bit strained, he stayed at arm's length from his brother.

A close look at these presidents' brothers suggests a pattern: when there were multiple sons—the half-dozen Eisenhowers, the Fords, Bushes, and Kennedys with four boys each—there was vigorous competition within the family, and while the boys might fight and bicker, there was no corrosive rivalry. The likely explanation is that when the mother's love and attention were spread among several, a sort of parity resulted; when it was divided between two, the competition was too intense. It is hard to live in a brother's shadow.

Little was heard from sisters, who were oddly underrepresented in these twelve families—of fifty-one children, only thirteen were girls, five of them Kennedys. None of them caused embarrassment to their big brothers in the White House. But the sisters were not competing with their brothers for that I'm-your-best-son attachment with their mothers; they were more likely to be seeking the attention and praise of their fathers.

And where were the future presidents' fathers as these bonds were being forged between mothers and sons? Taken as a group, which homogenizes sharply different men into one sample, these fathers tended to be rather remote presences, physically and/or emotionally. Some were weak or feckless figures, even outright failures, who couldn't have exercised great influence, either by example or by teaching, on their sons. And some were rarely there at all. Even the three most successful and powerful fathers, Joseph P. Kennedy and Prescott and George H. W. Bush, were absent a great deal in pursuit of their careers and lifestyles. But a striking number of the other fathers—those of Truman, Eisenhower, Reagan, Nixon, and Johnson—failed to achieve the success in life they sought so desperately.

Some of the wives turned to their sons to compensate for their weak husbands and the disappointment of dreams gone awry. Rebekah Johnson, Nelle Reagan, and, most of all, Virginia Clinton stand out as classic examples. Luci Baines Johnson, Rebekah's granddaughter, offers the candid appraisal that Rebekah's "character and characteristics were more laudable than her husband's," a tactful way of saying that Lyndon Johnson's mother married a man who fell far short of her lofty aspirations.

Circumstances similar to Rebekah's led to Nelle Reagan's eagerness to draw her more compatible son, Ronald, into her inner world of the finer things. And on a different level, Ida Eisenhower created a cheerful home, a place of smiles and optimism, to offset the gloom that engulfed her husband. ("My father," Eisenhower wrote, "lived in the shadows of his early failure.") The sons, as they grew up, felt the need to make up for the shortfalls of their fathers, to achieve more, to fulfill their mothers' dreams.

In *The Devil's Dictionary* Ambrose Bierce defines a bride as a woman whose brightest expectations lie behind her, a sardonic quip that is apt in most of these women's lives. But though their husbands failed to live up to their dreams, though their marriages were often hollow, these wives did not hector or belittle their husbands; they were not the emasculating shrew-wives portrayed in fiction. In a time when divorce was not acceptable, they married for better, for worse, forever. The wife would stay with her husband and starch his shirt collars and put a good supper on the table every night and never talk about the emptiness, the unspoken wishes, the unsatisfied desires. Not surprisingly, she would turn to her son, leaning on him psychologically to fill the void caused by the disengaged father.

Even now, in the view of psychotherapist Michael Gurian, author of *Mothers, Sons and Lovers,* "Fathers do not even know they are supposed to be involved. Fathers do alter somewhat their interaction with us as we enter adolescence, but not very much, and not in rituals that connect them to us and to our masculine spirituality. . . . We are still attached to our mothers, we are still yearning for our fathers." Gurian finds that even today, along with absences and emotional distance, there is "excessive passivity" among many fathers.

Historian David McCullough has written in defense of a strong attachment to mother: "Growing up as a mama's boy may not necessarily be detrimental, and the sort of person Mama is probably has as much as anything to do with the outcome." (A mother is not necessarily a force for good: Adolf Hitler was an egregious mama's boy.) Margaret Truman Daniel, author of many books, speaks of "the enormously strong intellectual and emotional bond between Dad and his mother—the sort of bond which, I have discovered in my delvings into presidential lore, has existed between an astonishing number of presidents and their mothers."

But these presidential mothers and fathers and their children shared a world that is receding into memory. They had a rootedness, a firm sense of where they came from. For the sons, it was the place that would leap to their minds in a hundred images whenever there was talk of boyhood and home. That place was almost always a small town, some not much more than a hamlet no one had heard of until a historic marker declared it "The Boyhood Home of . . ."

For such insignificant places to produce this nation's presidents is, at first thought, puzzling: how did they get from there to here? Ronald Reagan, as he entered politics, said, "You have to start with the small-town beginnings. You're part of everything that goes on. In a small town you can't stand on the sidelines and let somebody else do what needs doing; you can't coast along on someone else's opinions. That really is how I became an activist— there was a sense of urgency about getting involved." Biographer David Maraniss tied Clinton's rise to the presidency to his small-town origins: "Hope and Hot Springs nourished Clinton as few larger communities could have done." None of these modern presidents grew up in big cities, with their crowded isolation and without the bonding sense of community. Even the Kennedys and the Bushes lived in the exurban enclaves of Greenwich and Bronxville (close to the city, yet a world apart) and the summer paradises of their family compounds on the seacoasts of Maine and Cape Cod.

As presidents, all of them remembered their childhood as idyllic, even when times were hard, their homes cramped and short on creature comforts. There was so much to do: places to explore, books to read, family to play and fight with. When Harry Truman was asked if he grew bored on the farm and in small-town Independence, he was surprised at the notion: "Oh, my, no. I can't remember being bored, not once in my whole life."

Above all, it was the values they were taught that determined what these boys were made of. A child's character, the professionals say, is shaped by the age of five. That downloads a heavy burden on mothers, especially in bringing up the first child, which is usually on-the-job training—and seven of these twelve were firstborns. (Only Eisenhower came as far down the pecking order as third.) She must not only cope with the physical demands of this creature but also establish the signposts to guide him on the right path, for the old saw, "the child is father of the man," has real meaning. The late James David Barber, professor (retired) of political science at Duke University and author of *The Presidential Character*, established, in analyzing presidents' performance, that "in general, character has its main development in childhood, style in early adulthood." His thesis places character building squarely in the mother's court.

In one respect, parenting was easier in the first half of their century when values were not negotiable, before standards and attitudes began to shift, or drift in a kind of "situational ethics" fashion. "In those days," Truman said of his youth, "what was right was right, and what was wrong was wrong, and you didn't have to talk about it. You accepted it." Margaret Truman Daniel summed up the moral code that her grandmother, the stalwart Martha, laid down for her family: "Do the right thing, do it the best you can, never complain and never take advantage of others." In Martha's no-nonsense Missouri style, those few words would arm a child with guidelines to see him through any moral quandary.

Along with their inflexible standards of right and wrong, the mothers applied a measure of homespun psychology to encourage the boys to excel: they had a family name and a place in the community to live up to; they must set an example for younger brothers and sisters; they must show the world, their own close-at-hand world, that they were meant to be leaders. These sons were given to know that much was expected of them—because they were special. If the mother consciously recognized the role her son played in her own fulfillment, she would not have made that point.

Undergirding these mothers' reinforcement of values was a solid spirituality, an unshakable belief in a Divine Power that sustained and guided them through good times and bad. Most were mainstream Protestants (eight were Episcopalian or Baptist) and churchgoing was central to their lives, in part because the church was also the center of social life in small towns. Dorothy Bush and Dorothy Ford were leaders in their Episcopal parishes and Nelle Reagan engaged in a virtual street ministry for the Disciples of Christ (the Church of Christ). Devout Catholic Rose Kennedy, who attended Mass every day until she was almost one hundred years old, took the view that "God rewards people who [have] faith," which would surely put her in heaven's box seats. Perhaps the most individualistic was Ida Eisenhower, who was reared in the fundamentalist River Brethren sect and in her later years became a Jehovah's Witness.

These boys grew up in times before "affluence" became a common noun, in places where most people had to struggle to pay the rent and keep clothes on their backs. Yet again and again in presidential memoirs there are variations of the comment "We were poor but we didn't know it." When your neighbors are in the same boat, the hardships are accepted as just the way things are; everything about one's perception of life is relative to the people, the setting, and the times that are the yardstick. Then, too, great success tends to erase the tape of painful memories and early hard times. When these boys were growing up, their parents encouraged—even expected—their sons to take jobs. The work ethic permeated their lives, not as an exercise in character building but because their families needed the money. (An exception was Bobby Kennedy delivering magazines via Rolls-Royce.)

For those future presidents who endured real hardship, their mothers' drudgery and sacrifices in those pinched times left a deep impression on them. All were thoughtful gift-givers, eager to lavish on their mothers the things she had never been able to have. From their early days to "Hail to the Chief" in the Rose Garden, the president-sons never stopped being grateful to their mothers. And there were tributes for those who had not lived to share the joy. Each son had carried out Mother's subliminal message: he had made it; he had more than fulfilled her dreams for him; he had compensated for her disappointments. Even now, friends notice that Gerald Ford's voice quavers and George Bush's eyes mist over when they reminisce about their mothers.

When Harry Truman was over eighty and happily retired back in Independence, he expressed the feelings of all of these presidents in a note of sympathy to Dean Acheson, his former secretary of state, on the death of his mother. "There is no supporter like your mother," wrote the thirty-third president, who had so often been bludgeoned by criticism. "Right or wrong, from her viewpoint you are always right."

In 1930 Lyndon Johnson wrote his first editorial, a Mother's Day tribute, for the Southwest Texas State Teachers College newspaper: "The affection of all loved ones may be estranged, but Mother's love abides to the end. . . . There is no love on earth comparable to that of Mother." He enjoined his fellow students to "make our lives living tributes to the most potent and vital force for good in the world—the Mothers of men." He was only twenty-two at the time.

Simone de Beauvoir, the French feminist-philosopher, declared in her landmark work, *The Second Sex:* "A son will be a leader of men, a soldier, a creator; he will bend the world to his will, and his mother will share his immortal fame; he will give her the houses she has not constructed, the lands she has not explored, the books she has not read."

And justify her hopes; fulfill her dreams.

SELECTED READINGS

★

CHAPTER I: SARA DELANO ROOSEVELT

Asbell, Bernard. *The F.D.R. Memoirs: A Speculation on History*. Doubleday & Co., 1973.

Burns, James MacGregor. *FDR: The Lion and the Fox, 1882–1940*. Harcourt Brace & Co., 1956.

———. *Roosevelt: The Soldier of Freedom, 1940–1945*. Harcourt Brace Jovanovich, 1970.

Burt, Nathaniel. *First Families: The Making of an American Aristocracy*. Little, Brown, 1970.

Cook, Blanche Wiesen. *Eleanor Roosevelt, Vol. 1*. Viking, 1992.

Davis, Kenneth S. *FDR: The Beckoning of Destiny*. G. P. Putnam's Sons, 1972.

Flynn, John T. *Country Squire in the White House*. Doubleday Doran, 1955.

Fraser, Kennedy. *Ornament and Silence: Essays on Women's Lives*. Alfred A. Knopf, 1997.

Goodwin, Doris Kearns. *No Ordinary Time*. Simon & Schuster, 1994.

Graff, Robert, Robert Bennett Ginn, and Roger Butterfield. *FDR*. Harper & Row, 1963.

Gurko, Leo. *The Angry Decade*. Dodd, Mead, 1947.

Lash, Joseph P. *Eleanor and Franklin.* W. W. Norton, 1971.

Miller, Nathan. *FDR: An Intimate History.* Doubleday & Co., 1983.

National Park Service. *Home of Franklin D. Roosevelt.*

Perkins, Frances. *The Roosevelt I Knew.* Viking, 1946.

Robinson, Edgar Eugene. *The Roosevelt Leadership, 1933–1945.* J. B. Lippincott, 1955.

Roosevelt, Eleanor. *This Is My Story.* Garden City Publishing, 1937.

———.*This I Remember.* Harper & Brothers, 1949.

Roosevelt, Elliott, and James Brough. *An Untold Story: The Roosevelts of Hyde Park.* G. P. Putnam's Sons, 1973.

Roosevelt, Mrs. James, as told to Isabel Leighton and Gabrielle Forbush. *My Boy Franklin.* Ray Long & Richard R. Smith, 1933.

Schlesinger, Arthur M., Jr. *Franklin Roosevelt: Crisis of the Old Order, 1919–1933.* Houghton Mifflin, 1957.

———. *Franklin Roosevelt: The Coming of the New Deal.* Houghton Mifflin, 1959.

Sherwood, Robert E. *Roosevelt and Hopkins.* Harper & Brothers, 1948.

Tully, Grace. *F.D.R.: My Boss.* Charles Scribner's Sons, 1949.

Ward, Geoffrey C. *A First-Class Temperament.* Harper & Row, 1989.

———. *Before the Trumpet: Young Franklin Roosevelt.* Smithmark, 1994.

West, J. B., with Mary Lynn Kotz. *Upstairs at the White House: My Life with the First Ladies.* Coward, McCann, Geoghegan, 1973.

CHAPTER 2: MARTHA YOUNG TRUMAN

The American Experience. Public Broadcasting System, September 1997.

Berger, Meyer. "Mother Truman: Portrait of a Rebel." *New York Times Sunday Magazine,* June 23, 1946.

Ferrell, Robert H., ed. *Off the Record: The Private Papers of Harry S. Truman.* Harper & Row, 1980.

Jenkins, Roy. *Truman.* Harper & Row, 1986.

McCullough, David. *Truman.* Touchstone/Simon & Schuster, 1992.

Miller, Merle. *Plain Speaking: An Oral Biography of Harry S. Truman.* G. P. Putnam's Sons, 1974.

Miller, Richard Lawrence. *Truman: The Rise to Power.* McGraw-Hill, 1986.

Potts, Rev. Edward W. "The President's Mother." *The Christian Advocate,* July 1945.

Robbins, Jhan. *Bess & Harry: An American Love Story.* G. P. Putnam's Sons, 1980.

Steinberg, Alfred. *The Man from Missouri*. G. P. Putnam's Sons, 1962.

Stratton, Joanna L. *Pioneer Women: Voices from the Kansas Frontier*. Simon & Schuster, 1981.

Truman, Harry S. *Memoirs Vol. 1, Years of Decisions*. Doubleday, 1955.

Truman, Margaret. *Harry S. Truman*. William Morrow, 1973.

———, ed. *Where the Buck Stops: The Personal and Private Writings of Harry S. Truman*. Warner Books, 1989.

CHAPTER 3: IDA STOVER EISENHOWER

Ambrose, Stephen E. *Eisenhower: Soldier and President*. Simon & Schuster, 1990.

Biography. "Dwight David Eisenhower." Arts & Entertainment Network. October 7, 1997.

Brendon, Piers. *Ike: His Life & Times*. Harper & Row, 1986.

Childs, Marquis. *Eisenhower: Captive Hero*. Harcourt, Brace, 1958.

Davis, Kenneth S. *Dwight D. Eisenhower: Soldier of Democracy*. Doubleday, Doran, 1945.

Donovan, Robert J. *Eisenhower: The Inside Story*. Harper & Brothers, 1956.

Eisenhower, David. *Eisenhower at War, 1943–1945*. Random House, 1986.

Eisenhower, Dwight D. *At Ease: Stories I Tell to Friends*. Doubleday & Co., 1967.

Eisenhower, Julie Nixon. *Special People*. Simon & Schuster, 1977.

Eisenhower, Susan. *Mrs. Ike*. Farrar, Straus & Giroux, 1996.

Ferrell, Robert H., ed. *The Eisenhower Diaries*. W. W. Norton, 1981.

Jackson, Nettie Stover. *Oral History*. Dwight Eisenhower Library.

Krick, Robert K. "Stonewall Jackson's Deadly Calm." *American Heritage*, December, 1996.

McCann, Kevin. *Man from Abilene*. Doubleday & Co., 1952.

Miller, Francis Trevelyan. *Eisenhower: Man and Soldier*. John C. Winston, 1944.

Miller, Merle. *Ike the Soldier, As They Knew Him*. G. P. Putnam's Sons, 1987.

Montgomery, Ruth. *Hail to the Chiefs*. Coward-McCann, 1970.

Park, Lillian Rogers. *My 30 Years Backstairs at the White House*. Fleet Press, 1961.

Pusey, Merlo. *Eisenhower the President*. Macmillan, 1956.

Rovere, Richard H. *Affairs of State: The Eisenhower Years*. Farrar, Straus, 1956.

Webb, Walter Prescott. *The Great Plains*. University of Nebraska Press, 1931, 1981.

Women's National Press Club. *Who Says We Can't Cook!* McIver Art and Publications, 1955.

CHAPTER 4: ROSE FITZGERALD KENNEDY

Adams, Cindy, and Susan Crimp. *Iron Rose*. Dove Books, 1995.

Clymer, Adam. *Edward M. Kennedy: A Biography*. William Morrow, 1999.

Dallas, Rita, with Jeanire Ratcliffe. *The Kennedy Case*. G. P. Putnam's Sons, 1973.

David, Lester, and Irene David. *Bobby Kennedy: The Making of a Folk Hero*. Dodd, Mead, 1986.

Gibson, Barbara, with Caroline Latham. *Life with Rose Kennedy*. Simon & Schuster, 1971.

Gibson, Barbara, and Ted Schwarz. *Rose Kennedy and Her Family: The Best and Worst of Their Lives and Times*. Birch Lane Press/Carol Publishing Group, 1995.

Goodwin, Doris Kearns. *The Fitzgeralds and the Kennedys*. Simon & Schuster, 1987.

Hersh, Burton. *The Education of Edward Kennedy: A Family Biography*. William Morrow, 1972.

Higham, Charles. *Rose: The Life and Times of Rose Fitzgerald Kennedy*. Pocket Books, 1995.

Kennedy, Rose F. *Times of My Life*. Doubleday, 1974.

Ladies' Home Journal Special. *Rose Kennedy: The Legend and the Legacy*. March 1994.

Leamer, Laurence. *The Kennedy Women: The Saga of an American Family*. Villard Books, 1994.

Levin, Murray B. *Kennedy Campaigning*. Beacon Press, 1966.

Life. "Rose Kennedy at 100." March 1990.

Life. "The New Kennedy Generation," July 1997.

Lowe, Jacques. *Kennedy: A Time Remembered*. Quartet/Visual Arts, 1983.

Manchester, William. *The Death of a President*. Harper & Row, 1967.

Rainie, Harrison, and John Quinn. *Growing Up Kennedy: The Third Wave Comes of Age*. G. P. Putnam's Sons, 1983.

Reeves, Richard. *President Kennedy: Profile of Power*. Simon & Schuster, 1993.

Salinger, Pierre. *With Kennedy*. Doubleday, 1966.

Schlesinger, Arthur Jr. *Robert Kennedy and His Times*. Houghton Mifflin, 1978.

Sidey, Hugh. *John F. Kennedy, President*. Atheneum, 1963.

Sorensen, Theodore C. *Kennedy*. Harper & Row, 1965.

Whalen, Richard J. *The Founding Father: The Story of Joseph P. Kennedy*. Regnery Gateway, 1993.

White, Theodore H. *The Making of the President 1960*. Atheneum, 1961.

CHAPTER 5: REBEKAH BAINES JOHNSON

Barnes, Lorraine. "Lyndon's Mother." *Austin American*, November 19, 1955.

Baylor College Quarterly, Belton. February 1921.

Biography. "Lyndon Baines Johnson." Arts & Entertainment Network.

Caro, Robert A. *The Years of Lyndon Johnson: The Path to Power*. Alfred A. Knopf, 1982.

Carpenter, Elizabeth. "Rebekah Baines Johnson." *Houston Post*, June 20, 1954.

Goldman, Eric F. *The Tragedy of Lyndon Johnson*. Alfred A. Knopf, 1969.

James, Eleanor. *Forth from Her Portals: The First One Hundred Years in Belton*. University of Mary Hardin–Baylor, 1986.

Johnson, Lady Bird. *A White House Diary*. Holt, Rinehart & Winston, 1970.

Johnson, Lyndon Baines. *The Vantage Point*. Holt, Rinehart & Winston, 1971.

Johnson, Rebekah Baines. *A Family Album*. McGraw-Hill, 1965.

Johnson, Sam Houston. *My Brother Lyndon*. Cowles Book Co., 1969.

Kearns, Doris. *Lyndon Johnson and the American Dream*. Harper & Row, 1976.

Lyndon Baines Johnson Library. Oral histories: Ben Crider, Truman Fawcett, Wilma Fawcett, Eric Goldman, Jessie Hatcher, Sam Houston Johnson, Otto Lindig, Josepha Baines Saunders, Emmette S. Redford, Juanita Duggan Roberts.

McKee, Rose. "Senator Johnson's Mother." *Austin American*, September 13, 1958.

Montgomery, Ruth. *Mrs. LBJ*. Holt, Rinehart & Winston, 1964.

Russell, Jan Jarboe. *Lady Bird: A Biography of Mrs. Johnson*. A Lisa Drew Book/Scribner, 1999.

Sidey, Hugh. *A Very Personal Presidency: Lyndon Johnson in the White House*. Atheneum, 1968.

Steinberg, Alfred. *Sam Johnson's Boy*. Macmillan, 1970.

University of Mary Hardin–Baylor Sesquicentennial Album. Mary Hardin–Baylor University, 1995.

CHAPTER 6: HANNAH MILHOUS NIXON

Aitken, Jonathan. *Nixon: A Life*. Weidenfeld & Nicholson, 1993; Regnery, 1993.

Ambrose, Stephen E. *Nixon: The Education of a Politician, 1913–1962*. Simon & Schuster, 1987.

Brodie, Fawn M. *Richard Nixon: The Shaping of His Character*. W. W. Norton, 1981.

California State University, Fullerton. Oral histories. Jane Milhous Beeson, Ollie O. Burdg, Gladys Gauldin, Janet L. Goeske, Muriel Kenny, Helen S. Letts, Olive and Oscar Marshburn, Cecil Pickering, Mary G. Skidmore, Paul Smith, Ada Ware Sutton, Merle West, Floyd Wildermuth.

Costello, William. *The Facts About Nixon*. Viking, 1960.

Crowley, Monica. *Nixon Off the Record*. Random House, 1996.

Eisenhower, Julie Nixon. *Pat Nixon: The Untold Story*. Simon & Schuster, 1986.

Graham, the Rev. Dr. Billy. *Just As I Am*. HarperCollins, 1997.

Keogh, James. *This Is Nixon*. G. P. Putnam's Sons, 1956.

Kornitzer, Bela. *The Real Nixon*. Rand McNally, 1960.

Mazlish, Bruce. *In Search of Nixon*. Basic Books, 1972.

Mazo, Earl. *Richard Nixon: A Political and Personal Portrait*. Harper & Brothers, 1959.

———, and Stephen Hess. *Nixon: A Political Portrait*. Popular Library, 1968.

Morris, Roger. *Richard Milhous Nixon*. Henry Holt, 1990.

Nixon, Richard. *The Memoirs of Richard Nixon*. Grosset & Dunlap, 1978.

Schulte, Renee K., ed. *The Young Nixon: An Oral Inquiry*. Oral History Program, California State University, 1978.

Strober, Gerald S., and Deborah Hart Strober. *Richard Nixon: An Oral History of His Presidency*. HarperCollins, 1994.

Toledano, Ralph de. *Nixon*. Henry Holt, 1956.

White, Theodore H. *The Making of the President 1972*. Atheneum, 1973.

———. *Breach of Faith: The Fall of Richard Nixon*. Atheneum, Reader's Digest Press, 1975.

Wills, Garry. *Nixon Agonistes*. Houghton Mifflin, 1970.

Witcover, Jules. *The Resurrection of Richard Nixon*. G. P. Putnam's Sons, 1970.

CHAPTER 7: DOROTHY GARDENER FORD

The Ancestry of Gerald R. Ford. Gerald Ford Library.

Cannon, James. *Time and Chance*. HarperCollins, 1994.

Ford, Betty, with Chris Chase. *The Times of My Life*. Harper & Row, 1978.

———. *Betty: A Glad Awakening*. Doubleday, 1987.

Ford, Gerald R. *A Time to Heal: The Autobiography of Gerald R. Ford*. Harper & Row and The Reader's Digest Association, 1979.

Gerald R. Ford Library. Gerald R. Ford Vice Presidential Papers, Box 88, Folder "Parents' Divorce Case."

Reeves, Richard. *A Ford, Not a Lincoln*. Harcourt Brace Jovanovich, 1975.

terHorst, Jerald. *Gerald Ford and the Future of the Presidency*. The Third Press, 1974.

Vestal, Bud. *Jerry Ford, Up Close*. Coward, McCann & Geoghegan, 1974.

CHAPTER 8: LILLIAN GORDY CARTER

Bourne, Peter, G. *Jimmy Carter: A Comprehensive Biography from Plains to Postpresidency*. Scribner's, 1997.

Carter, Jimmy. *Why Not the Best?* Bantam Books, 1976.

———. *Always a Reckoning*. Times Books, 1995.

———. *Living Faith*. Times Books, 1996.

———. *The Virtues of Aging*. Ballantine, 1998.

Carter, Lillian, and Gloria Carter Spann. *Away from Home*. Simon & Schuster, 1977.

Carter, Rosalynn. *First Lady from Plains*. Houghton Mifflin, 1984.

Glad, Betty. *Jimmy Carter: In Search of the Great White House*. W. W. Norton, 1980.

Schram, Martin. *Running for President 1976: The Carter Campaign*. Stein & Day, 1977.

Stroud, Kandy. *How Jimmy Won*. William Morrow, 1977.

Tartan, Beth (Elizabeth Hedgecock Sparks). *Miss Lillian and Her Friends*. Modern American Library, 1977.

U.S. Government Printing Office. *The Presidential Campaign 1976: Vol. 1, Parts 1 and 2*. 1978.

Wooten, James. *Dasher: The Roots and the Rising of Jimmy Carter*. Weidenfeld & Nicolson, 1978.

CHAPTER 9: NELLE WILSON REAGAN

Barrett, Laurence. *Gambling with History: Reagan in the White House*. Doubleday, 1983.

Biography. "Ronald Reagan." Arts & Entertainment Network. April 11, 1997.

Boyarsky, Bill. *The Rise of Ronald Reagan*. Random House, 1968.

Cannon, Lou. *Reagan*. Simon & Schuster, 1982.

———. *President Reagan: The Role of a Lifetime*. Simon & Schuster, 1991.

Davis, Patti. *The Way I See It*. G. P. Putnam's Sons, 1992.

D'Souza, Dinesh. *Ronald Reagan: How an Ordinary Man Became an Extraordinary Leader*. Free Press, 1997.

Dugger, Ronnie. *On Reagan: The Man & His Presidency*. McGraw-Hill, 1983.

Edwards, Anne. *Early Reagan: The Rise to Power*. William Morrow, 1987.

Korda, Michael. "Reagan." *The New Yorker*, October 6, 1997.

Leamer, Laurence. *Make-Believe: The Story of Nancy & Ronald Reagan*. Harper & Row, 1983.

Leighton, Frances Spatz. *The Search for the Real Nancy Reagan*. Macmillan, 1987.

Morris, Edmund. *Dutch: A Memoir*. Random House, 1999.

Public Broadcasting System. *The American Experience*, 1998.

Reagan, Maureen. *First Father, First Daughter*. Little, Brown, 1989.

Reagan, Michael, with Joe Hyams. *On the Outside Looking In*. Kensington, 1988.

Reagan, Nancy, with William Novak. *My Turn*. Random House, 1989.
Reagan, Ronald, with Richard G. Hubler. *Where's the Rest of Me?* Duell, Sloan & Pearce, 1965.
———. *An American Life*. Simon & Schuster, 1990.

CHAPTER 10: DOROTHY WALKER BUSH

Bush, Barbara. *Barbara Bush: A Memoir*. A Lisa Drew Books/Charles Scribner's Sons, 1994.
Bush, George, with Victor Gold. *Looking Forward*. Doubleday, 1987.
———. *All the Best, George Bush: My Life in Letters and Other Writings*. A Lisa Drew Book/Charles Scribner's Sons, 1999.
Bush, George W. *A Charge to Keep*. William Morrow, 1999.
Duffy, Michael, and Dan Goodgame. *Marching in Place: The Status Quo Presidency of George Bush*. Simon & Schuster, 1992.
Graham, the Rev. Dr. Billy. *Just As I Am*. HarperCollins, 1997.
Green, Fitzhugh. *George Bush: An Intimate Portrait*. Hippocrene Books, 1991.
King, Nicholas. *George Bush: A Biography*. Dodd, Mead, 1980.
Levin, Phyllis. *Abigail Adams*. St. Martin's Press, 1987.
Minutaglio, Bill. *First Son*. Times Books, 1999.
Parmet, Herbert S. *George Bush: The Life of a Lone Star Yankee*. A Lisa Drew Book/Charles Scribner's Sons, 1997.
Radcliffe, Donnie. *Simply Barbara Bush*. Warner Books, 1989.
Time magazine editors. *The Winning of the White House—1988*. A Time Book, 1988.
Walker, Mary. *Greenwich Library Oral History Project*, 1991.
Wead, Doug. *George Bush: Man of Integrity*. Harvest House, 1988.

CHAPTER 11: VIRGINIA CLINTON KELLEY

Allen, Charles F., and Jonathan Portis. *The Life and Career of Bill Clinton: The Comeback Kid*. Birch Lane Press, 1993.
Brummett, John, *High Wire: The Education of Bill Clinton*. Hyperion, 1994.
Cassidy, John. "The Political Scene." *The New Yorker*, September 21, 1998.
Clinton, Roger, with Jim Moore. *Growing Up Clinton*. Summit, 1995.
Fick, Paul M. *The Dysfunctional President*. Birch Lane Press, 1995.
Gallen, David. *Bill Clinton As They Know Him: An Oral Biography*. Gallen, 1994.

Jackson County (Missouri) Court: marriage license, William J. Blythe and Wanetta Alexander, May 3, 1941, document 834-32; birth certificate, Sharon Lee Blythe, May 11, 1941, registrar number 2605; divorce petition filed by Wanetta Blythe March 28, 1944, document 98, 687.

Kelley, Virginia Clinton, with James Morgan. *Leading with My Heart: My Life*. Simon & Schuster, 1994.

King, Norman. *Hillary: Her True Story*. Birch Lane Press, 1993.

Levin, Jerome D. *The Clinton Syndrome*. Prima, 1998.

Levin, Robert E. *Bill Clinton: The Inside Story*. S.P.I. Books/Shapolski, 1992.

Maraniss, David. *First in His Class*. Touchstone Books/Simon & Schuster, 1996.

———. *The Clinton Enigma*. Simon & Schuster, 1998.

Morris, Roger. *Partners in Power*. Henry Holt/A John Macrae Book, 1996.

Murdock, Maureen. *Fathers' Daughters: Transforming the Father-Daughter Relationship*. Fawcett Columbine, 1996.

People. "Blythe Spirit." September 13, 1993.

Radcliffe, Donnie. *Hillary Rodham Clinton: A First Lady for Our Time*. Warner Books, 1993.

Renshon, Stanley A. *High Hopes*. New York University Press, 1996.

Walker, Martin. *The President We Deserve*. Crown, 1996.

Warner, Judith. *Hillary Clinton: The Inside Story*. Signet, 1993.

CHAPTER 13: PASSIONATE ATTACHMENTS

Barber, James David. *The Presidential Character*. Prentice-Hall, 1972, 1992.

Bernard, Jesse. *The Future of Motherhood*. Penguin, 1975.

Brockett, L. P. *Life & Times of Abraham Lincoln*. Bradley & Co., 1865.

Brodie, Fawn. *Thomas Jefferson: An Intimate History*. Norris, 1974.

Burlingame, Michael. *The Inner World of Abraham Lincoln*. University of Illinois Press, 1994.

Ellis, Joseph J. *American Sphinx: The Character of Thomas Jefferson*, Alfred A. Knopf, 1997.

Forcey, Linda R. *Mothers of Sons*. Praeger, 1987.

Freeman, Douglas Southall (abridged by Richard Harwell). *Washington*. Charles Scribner's Sons, 1968.

Freud, Sigmund (trans., ed. A. A. Brill). *The Basic Writings of Sigmund Freud*. Modern Library/Random House, 1938.

Gurian, Michael. *Mothers, Sons & Lovers: How a Man's Relationship with His Mother Affects the Rest of His Life*. Shambhala, 1994.

Kagan, Jerome. *The Nature of the Child*. Basic Books, 1984.

Kaplan, Dr. Louise J. *Oneness and Separateness*. Touchstone, 1978.

Klein, Carole. *Mothers and Sons*. Houghton Mifflin, 1984.

Malone, Dumas. *Jefferson the Virginian*. Little, Brown, 1948.

McCullough, David. "Mama's Boys." *Psychology Today*, October 1983.

Neimark, Jill. "Twins: Nature's Clones." *Psychology Today*, July–August 1997.

Olsen, Paul. *Sons and Mothers*. Fawcett Crest, 1981.

Roberts, Paul. "The Psychology of Parenting." *Psychology Today*, May–June 1996.

Silverstein, Olga, and Beth Rashbaum. *The Courage to Raise Good Men*. Viking, 1994.

Tocqueville, Alexis de (ed. Andrew Hacker, trans. Henry Reeve). *Democracy in America*. Washington Square Press, 1964.

Wilson, Dorothy Clarke. *Lady Washington: The Story of America's First First Lady*. Doubleday, 1984.

BACKGROUND

Bassett, Margaret. *Profiles of American Presidents & Their Wives*. Bond Wheelwright, 1969.

Ben Cramer, Richard. *What It Takes: The Way to the White House*. Random House, 1992.

Burns, James MacGregor. *Presidential Government: The Crucible of Leadership*. Avon Books, 1965.

Frank, Sid, and Arden Davis Melick. *The Presidents: Tidbits & Trivia*. Hammond, 1980.

Freidel, Frank, ed. *Our Country's Presidents*, 8th ed. National Geographic Society, 1979.

Heckler-Feltz, Cheryl. *Heart & Soul of the Nation*. Doubleday, 1997.

Iremonger, Lucille. *The Fiery Chariot: British Prime Ministers and the Search for Love*. Secker & Warburg, 1970.

Johnson, Walter. *1600 Pennsylvania Avenue: Presidents and the People 1929–1959*. Little, Brown, 1960.

Kane, Joseph Nathan. *Facts About the Presidents*. Pocket Books, 1993.

Moos, Malcolm, and Steven Hess. *Hats in the Ring: The Making of Presidential Candidates*. Random House, 1960.

Nash, J. Madeleine. "Special Report: Fertile Minds." *Time*, February 3, 1997.

National Geographic. "The Territorial Growth of the United States." September 1987.

Neustadt, Richard E. *Presidential Power: The Politics of Leadership*. John Wiley & Sons, 1960.

Rossiter, Clinton. *The American Presidency*. Mentor Books, New American Library, 1960.

Sadler, Christine. *America's First Ladies*. Macfadden, 1963.

Schlesinger, Arthur M. *Political and Social Growth of the United States 1852–1933*. Macmillan, 1934.

Smiley, Jane, Andrew J. Cherlin, Peggy Orenstein. "Mothers Can't Win." *New York Times Magazine*, April 5, 1998.

Taylor, Tim. *The Book of Presidents*. Arno Press, 1972.

ACKNOWLEDGMENTS

★

$\mathcal{F}$OR a book with such a long gestation as this one, it is impossible for me to name all who have been helpful along the way, but some must be specially thanked.

First, the former presidents and their family members (who are named in the introduction) generously gave me their time to talk about these women who meant so much in their lives. They relived old memories, spun tales that had not been told before, and shared personal insights in interviews that are the cornerstone of this book. Six mothers—Sara Roosevelt, Rose Kennedy, Rebekah Johnson, Lillian Carter, Virginia Clinton Kelley and Barbara Bush—wrote their own life stories, which were invaluable, for mothers see their offsprings' emerging character and personality more clearly than do the children themselves. The memoirs of seven of these presidents added evocative details about their boyhood years, and six daughters-in-law wrote books enriched by their unique relationships with both the husband and his mother.

The presidential libraries, a national treasure, yielded warm letters, old clippings, rich oral histories, and early photographs from family albums; National Park Service guides made boyhood homes come alive for me. Richard Nixon's southern California roots were illuminated by Joseph Dmohowski, librarian of the special Nixon collection at Whittier College; and the outstanding collection of oral histories and old photographs at California State University at Fullerton were opened up to me by Kathy Frizee and Gail Gutierrez. The chapter on Nelle Reagan benefited greatly from the kindness of Lorraine Makler Wagner, who recounted her friendships with both Nelle and her "Dutch" and shared with me touching, never-published letters she had received from Nelle over a number of years.

In writing this book I drew on scores of biographies, but I am particularly indebted to three authors: Lou Cannon, who covered Reagan from his entry into politics and wrote two richly detailed biographies; James Cannon, who was an adviser to Vice President Ford, introduced a wealth of previously unmined material about the early background of our only unelected president; and Doris Kearns Goodwin, whose brilliant works on three of the modern presidents were intimidating to one following in her wake. Though many writers have dined from the same available smorgasbord of information, I have tried to give credit in the text to those who published new findings or fresh direct quotations.

First among the individuals to whom I am grateful is my son, Christopher (Kip to all) Levy, who was enthusiastic and helpful from the first dip into research to the last dot-com. Cherie Burns, my agent, nudged the book out of my head and onto these pages, and the skillful editing of Henry Ferris vastly improved the product.

My colleagues at *Time* magazine were supportive from the beginning, steering me to useful sources and contributing ideas. In particular, Hugh Sidey was a fount of anecdotes and observations drawn from his years of writing about a succession of nine presidents; Anne Hopkins probed the psychology of sons and mothers; and Mary Cronin bucked me up whenever I was fading.

In Washington Michael Duffy extended the courtesy of the

Time bureau, in which I had first begun thinking about this subject; Don Collins Jr. and Brian Doyle rescued me when my computer turned hostile; Christopher Ogden's advice and Lissa August's input were much appreciated. Encouragement, recollections, and opinions were volunteered by veteran White House correspondents Frances Lewine, Helen Thomas (herself a White House legend), Sid Davis, Robert Clark, and Al Spivak. Two of the best-ever White House press secretaries, Elizabeth Carpenter, the genius behind Lady Bird Johnson, and Joe Laitin, who served a record number of presidents and cabinet members, offered their insiders' insights. David Hawpe, longtime editor of the *Louisville Courier-Journal,* was among the first to spur me on; producer and author Clare Crawford-Mason's suggestions and vision were an inspiration; and James O'Shea Wade provided valuable guidance.

Some who enlightened me with their personal knowledge of the presidents' boyhood years were Harry Jeffrey, professor of history at Fullerton, who brought together a group of Nixon historians for a lively discussion; Bob Gibler, president of the Lee County (Illinois) Historical Society; Carolyn Yeldell Staley, the close friend and neighbor of classmate Bill Clinton and his mother; and James Morgan, Virginia Clinton Kelley's coauthor.

And my fond thanks to June Arey, my friend since the fifth grade, and the evergreen Eleanor Lambert, who were always there to keep me afloat.

To these and so many others I am eternally grateful.

INDEX

★

religious beliefs of, 254–55, 437
second marriage of, 233–42, 243, 255
Ford, Gerald R., 228–57
biological father of, 228–33, 238, 240–46, 253
birth name of, 228, 232, 233, 234
birth of, 232
childhood of, 233–42
congressional campaign of (1948), 249–50, 256
as congressman, 237, 252, 256
education of, 235, 239, 240, 241, 242–46, 247
football played by, 235, 239, 240, 241, 242, 243, 375
as lawyer, 247, 248
legal studies of, 235, 239, 242, 245–46, 247, 250–51
memoirs of, 234–35, 241, 251–52
military service of, 245, 247–48
personality of, 238, 239–40, 246
physical appearance of, 230, 246–47
political career of, 237, 249–50, 252–53
as president, 246, 251–52, 256
presidential campaign of (1976), 251
reading by, 242–43
RN pardoned by, 251
as vice president, 251–53, 256
Ford, Gerald R., Sr.:
background of, 233–34
business ventures of, 235, 243, 345
GF's relationship with, 234–35, 256–57
personality of, 234, 249, 252
Ford, James, 235–36, 237, 248
Ford, Susan, 250, 257
Ford, Tom, 235–36, 255–56
Foster, Stephen, 47
Freeman, Douglas Southall, 420, 422
Freud, Sigmund, 1, 428–29
Friendly Persuasion (West), 200
Fulbright, J. William, 388–89

Gardner, Adele, 232
Gardner, Levi, 229, 230, 232, 233
Garel, Byrd, 239–40
Garment, Leonard, 396
Gates family, 63–64
Gelb, Bruce, 351
George, Mary, 209, 210
George VI, King of England, 36
Georgetown University, 380–81, 387–88, 389
Gliddon, Stella, 161
Goldwater, Barry, 322
Good, Milton, 90, 91
Goodwin, Doris Kearns, 137, 168, 169, 190
Gorbachev, Mikhail, 301
Gordy, James Jackson "Jim Jack," 259–60, 278, 284, 285, 366

Gordy family, 259
Gore, Al, 414
Graham, Billy, 189, 335, 341–42
Graham, Wallace H., 64–65
Grandview, Mo., 41, 42–43, 51, 58–61, 63
Great Depression, 235, 243, 244–45, 268, 297, 319–22, 337, 345, 371, 392
Great Society, 180, 190
Green, Fitzhugh, 343
Greenwich Country Day School, 350
Grisham, Buddy, 368, 369
Groton school, 1, 9–10
Gurian, Michael, 434

Hamer, Dean, 299
Harding, Gladys, 97
Harding, Warren G., 213
Harriman, Averell, 334
Hartington, William John Robert Cavendish, Marquess of, 146
Harvard University, 1, 2, 10–11, 16, 17, 29–30, 121, 124, 129, 137
Hatcher, Jessie, 157–58, 174
Hayes, Rutherford B., 66
Hemingway, Ernest, 34, 228
Hickok, Wild Bill, 88, 166
Hiss, Alger, 196
Hitler, Adolf, 24, 434
Hoover, Herbert, 319–20
Hot Springs, Ark., 369–71, 375, 382, 398–401, 435
Hot Springs High, 375
Hughes, Howard, 432
Hyde Park estate, 2, 15, 16, 21, 23, 27, 28, 29, 30, 32–33, 34, 35, 43

Ickes, Harold, 34
"If" (Kipling), 238
Independence, Mo., 45–46, 47, 50, 61, 63
India, 270, 279–83, 291
Indians, American, 53, 115, 167

Jackson, Andrew, 66, 363, 414
Jackson, Nettie Stover, 80
Jackson, Thomas J. "Stonewall," 85
Jackson, Wes, 79
Jacobson, Eddie, 63, 70
Jedrzejowska, Jadwiga, 333
Jefferson, Jane Randolph, 422–23
Jefferson, Thomas, 420, 422–23
Jehovah's Witnesses, 83, 437
Johnson, Alvan, 278
Johnson, Claudia Taylor "Lady Bird":
education of, 182
as First Lady, 182, 190–91
LBJ as viewed by, 159, 168, 173, 178, 183, 191
LBJ's courtship of, 182–84